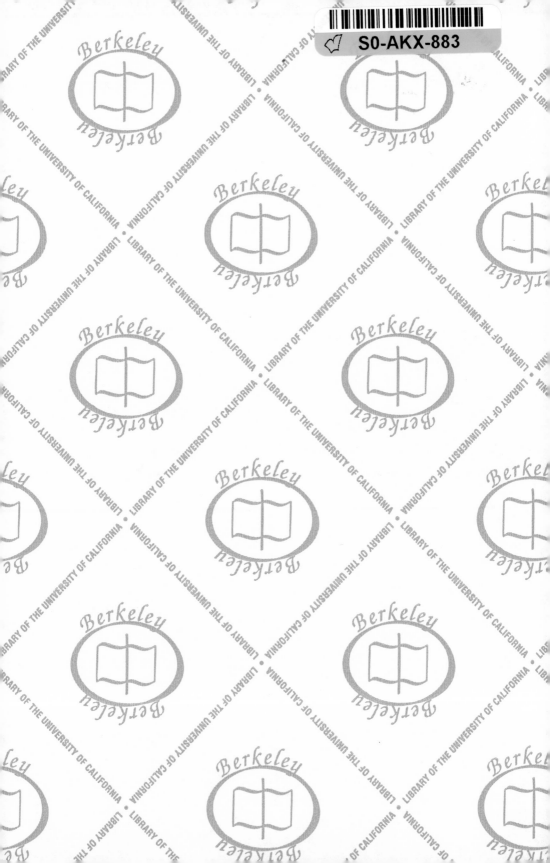

GENERAL RELATIVITY AND GRAVITATIONAL PHYSICS

Related Titles from AIP Conference Proceedings

To learn more about these titles, or the AIP Conference Proceedings Series, please visit the
webpage **http://proceedings.aip.org/proceedings**

GENERAL RELATIVITY AND GRAVITATIONAL PHYSICS

16th SIGRAV Conference on General Relativity
and Gravitational Physics

Vietri sul Mare, Italy   13 – 16 September 2004

EDITORS
Giampiero Esposito
Instituto Nazionale di Fisica Nucleare - Sezione di Napoli
Napoli, Italy

Gaetano Lambiase
Università degli Studi di Salerno
Salerno, Italy

Giuseppe Marmo
Università di Napoli "Federico II"
Napoli, Italy

Gaetano Scarpetta
Gaetano Vilasi
Università degli Studi di Salerno
Salerno, Italy

SPONSORING ORGANIZATIONS
Salerno University, Italy
Naples University, Italy
Italian Society of General Relativity and Gravitational Physics - (SIGRAV) - Torino, Italy
Instituto Nazionale di Fisica Nucleare - (INFN) - Roma, Italy

AMERICAN
INSTITUTE
OF PHYSICS

Melville, New York, 2005
AIP CONFERENCE PROCEEDINGS ■ VOLUME 751

Editors:

Giampiero Esposito
INFN - Sezione di Napoli
Complesso Universitario di Monte Sant'Angelo
Via Cinthia
I-80126 Napoli
ITALY
E-mail: giampiero.esposito@na.infn.it

Gaetano Lambiase
Gaetano Scarpetta
Gaetano Vilasi

Dipartimento di Fisica "E.R. Caianiello"
Università degli Studi di Salerno
Via S. Allende
I-84081 Baronissi (SA)
ITALY

E-mail: lambiase@sa.infn.it
          scarpetta@sa.infn.it
          vilasi@sa.infn.it

Giuseppe Marmo
Università di Napoli "Federico II"
Dipartimento di Scienze Fisiche
Complesso Universitario di Monte Sant'Angelo
Via Cinthia
I-80126 Napoli
ITALY
E-mail: marmo@na.infn.it

L.C. Catalog Card No. 2005921827
ISBN 0-7354-0236-1
ISSN 0094-243X
Printed in the United States of America

CONTENTS

SIGRAV PRIZES

PLENARY SESSION

PREFACE

The present volume contains the proceedings of the 16th SIGRAV Conference held in Vietri sul Mare (SA), September 13-16, 2004. The SIGRAV Society was founded in 1990 to provide a forum for discussions on research advances in gravitation and general relativity, emphasizing their mathematical, theoretical and experimental applications and predictions. The main aim of these meetings is to bring together Italian and international leaders on gravitational physics, as well as bright and eager beginners, and to make it easier to exchange ideas for a deeper understanding of more recent theories and experiments on gravity. General relativistic theories have become, indeed, the basic models for new fields of research, in particular in relativistic astrophysics. These new fields encompass important experiments and observations that represent, on one hand, a frontier on which Einstein's theory of gravity can be tested, while on the other hand they provide new insight into gravitational physics.

The Italian Conferences on General Relativity and Gravitational Physics have been organized since 1975. The first was indeed held in Padua in 1975, the second one in Ferrara in 1976. In the following years, the (SIGRAV) Conferences have been organized every two years: Turin in 1978, Pavia in 1980, Catania in 1982, Florence in 1984, Rapallo in 1986, Cavalese in 1988, Capri in 1990, Bardonecchia in 1992, Trieste in 1994, Rome in 1996, Monopoli (Bari) in 1998, Genoa in 2000, Frascati (Rome) in 2002.

The SIGRAV Society also organizes Schools, which are primarily addressed to PhD students in physics and mathematics with interest in general relativity, astrophysics, experimental gravity and the modern quantum theories of gravitation. Every year a particular topic is chosen and Lecturers are invited to deliver a number of main courses, collectively covering various aspects of the chosen theme.

Following the tradition of previous years, the scientific program of the SIGRAV04 Conference has been structured in a number of invited lectures (plenary talks) reviewing the status of current theories and experiments and observations, and parallel sessions based on short presentations of participants' research. The frontier arguments presented at this meeting, as well as in the previous editions, testify to the spirit of the SIGRAV Conferences during the years: to follow the most important developments on general relativity and gravitation, becoming an important firm point for present and future research. We feel confident that the reader will enjoy studying papers on such a huge variety of topics, e.g. higher order theories of gravity, gravitomagnetism, the C metric, cosmic censorship hypothesis, gravitational waves and holography in quantum gravity, among the others.

The SIGRAV04 Award ceremony was held in the "Palazzo della Provincia", in Salerno. The prestigious Amaldi Medal was awarded to Professor Roger Penrose. Roger Penrose has been Rouse Ball Professor of Mathematics at Oxford University, England. Most of his work is devoted to relativity theory and quantum physics, i.e. space-time singularities, conformal infinity, spinor geometry, twistor theory, foundations of quantum mechanics, but he is also fascinated with a field of geometry known as tessellation, the covering of a surface with tiles of prescribed shapes. During his brilliant career, Professor Penrose has received many prizes, including, in particular, the Cambridge University Adams Prize Essay in 1966, the prestigious Wolf Prize for Physics in 1988, that he shared with Stephen Hawking, and the Albert Einstein Prize and Medal of the Albert Einstein Society in 1990.

The SIGRAV Prizes to young researchers with an outstanding scientific record was awarded to Carlo Angelantonj for his studies on "Open Strings and Supersymmetry Breaking" and Giovanni Miniutti for his studies on "Flux and energy modulation of iron emission due to relativistic effects in NGC 3516".

We wish to acknowledge the financial support from the Rector of Salerno University, Professor Raimondo Pasquino, Rector of Naples University Federico II, Professor Guido Trombetti, Dipartimento di Fisica E. R. Caianiello of Salerno University, Dipartimento di Matematica e Informatica of Salerno University, Dipartimento di Scienze Fisiche of Naples University, INFN, SIGRAV. The Conference was also sponsored by Regione Campania, Provincia di Salerno and Ente Cassa di Risparmio di Firenze.

We also thank Canada-France-Hawaii Telescope and Coelum Astronomia for permission to use the image of NGC 891 in our poster.

Last, but not least, we should acknowledge with warm thanks the enormous efforts of our Secretary, Stefania Russo, who together with her colleagues, Domenico Borrelli and Vincenzo Di Marino, have gone well beyond the call of duty in taking care of any conceivable step.

The Editors

ACHIEVEMENTS OF THE RECIPIENTS

Open Strings and Supersymmetry Breaking
C. Angelantonj

Carlo Angelantonj obtained his Laurea degree from the University of Aquila in 1994, working under the supervision of the late Professor F. Occhionero. His Phd., again from the University of Aquila, was instead supervised at the University of Rome "Tor Vergata", jointly, by Professor M. Bianchi and Professor A. Sagnotti. His Ph.D. Thesis contains the first instance of a chiral four-dimensional open-string vacuum and the first systematic scanning of rational open-string models in six-dimensions, two results that have had a wide impact on the field.

He was then Research Fellow at the École Polytechnique, "Marie Curie Fellow" first at École Normale Supérieure and then again at the École Polytechnique, CERN Fellow and Alexander von Humboldt Fellow in Berlin and in Munich. Dr. Angelantonj recently joined the Theoretical Physics Department of the University of Torino.

Clearly the best student formed in the string group of the University of Rome "Tor Vergata" for over a decade, his research has centered from the beginning on the possible mechanisms to attain the breaking of supersymmetry in the presence of open strings and on the key properties of the resulting models. He has given a number of contributions to these aspects of String Theory, both on his own and in a number of collaborations. To cite but a few, he was the first to investigate the non-tachyonic, and thus classically stable, compactifications of the non-supersymmetric 0B' string theory, and clarified the role of the quantized NS-NS Bab tensor in generating several tensor multiplets in orbifold compactifications. More recently he has been focusing on mechanisms to lower the value of the cosmological constant induced by the breaking of supersymmetry.

Flux and Energy Modulation of Iron Emission due to Relativistic Effects in NGC 3516G
G. Miniutti

After graduating in Physics, magna cum laude, in 1999 at the University of Rome "La Sapienza" Giovanni Miniutti started the PhD program at the same University under the supervision of Professor V. Ferrari.

His research program focused on the study of gravitational waves emitted by binary systems consisting of compact objects, by pulsating neutron stars and by newly born, hot proto-neutron stars. He got the PhD in 2003, defending the thesis "Neutron star Oscillations in General Relativity and emission of Gravitational Waves".

He is currently working at the Institute of Astronomy of the University of Cambridge as a Post-Doctoral Research Associate in collaboration with Professor A.C. Fabian. His work mainly focuses on data analysis and theoretical modelling of X-ray astronomical observations of accretion processes involving both stellar-mass black holes in binary systems in the Galaxy, and supermassive (millions or billions of solar masses) black holes sitting at the center of active galaxies. One of his main contributions in this research consists in including general relativistic effects in the theoretical models constructed to explain the spectral properties and the temporal variability patterns that are observed in the X-ray emission of these sources.

He has shown that X-ray data from several active galaxies such as MCG-6-30-15 and 1H 0707-495, and from the Galactic stellar-mass black hole binary XTE J1650-500 can be interpreted if the effects induced by special and general relativity on the X-ray emission are properly taken into account. In particular, the spectral variability in these objects can be explained in the framework of a model he proposed with Professor A.C. Fabian, introducing the idea that light bending by the strong gravitational field of the central black hole is essential if X-ray emission from the innermost regions of the accretion flow has to be modelled.

More recently, he has given an important contribution to the study of the data collected by the X-ray satellite XMM-Newton on the active galaxy NGC 3516. In his work with Dr. Iwasawa and Professor A.C. Fabian, he introduced a new data analysis technique devised to search for the variability of faint X-ray emission lines in the spectra of active galaxies with the potential of revealing the motion of the emitting gas in the immediate vicinity of the central black hole, probing the nature of the accretion flow and the spacetime itself in a regime in which General Relativity plays a crucial role. Indeed, his work revealed for the first time a modulation of the iron emission line in NGC 3516 both in flux and energy. The modulation is consistent with special relativistic beaming and Doppler/gravitational energy shifts as the emitting matter orbits the central black hole at a distance of about 5 Schwarzschild radii. About four cycles of modulation were found allowing to estimate the distance from the black hole and the orbital period which allowed to estimate the black hole mass in NGC 3516 in the range of 10 to 50 million of solar masses. This is the first time such an estimate is made by using X-rays from the innermost regions of the accretion flow around a black hole and the result opens a new promising window to our understanding of the spacetime geometry and gas motion in the strong field regime of general relativity. These results were presented by Giovanni Miniutti and Dr. Iwasawa in a press conference on September 9, 2004 in New Orleans at the meeting of the High Energy Astrophysics Division of the American Astronomical Society.

SIGRAV PRIZES

Open Strings and Supersymmetry Breaking

Carlo Angelantonj

*Department für Physik, Ludwig-Maximilians-Universiät, München
Theresienstr. 37, D80333 München*

Abstract. We review several mechanisms for supersymmetry breaking in orientifold models. In particular, we focus on non-supersymmetric open-string realisations that correspond to consistent flat-space solutions of the classical equations of motion. In these models, the one-loop vacuum energy is typically fixed by the size of the compact extra dimensions, and can thus be tuned to extremely small values if enough extra dimensions are large.

A GLIMPSE AT ORIENTIFOLD CONSTRUCTIONS

Type-I models have become the subject of an intense activity during the last few years, since their perturbative definition offers interesting new possibilities for low-energy phenomenology. Their consistency and a number of their most amusing features may be traced back to the relation to suitable "parent" models of oriented closed strings [1, 2], from which their spectra can be derived. In this procedure, a special role is played by "tadpole conditions" for Ramond–Ramond (R-R) and Neveu–Schwarz–Neveu–Schwarz (NS-NS) massless states. One of the most amusing features of orientifold constructions is the different origin of gravitational and gauge interactions. Although this observation might seem in contrast with the old dream of unifying geometry and matter, it actually allows for more generic vacuum configurations with interesting implications. On the one hand, gravity originates from closed strings, and as such permeates the whole ten-dimensional space-time. On the other hand, gauge interactions are associated to open strings whose free ends live on Dp-branes [3], $(p+1)$-dimensional hyper-manifolds embedded in the ten-dimensional bulk. In particular, if we denote by x^μ (y^a) the coordinates along (transverse to) the world-volume of D-branes, this simple remark implies for example that the full metric tensor and the space-time gauge field are of the form

$$g_{MN} = g_{MN}(x^\rho, y^a), \qquad A_\mu = A_\mu(x^\rho), \tag{1}$$

and, as we shall see, this fact has dramatic consequences in orientifold constructions.

Given this geometric description of orientifold constructions, one can associate a more physical interpretation to tadpole conditions. While R-R tadpoles should be regarded as *global* neutrality conditions for R-R charges [4], NS-NS tadpoles ensure that the configurations of D-branes and O-planes (a sort of rigid mirrors that revert the orientation of closed and open strings) be *globally* massless. In supersymmetric models these two conditions are related by supersymmetry transformations, and thus the vanishing of R-R tadpoles naturally implies that the NS-NS ones vanish as well. However, it is worth stressing here that they have completely different consequences: while the R-R tadpole

CP751, *General Relativity and Gravitational Physics, 16th SIGRAV Conference*, edited by G. Esposito et al.

conditions are required by gauge invariance and their violation is linked to the emergence of irreducible gauge and gravitational anomalies [5, 6], NS-NS tadpoles are not associated to any inconsistency, and thus in principle may be relaxed. In doing so, however, one perturbs the background geometry [7], with the end result that full fledged string-theory calculations are more difficult [8].

NON TACHYONIC TYPE-0 VACUA IN VARIOUS DIMENSIONS

Looking for non-supersymmetric vacua, type-0 strings [9] offer a natural arena. Typically, these theories include in their spectrum tachyonic modes (both in the closed and in the open sector) that induce instabilities in the vacuum. However, it was shown in [10, 11] that orientifold constructions allow for different choices of projections and in particular in [12] a tachyon-free descendant of type-0B theory was built. Together with non-tachyonic non-supersymmetric heterotic strings [13], this is a notable example of a classically stable string vacuum without supersymmetry.

A natural question is then whether these properties will survive upon compactification on non-trivial manifolds. While surprises are not expected in simple toroidal reductions, new interesting features emerge when orbifolds are considered. Since the ten-dimensional parent strings are not supersymmetric to begin with, one is now entitled to use both supersymmetry-preserving [14] and supersymmetry-breaking twists [15]. In the case of $T^4/\mathbb{Z}_N$ and $T^6/\mathbb{Z}_N$ compactifications, an exhaustive analysis has revealed that, for generic N, aside from the surviving untwisted tachyon of the parent closed oriented theory, new complex twisted tachyons are typically present in the spectrum of light excitations [14, 15]. As a result, a generic orientifold projection can only make the twisted tachyons real, and thus classical instabilities are always present. The only exception is given by the supersymmetry-preserving $T^4/\mathbb{Z}_2$, $T^6/\mathbb{Z}_3$ and $T^6/\mathbb{Z}_2 \times \mathbb{Z}_2$ and supersymmetry-breaking $T^6/\mathbb{Z}_2$ cases, where no tachyons emerge in the twisted sectors. One can then properly deform the orientifold projection as in [12] and build new lower-dimensional vacua free from untwisted and twisted tachyonic modes [14]. One should stress here that, although these vacua are tachyon-free, the impossibility of cancelling their NS-NS tadpoles and the generation of a one-loop cosmological constant unavoidably destabilise the classical vacuum.

Type-0 theories and their D-branes have also triggered some activity in the context of gauge/gravity dualities [16] for non-supersymmetric non-conformal theories. While in oriented type-0B strings the presence of the tachyon complicates the gravity description and leads to instabilities in the dual strongly coupled field theory, a sensible holographic description of non-supersymmetric gauge theories was proposed in [17]. Resorting to the non-tachyonic orientifolds of [12], the gauge theory in the open-string sector turns out to be conformal in the planar large-N limit, and therefore several results can be argued from the parent $\mathcal{N} = 4$ theory. In particular, the leading (planar) geometry remains $AdS_5 \times S^5$ with a constant dilaton field. For finite N, however, the gauge theory is no longer conformal and new features are expected in the gravity description. Indeed, the lack of conformal invariance translates on the string side into the presence of a non-vanishing dilaton tadpole at the disk level that induces deviations from the $AdS_5 \times S^5$

4

geometry. In addition, the dilaton starts running consistently with the RG-determined behavior for the gauge coupling on the field theory side. Moreover, the back-reaction in the classical geometry predicts a quark-anti-quark potential interpolating between a (logarithmically running) Coulomb phase and a confining phase in the IR, as expected from the field theory analysis [17]. More recently, this study has been extended to obtain exact results in non-supersymmetric gauge theories [18].

SPONTANEOUS SUPERSYMMETRY BREAKING

Compared to explicit breakings of supersymmetry by suitable choices of modular invariant partition functions and/or compactification manifolds, the spontaneous breaking has more appealing properties since at times it makes it possible to attain a better control on radiative corrections of masses and couplings.

Independently of the specific mechanism at work, whenever supersymmetry is spontaneously broken at a scale M_{sb} (depending on the specific model one is considering), the (tree-level) mass splitting in a given super multiplet and the (one-loop induced) cosmological constant are typically both determined by M_{sb}:

$$\Lambda \sim M_{sb}^4, \qquad \delta M \sim M_{sb}. \tag{2}$$

One of the outstanding problems in string theory, as in any quantum theory including gravity, is to understand how a small cosmological constant can be accompanied by the generation of appropriate gaugino masses. This problem became much more severe after recent observations suggested a non-vanishing vacuum energy corresponding to a new energy scale, far smaller than every other scale in the physics of fundamental interactions, $\Lambda_{obs} = E_\Lambda^4$, with $E_\Lambda \sim 10^{-4}$ eV. Similarly, experiments in Particle Physics suggest that gauginos should be heavier than a few TeV. Given the expressions in eq. (2), two distinct mechanisms for breaking supersymmetry should be combined to successfully disentangle the cosmological constant scale from the gaugino mass scale.

Type-II string models with a cosmological constant possibly vanishing in perturbation theory were studied in refs. [19, 20]. Their main feature is a Fermi–Bose degenerate spectrum, leading to an automatic vanishing of one-loop vacuum energy. Aside from the question of higher-loop corrections [21], their main drawback is that the non-Abelian gauge sector, appearing at particular singular points of the compactification manifold, or on appropriate D-brane collections, is always supersymmetric [22, 23]. Thus, it is questionable whether such constructions can accommodate gauge degrees of freedom with large supersymmetry-breaking mass splittings without spoiling the vanishing of the vacuum energy.

Among the mechanisms to break supersymmetry in the gauge sector, intersecting brane models [24], T-dual to magnetic field backgrounds [25], have proved to be a natural setting to realise Standard-Model-like patterns of gauge symmetries and matter fields within string theory (see [26] for a review and references therein). However, they are generally plagued by tachyonic instabilities, and only in a few cases the fate of these unstable configurations has been studied in some detail [27]. Moreover, in standard realisations it seems quite difficult to give masses to the superpartners of the

$SU(3) \times SU(2) \times U(1)$ gauge bosons, and typically a sizeable vacuum energy determined by the intersection angles (or by the strength of the background magnetic fields) is generated already at the disk level [25]. These constructions can be extended to the case of magnetic backgrounds for the gravi-photon, corresponding to strings fluctuating in Melvin spaces. In this case the breaking of supersymmetry is felt also by the closed-string sector, and a T-dual description would include orientifold planes at generic angles [28]. Although this scenario has not found a direct application in string phenomenology, it provides an interesting example of conformal field theories with non-compact target spaces associated to orbifolds with irrational twists, and also exhibits some amusing arithmetic properties [29].

The Scherk–Schwarz mechanism provides an elegant realisation of supersymmetry breaking by compactification in field theory [30]. In the simplest case of circle compactification, it amounts to allowing the higher dimensional fields to be periodic around the circle up to an R-symmetry transformation. The Kaluza–Klein momenta of the various fields are correspondingly shifted proportionally to their R charges, and modular invariance dictates the extension of this mechanism to the full perturbative spectrum in models of oriented closed strings [31]. As a result, the gravitini get masses inversely proportional to the compactification radius

$$m_{3/2} \sim R^{-1}, \tag{3}$$

while the breaking of supersymmetry is accompanied by a one-loop vacuum energy that reproduces the behavior familiar from field theory

$$\Lambda_{SS} \sim R^{-4}, \tag{4}$$

aside from additional terms exponentially suppressed with R originating from the twisted sector [32]. When open strings are present, one has to distinguish between the two cases of Scherk–Schwarz deformations transverse or longitudinal to the world-volume of the branes [33]. In fact, as a result of (1) in the former case the open-string fields do not depend on the coordinates of the extra dimension, and therefore are not affected by the deformation. In this scenario, termed in [33] "M-theory breaking", the D-brane excitations stay supersymmetric (at least to lowest order) and the gaugino mass is identically vanishing, $m_{1/2} = 0$. In the latter case, instead, the R charges determine the masses of the fields and $m_{1/2} \sim R^{-1}$.

Finally, Brane Supersymmetry Breaking [34] is a purely stringy mechanism that, to lowest order, affects only the open-string excitations. The supersymmetric bulk is coupled to non-supersymmetric branes, where the mass splitting is set by the string scale itself, $m_{1/2} \sim M_s$. However, Brane Supersymmetry Breaking yields a non-vanishing contribution to the cosmological constant already at the disk level, by virtue of the impossibility of cancelling NS-NS tadpoles, with $\Lambda \sim M_s^4$. In a number of interesting examples, brane supersymmetry breaking provides a solution [35] to an old puzzle in the construction of open-string models, where some tadpole conditions were long known to allow apparently no consistent solution [36]. At the level of the low-energy effective action, the apparently broken supersymmetry is linearly realised in the open-string sector, and the singlet spinor, always present in these models, plays the role of the goldstino [37].

SCALES IN ORIENTIFOLD MODELS

We have long been accustomed to accepting the fate that typical string effects are confined to very high energies. This is directly implied by the often implicit identification of the string scale $M_s = \ell_s^{-1}$ with the Planck scale M_{Pl}, motivated by the experience with the weakly coupled heterotic strings, where gravitational and gauge interactions are associated to excitations of closed strings only. It is amazing that the different nature of gravity and gauge forces in orientifold models, together with the built-in observation that they generally propagate in different space-time directions, as recalled in eq. (1), makes it possible to decouple ℓ_s from ℓ_{Pl}. Indeed, the schematic low-energy effective action of a generic orientifold model in the presence of a Dp brane reads as

$$S = \int [d^{10}x] \frac{1}{\ell_s^8 g_s^2} R + \int [d^{p+1}x] \frac{1}{\ell_s^{p-3} g_s} F^2, \tag{5}$$

with g_s the string coupling constant, and R and F^2 the familiar Ricci scalar and the kinetic term for gauge fields. Upon compactification to four dimensions, the Planck length and the gauge coupling constant

$$\frac{1}{\ell_{Pl}^2} = \frac{V_\parallel V_\perp}{\ell_s^8 g_s^2}, \qquad \frac{1}{g_{YM}^2} = \frac{V_\parallel}{\ell_s^{p-3} g_s}, \tag{6}$$

are related to the volumes $V_\parallel$ and $V_\perp$ parallel and orthogonal to the branes as well as to the string length ℓ_s and to the string coupling constant g_s. Combining eqs. (6), one can arrive at the key expressions

$$M_{Pl}^2 = \frac{1}{g_{YM}^4 v_\parallel} M_s^{2+n} R_\perp^n, \qquad g_s = g_{YM}^2 v_\parallel, \tag{7}$$

with $n = 9 - p$ the number of dimensions, all of size $R_\perp$, orthogonal to the branes and $v_\parallel = V_\parallel / \ell_s^{p-3}$ the longitudinal volume measured in string units. It should then be clear that in a weakly coupled type-I string $v_\parallel \sim 1$, while the string scale and the size of the transverse directions are completely undetermined, though correlated through (7). For instance, for $M_s \sim 1$ TeV the size of the extra dimensions $R_\perp$ can in principle vary from 10^8 km, to 0.1 mm down to 0.1 fm for $n = 1, 2$ or 6 transverse directions [38]. Aside from the $n = 1$ case, clearly ruled out, all other cases are actually consistent with observations.

SUPPRESSING THE COSMOLOGICAL CONSTANT

It is very suggestive, if not a simple numerical coincidence, that for $n = 2$ the size of the transverse dimensions is of the same order of magnitude as the observed cosmological constant scale, $R_\perp \sim E_\Lambda^{-1}$. In fact, as we have seen, the behavior $\Lambda \sim R_\perp^{-4}$ is typical of models where supersymmetry is broken by Scherk–Schwarz deformations. However, in this case mass splittings are at most of order $R_\perp^{-1}$, and hence too small by several orders of magnitude. Similarly, in models featuring Brane Supersymmetry Breaking the gaugino mass, determined by the string scale itself, can be naturally tuned to a few TeV, consistently with data from particle accelerators. It would thus be tempting to combine Scherk–Schwarz reductions transverse to the branes with brane supersymmetry breaking in order to disentangle the cosmological constant scale and the gaugino mass scale, and to tune them to experimentally acceptable values. However, a naive combination of these two effects would appear to spoil the value of the vacuum energy, since

$$\Lambda(R) \sim (n_B^c - n_F^c)\frac{1}{R^4} + c_1(n_B^o - n_F^o)M_s^4 + c_2(n_B^o - n_F^o)\frac{M_s^{6-n}}{R^{n-2}} + \mathcal{O}\left(e^{-M_s^2 R^2}\right) \quad (8)$$

would receive contributions of order $M_s^4 \sim (\text{few TeV})^4$ from the brane supersymmetry breaking mechanism, unless the open-string spectrum is Fermi–Bose degenerate. In one possible realisation, this degeneracy might be thought of as emerging from a supersymmetric theory where the super partners have been displaced appropriately in position space.

In order to achieve such a Fermi–Bose degenerate open-string spectrum one turns on discrete values for the NS-NS B_{ab} along the compact directions. The presence of such a background, or of similar discrete vacuum expectation values of fields projected out by the world-sheet parity [40], reverts the nature of some of the orientifold planes [41], and therefore fewer numbers of D-branes are required to cancel tadpoles. As a result, in the open-string sector the rank of the gauge group is reduced proportionally to the rank of the non-vanishing B_{ab}, and moreover symplectic groups can be continuously connected to orthogonal ones [41]. In orbifold compactifications these phenomena are accompanied by a modified structure of fixed points, that in the presence of a quantised B_{ab} arrange themselves into multiplets and induce different projections in the twisted sector [42]. As a result, in six-dimensional $\mathcal{N} = (1,0)$ compactifications variable numbers of tensor multiplets are present in the closed unoriented spectrum, thus providing a geometrical description of the rational constructions first presented in [43]. The presence of these additional tensors is actually crucial for obtaining consistent vacuum configurations, since they play a significant role in a generalised Green–Schwarz mechanism for anomaly cancellation [44]. Discrete values for the NS-NS B-field are of crucial importance also in compactifications on magnetised backgrounds, where odd numbers of families of chiral matter are then allowed [45].

The simplest string theory construction with a Fermi–Bose degenerate spectrum [46] corresponds to the M-theory breaking model of [33] compactified on an additional T^2 that is permeated with a quantised B_{ab}. To be more concrete, the one-loop torus, Klein, annulus, and Möbius-strip amplitudes

$$\mathscr{T} = \frac{1}{2}\left[(V_8\bar{V}_8 + S_8\bar{S}_8)\Gamma_{m,2n}^{(1,1)} + (O_8\bar{O}_8 + C_8\bar{C}_8)\Gamma_{m,2n+1}^{(1,1)}\right.$$
$$\left. -(V_8\bar{S}_8 + S_8\bar{V}_8)\Gamma_{m+\frac{1}{2},2n}^{(1,1)} - (O_8\bar{C}_8 + C_8\bar{O}_8)\Gamma_{m+\frac{1}{2},2n+1}^{(1,1)}\right]\Gamma^{(2,2)}(B), \quad (9)$$

$$\mathscr{K} = \frac{1}{2}\left[(V_8 - S_8)W_{2n}^{(1)} + (O_8 - C_8)W_{2n+1}^{(1)}\right]W_{(2n^7,2n^8)}^{(2)}, \quad (10)$$

$$\mathscr{A} = \frac{1}{2}\left(N_D^2 + N_{\bar{D}}^2\right)(V_8 - S_8)W_n^{(1)}W_{(n^7,n^8)}^{(2)} + N_D N_{\bar{D}}(O_8 - C_8)W_{n+\frac{1}{2}}^{(1)}W_{(n^7+\frac{1}{2},n^8)}^{(2)}, \quad (11)$$

8

and

$$\mathcal{M} = -\tfrac{1}{2}(V_8 + (-1)^n S_8)W_n^{(1)}\left[(N_D + N_{\bar{D}})W_{(n^7,2n^8+1)}^{(2)}\right.$$
$$\left. -(N_D - N_{\bar{D}})(-1)^{n^7}W_{(n^7,2n^8)}^{(2)}\right], \tag{12}$$

encode all the information about the geometry and the spectrum of the vacuum configuration (see [46] for details). Supersymmetry is clearly broken *à la* Scherk–Schwarz in the bulk, while the D-brane massless excitations comprise the Kaluza–Klein reduction of ten-dimensional gauge bosons with gauge group $USp(8) \times SO(8)$, together with fermions in the $(28,1) + (1,36)$ representations, as a result of Brane Supersymmetry Breaking. The two gauge-group factors, and thus the two sets of branes, are distributed at different points along the compact directions where suitable orientifold planes are located. This open-string spectrum is clearly Fermi–Bose degenerate, although it is not supersymmetric.

The one-loop vacuum energy is then

$$\Lambda(R) = \int_{\mathscr{F}} \frac{d^2\tau}{\tau_2^{9/2}} \frac{\mathscr{T}(R)}{|\eta|^{10}} + \int_0^\infty \frac{d\tau_2}{\tau_2^{9/2}} \frac{\mathscr{K}(R)}{\eta^5}$$
$$+ \int_0^\infty \frac{d\tau_2}{\tau_2^{9/2}} \frac{\mathscr{A}(R)}{\eta^5} + \int_0^\infty \frac{d\tau_2}{\tau_2^{9/2}} \frac{\mathscr{M}(R)}{\eta^5}, \tag{13}$$

and, aside from the power-low behavior originating from the torus amplitude, that after a four-dimensional reduction on a spectator T^3 yields the announced R^{-4} term in (8), receives only exponentially suppressed contributions from $\mathscr{K}$, $\mathscr{A}$ and $\mathscr{M}$ provided the radii are appropriately correlated [46].

This is the simplest instance of a class of non-supersymmetric orientifolds with two large transverse dimensions and a naturally small cosmological constant. In the large-radius limit supersymmetry is restored in the bulk, while the D-brane spectra stay non-supersymmetric, but exhibit Fermi–Bose degeneracy at all massive string levels. In loose terms, the model contains two "mirror worlds", and the degeneracy results from an interchange of the ordinary superpartners on the two branes.

TAMING HIGHER-ORDER CORRECTIONS

Having found a model with a Fermi–Bose degenerate massless spectrum in the open-string sector does not suffice, however, to guarantee that higher-loop corrections do not induce sizeable contributions to the vacuum energy. In orientifold models these originate by diagrams with increasing numbers of handles, crosscaps and holes, and are in general quite hard to compute directly (see, for instance, [47]). Despite these technical difficulties it is at times possible to get a flavor of the qualitative behavior of higher-genus vacuum amplitudes and to discriminate between those that are vanishing identically and those that are not.

Let us consider for example the model presented in [48]. It consists essentially in a discrete deformation of the open-string sector allowed by the two-dimensional CFT constraints [10], in such a way that a fully supersymmetric bulk is accompanied by non-supersymmetric branes with gauge group $SO(8) \times USp(8)$ and fermions in the $(36, 1) + (1, 28)$. The relevant transverse-channel amplitudes for our discussion are (see [48] for more details)

$$\tilde{\mathscr{K}} = \frac{2^4}{2}(V_8 - S_8)(O_4O_4 + V_4V_4), \qquad (14)$$

$$\tilde{\mathscr{A}} = \frac{2^{-4}}{2} \left\{ (N+M)^2 (V_8 - S_8)(O_4O_4 + V_4O_4 + S_4S_4 + C_4S_4) \right. $$
$$\left. + \left[(N-M)^2 V_8 - (-N+M)^2 S_8 \right] (V_4V_4 + O_4V_4 + C_4C_4 + S_4C_4) \right\}, \quad (15)$$

and

$$\tilde{\mathscr{M}} = -\left\{ (N+M)(\hat{V}_8 - \hat{S}_8)\hat{O}_4\hat{O}_4 + \left[(N-M)\hat{V}_8 - (-N+M)\hat{S}_8 \right] \hat{V}_4\hat{V}_4 \right\}. \qquad (16)$$

It is then clear that for $N = M$ (both equal to eight as a result of tadpole conditions) these one-loop amplitudes vanish identically, even in the presence of supersymmetry breaking, as a result of the Jacobi identity $V_8 \equiv S_8$.

Moving to higher genus, the amplitudes associated with closed Riemann surfaces, both oriented and unoriented, are expected to vanish, since the closed-string sector is not affected by the deformation and hence have the same properties as the type-I superstring. However, more care is needed when surfaces with boundary are considered. Let us specialise to surfaces with two crosscaps and one boundary. Similarly to the one-loop case, there is a particular choice for the period matrix $\Omega_{\alpha\beta}$ for which this surface describes a tree-level three-closed-string interaction diagram, weighted by the product of disc (B_i) and cross-cap (Γ_i) one-point functions of closed states, that can be read from the transverse-channel Klein-bottle, annulus and Möbius-strip amplitudes. More precisely, the expression for the amplitude would be

$$\mathscr{R}_{[0,1,2]} = \sum_{i,j,k} \Gamma_i \Gamma_j B_k \mathscr{N}_{ij}{}^k \mathscr{V}_{ij}{}^k(\Omega_{\alpha\beta}), \qquad (17)$$

where the $\mathscr{N}_{ij}{}^k$ are the fusion rule coefficients and $\mathscr{V}_{ij}{}^k(\Omega_{\alpha\beta})$ are complicated functions of the period matrix encoding the three-point interaction among states i, j and k. This amplitude is expected to vanish in the supersymmetric (undeformed) case, and this requirement imposes some relations among the functions $\mathscr{V}_{ij}{}^k(\Omega_{\alpha\beta})$ that we have not defined explicitly. For instance, for the undeformed supersymmetric version of the model in (15) and (16), whose open-string amplitudes are now

$$\tilde{\mathscr{A}} = \frac{2^{-4}}{2} \left[(N+M)^2 (V_8 - S_8)(O_4O_4 + V_4O_4 + S_4S_4 + C_4S_4) \right. $$
$$\left. + (N-M)^2 (V_8 - S_8)(V_4V_4 + O_4V_4 + C_4C_4 + S_4C_4) \right] \qquad (18)$$

and

$$\tilde{\mathscr{M}} = -(N+M)(\hat{V}_8 - \hat{S}_8)\hat{O}_4\hat{O}_4 - (N-M)(\hat{V}_8 - \hat{S}_8)\hat{V}_4\hat{V}_4, \qquad (19)$$

the genus three-half amplitude takes the form

$$\mathcal{R}_{[0,1,2]} = \frac{1}{4}(N+M)\left[\mathcal{V}_{111} + 3\mathcal{V}_{133} + \mathcal{V}_{122} + \mathcal{V}_{144} + 2\mathcal{V}_{234}\right]$$
$$+ \frac{1}{4}(N-M)\left[\mathcal{V}_{122} + \mathcal{V}_{144} + 2\mathcal{V}_{234}\right], \tag{20}$$

where the relative numerical coefficients of the $\mathcal{V}$ take into account the combinatorics of diagrams with given external states. The indices $1,2,3,4$ refer to the four characters $V_8 O_4 O_4$, $V_8 V_4 V_4$, $-S_8 O_4 O_4$ and $-S_8 V_4 V_4$, that identify the only states with a non-vanishing Γ_i, as one can read from eq. (14), and their non-vanishing fusion rule coefficients, all equal to one, are $\mathcal{N}_{111}$, $\mathcal{N}_{122}$, $\mathcal{N}_{133}$, $\mathcal{N}_{144}$ and $\mathcal{N}_{234}$. Finally, both in $\mathcal{V}$ and in $\mathcal{N}$ we have lowered the indices by using the diagonal metric δ_{kl}, since all characters in this model are self-conjugate.

For a supersymmetric theory this amplitude is expected to vanish independently of brane locations, and thus the condition $\mathcal{R}_{[0,1,2]} = 0$ amounts to the two constraints

$$\mathcal{V}_{111} + 3\mathcal{V}_{133} = 0,$$
$$\mathcal{V}_{122} + \mathcal{V}_{144} + 2\mathcal{V}_{234} = 0. \tag{21}$$

Turning to the non-supersymmetric open sector in eqs. (15) and (16), one finds instead

$$\mathcal{R}_{[0,1,2]} = \frac{1}{4}(N+M)\left[\mathcal{V}_{111} + 3\mathcal{V}_{133} + \mathcal{V}_{122} + \mathcal{V}_{144} + 2\mathcal{V}_{234}\right]$$
$$- \frac{1}{2}(N-M)\left[\mathcal{V}_{122} - \mathcal{V}_{144}\right], \tag{22}$$

since now $B_4 = -N + M$ has a reversed sign. Using eq. (21) the non-vanishing contribution to the genus three-half vacuum energy would be

$$\mathcal{R}_{[0,1,2]} = -\frac{1}{2}(N-M)\left[\mathcal{V}_{122} - \mathcal{V}_{144}\right] \tag{23}$$

that however vanishes for our choice $N = M$.

Similar considerations hold for other genus-g surfaces with boundary. In all cases it can be shown that all potentially non-vanishing contributions are multiplied by the breaking coefficients $N - M$, and hence are zero for $M = N$. As a result, no contributions to the vacuum energy are generated at any order in perturbation theory. Unfortunately, this configuration of orientifold planes and D-branes is unstable, and in the true vacuum already the one-loop amplitudes contribute to Λ.

ACKNOWLEDGMENTS

I would like to thank Professor Eugenio Coccia and the Board of the Italian Society of General Relativity and Gravitation for selecting me for the 2004 SIGRAV prize. It is amusing to notice a nice coincidence of dates: this 2004 prize came ten years after my supervisors had been awarded the first SIGRAV prize for the "systematics of open-string constructions", and, at the same time, I had started my PhD under their supervision. It is a real pleasure to acknowledge Adi Armoni, Massimo Bianchi, Ralph Blumenhagen, Matteo Cardella, Giuseppe D'Appollonio, Riccardo D'Auria, Emilian Dudas, Sergio

Ferrara, Kristin Förger, Matthias Gaberdiel, Jihad Mourad, Gianfranco Pradisi, Yassen S. Stanev, Mario Trigiante and especially Ignatios Antoniadis and Augusto Sagnotti for their friendship and for long-lasting stimulating collaborations that have contributed significantly to shape my understanding of string theory. Ideally, I would like to share the SIGRAV prize with them all.

REFERENCES

1. A. Sagnotti, "Open Strings And Their Symmetry Groups," Cargèse Summer Institute on Non-Perturbative Methods in Field Theory, Cargèse, France, 1987 [arXiv:hep-th/0208020]; G. Pradisi and A. Sagnotti, "Open String Orbifolds," Phys. Lett. B216 (1989) 59; M. Bianchi and A. Sagnotti, "On The Systematics Of Open String Theories," Phys. Lett. B247 (1990) 517; "Twist Symmetry And Open String Wilson Lines," Nucl. Phys. B361 (1991) 519; P. Horava, "Strings On World Sheet Orbifolds," Nucl. Phys. B327 (1989) 461.
2. E. Dudas, "Theory and phenomenology of type I strings and M-theory," Class. Quantum Grav. 17 (2000) R41 [arXiv:hep-ph/0006190]; C. Angelantonj and A. Sagnotti, "Open strings," Phys. Rep. 371 (2002) 1 [Erratum-ibid. 376 (2003) 339] [arXiv:hep-th/0204089].
3. J. Dai, R. G. Leigh and J. Polchinski, "New Connections Between String Theories," Mod. Phys. Lett. A4 (1989) 2073; R. G. Leigh, "Dirac–Born–Infeld Action From Dirichlet Sigma Model," Mod. Phys. Lett. A4 (1989) 2767.
4. J. Polchinski, "Dirichlet Branes and Ramond–Ramond Charges," Phys. Rev. Lett. 75 (1995) 4724 [arXiv:hep-th/9510017].
5. J. Polchinski and Y. Cai, "Consistency Of Open Superstring Theories," Nucl. Phys. B296 (1988) 91.
6. M. Bianchi and J. F. Morales, "Anomalies and tadpoles," JHEP 0003 (2000) 030 [arXiv:hep-th/0002149].
7. W. Fischler and L. Susskind, "Dilaton Tadpoles, String Condensates And Scale Invariance," Phys. Lett. B171 (1986) 383; "Dilaton Tadpoles, String Condensates And Scale Invariance. 2," Phys. Lett. B173 (1986) 262.
8. E. Dudas, G. Pradisi, M. Nicolosi and A. Sagnotti, "On tadpoles and vacuum redefinitions in string theory," arXiv:hep-th/0410101.
9. L. J. Dixon and J. A. Harvey, "String Theories In Ten-Dimensions Without Space-Time Supersymmetry," Nucl. Phys. B274 (1986) 93; N. Seiberg and E. Witten, "Spin Structures In String Theory," Nucl. Phys. B276 (1986) 272.
10. D. Fioravanti, G. Pradisi and A. Sagnotti, "Sewing constraints and nonorientable open strings," Phys. Lett. B321 (1994) 349 [arXiv:hep-th/9311183]; G. Pradisi, A. Sagnotti and Y. S. Stanev, "Planar duality in SU(2) WZW models," Phys. Lett. B354 (1995) 279 [arXiv:hep-th/9503207]; "The Open descendants of nondiagonal SU(2) WZW models," Phys. Lett. B356 (1995) 230 [arXiv:hep-th/9506014]; "Completeness Conditions for Boundary Operators in 2D Conformal Field Theory," Phys. Lett. B381 (1996) 97 [arXiv:hep-th/9603097].
11. L. R. Huiszoon, A. N. Schellekens and N. Sousa, "Klein bottles and simple currents," Phys. Lett. B470 (1999) 95 [arXiv:hep-th/9909114]. L. R. Huiszoon and A. N. Schellekens, "Crosscaps, boundaries and T-duality," Nucl. Phys. B584 (2000) 705 [arXiv:hep-th/0004100]. J. Fuchs, L. R. Huiszoon, A. N. Schellekens, C. Schweigert and J. Walcher, "Boundaries, crosscaps and simple currents," Phys. Lett. B495 (2000) 427 [arXiv:hep-th/0007174].
12. A. Sagnotti, "Some properties of open string theories," arXiv:hep-th/9509080; "Surprises in open-string perturbation theory," Nucl. Phys. Proc. Suppl. 56B (1997) 332 [arXiv:hep-th/9702093].
13. L. Alvarez-Gaume, P. H. Ginsparg, G. W. Moore and C. Vafa, "An $O(16) \times O(16)$ Heterotic String," Phys. Lett. B171 (1986) 155.
14. C. Angelantonj, "Non-tachyonic open descendants of the 0B string theory," Phys. Lett. B444 (1998) 309 [arXiv:hep-th/9810214]; R. Blumenhagen, A. Font and D. Lust, "Non-supersymmetric gauge theories from D-branes in type-0 string theory," Nucl. Phys. B560 (1999) 66 [arXiv:hep-th/9906101]; K. Förger, "On non-tachyonic $Z_N \times Z_M$ orientifolds of type-0B string theory," Phys. Lett. B469 (1999) 113 [arXiv:hep-th/9909010].

15. R. Blumenhagen and A. Kumar, "A note on orientifolds and dualities of type-0B string theory," Phys. Lett. B464 (1999) 46 [arXiv:hep-th/9906234].
16. J. M. Maldacena, "The large N limit of superconformal field theories and supergravity," Adv. Theor. Math. Phys. 2 (1998) 231 [Int. J. Theor. Phys. 38 (1999) 1113] [arXiv:hep-th/9711200]; S. S. Gubser, I. R. Klebanov and A. M. Polyakov, "Gauge theory correlators from non-critical string theory," Phys. Lett. B428 (1998) 105 [arXiv:hep-th/9802109]; E. Witten, "Anti-de Sitter space and holography," Adv. Theor. Math. Phys. 2 (1998) 253 [arXiv:hep-th/9802150]. See also O. Aharony, S. S. Gubser, J. M. Maldacena, H. Ooguri and Y. Oz, "Large N field theories, string theory and gravity," Phys. Rep. 323 (2000) 183 [arXiv:hep-th/9905111] for a review.
17. C. Angelantonj and A. Armoni, "Non-tachyonic type 0B orientifolds, non-supersymmetric gauge theories and cosmological RG flow," Nucl. Phys. B578 (2000) 239 [arXiv:hep-th/9912257]. C. Angelantonj and A. Armoni, "RG flow, Wilson loops and the dilaton tadpole," Phys. Lett. B482 (2000) 329 [arXiv:hep-th/0003050].
18. For a review, see A. Armoni, M. Shifman and G. Veneziano, "From super-Yang-Mills theory to QCD: Planar equivalence and its implications," arXiv:hep-th/0403071.
19. S. Kachru, J. Kumar and E. Silverstein, "Vacuum energy cancellation in a non-supersymmetric string," Phys. Rev. D59 (1999) 106004[arXiv:hep-th/9807076].
20. J. A. Harvey, "String duality and non-supersymmetric strings," Phys. Rev. D59 (1999) 026002 [arXiv:hep-th/9807213].
21. S. Kachru and E. Silverstein, "On vanishing two loop cosmological constants in non-supersymmetric strings," JHEP 9901 (1999) 004 [arXiv:hep-th/9810129]; R. Iengo and C. J. Zhu, "Evidence for non-vanishing cosmological constant in non-SUSY superstring models," JHEP 0004 (2000) 028 [arXiv:hep-th/9912074].
22. R. Blumenhagen and L. Görlich, "Orientifolds of non-supersymmetric, asymmetric orbifolds," Nucl. Phys. B551 (1999) 601 [arXiv:hep-th/9812158].
23. C. Angelantonj, I. Antoniadis and K. Forger, "Non-supersymmetric type-I strings with zero vacuum energy," Nucl. Phys. B555 (1999) 116 [arXiv:hep-th/9904092].
24. M. Berkooz, M. R. Douglas and R. G. Leigh, "Branes intersecting at angles," Nucl. Phys. B480 (1996) 265 [arXiv:hep-th/9606139].
25. C. Bachas, "A Way to break supersymmetry," arXiv:hep-th/9503030. R. Blumenhagen, L. Görlich, B. Körs and D. Lüst, "Noncommutative compactifications of type-I strings on tori with magnetic background flux," JHEP 0010 (2000) 006 [arXiv:hep-th/0007024]; C. Angelantonj, I. Antoniadis, E. Dudas and A. Sagnotti, "Type-I strings on magnetised orbifolds and brane transmutation," Phys. Lett. B489 (2000) 223 [arXiv:hep-th/0007090].
26. A. M. Uranga, "Chiral four-dimensional string compactifications with intersecting D-branes," Class. Quantum Grav. 20 (2003) S373 [arXiv:hep-th/0301032]; D. Cremades, L. E. Ibanez and F. Marchesano, "More about the standard model at intersecting branes," arXiv:hep-ph/0212048; D. Lüst, "Intersecting brane worlds: A path to the standard model?," Class. Quantum Grav. 21 (2004) S1399 [arXiv:hep-th/0401156].
27. C. Angelantonj, R. Blumenhagen and M. R. Gaberdiel, "Asymmetric orientifolds, brane supersymmetry breaking and non-BPS branes," Nucl. Phys. B589 (2000) 545 [arXiv:hep-th/0006033].
28. C. Angelantonj, E. Dudas and J. Mourad, "Orientifolds of string theory Melvin backgrounds," Nucl. Phys. B637 (2002) 59 [arXiv:hep-th/0205096]; C. Angelantonj, "Rotating D-branes and O-planes," Fortsch. Phys. 51 (2003) 646 [arXiv:hep-th/0212066].
29. D. Kutasov, J. Marklof and G. W. Moore, "Melvin models and diophantine approximation," arXiv:hep-th/0407150.
30. J. Scherk and J. H. Schwarz, "How To Get Masses From Extra Dimensions," Nucl. Phys. B153 (1979) 6; "Spontaneous Breaking Of Supersymmetry Through Dimensional Reduction," Phys. Lett. B82 (1979) 60.
31. R. Rohm, "Spontaneous Supersymmetry Breaking In Supersymmetric String Theories," Nucl. Phys. B237 (1984) 553; C. Kounnas and M. Porrati, "Spontaneous Supersymmetry Breaking In String Theory," Nucl. Phys. B310 (1988) 355; S. Ferrara, C. Kounnas, M. Porrati and F. Zwirner, "Superstrings With Spontaneously Broken Supersymmetry And Their Effective Theories," Nucl. Phys. B318 (1989) 75.
32. H. Itoyama and T. R. Taylor, "Supersymmetry Restoration In The Compactified $O(16) \times O(16)'$ Heterotic String Theory," Phys. Lett. B186 (1987) 129.

33. I. Antoniadis, E. Dudas and A. Sagnotti, "Supersymmetry breaking, open strings and M-theory," Nucl. Phys. B544 (1999) 469 [arXiv:hep-th/9807011]; I. Antoniadis, G. D'Appollonio, E. Dudas and A. Sagnotti, "Partial breaking of supersymmetry, open strings and M-theory," Nucl. Phys. B553 (1999) 133 [arXiv:hep-th/9812118].

34. S. Sugimoto, "Anomaly cancellations in type I D9-$\overline{\text{D9}}$ system and the USp(32) string theory", Prog. Theor. Phys. 102 (1999) 685 [arXiv:hep-th/9905159]; I. Antoniadis, E. Dudas and A. Sagnotti, "Brane supersymmetry breaking," Phys. Lett. B464 (1999) 38 [arXiv:hep-th/9908023]. For a review of the main features of BSB and its field theory implications, see C. Angelantonj, "Aspects of supersymmetry breaking in open-string models," Fortsch. Phys. 50 (2002) 735.

35. C. Angelantonj, I. Antoniadis, G. D'Appollonio, E. Dudas and A. Sagnotti, "Type I vacua with brane supersymmetry breaking," Nucl. Phys. B572 (2000) 36 [arXiv:hep-th/9911081].

36. M. Bianchi, PhD Thesis (1992); A. Sagnotti, "Anomaly cancellations and open string theories," arXiv:hep-th/9302099; G. Zwart, "Four-dimensional $N = 1$ $Z_N \times Z_M$ orientifolds," Nucl. Phys. B526 (1998) 378 [arXiv:hep-th/9708040].

37. E. Dudas and J. Mourad, "Consistent gravitino couplings in non-supersymmetric strings," Phys. Lett. B514 (2001) 173 [arXiv:hep-th/0012071]; G. Pradisi and F. Riccioni, "Geometric couplings and brane supersymmetry breaking," Nucl. Phys. B615 (2001) 33 [arXiv:hep-th/0107090].

38. I. Antoniadis, "A Possible New Dimension at a Few TeV," Phys. Lett. B246 (1990) 377; N. Arkani-Hamed, S. Dimopoulos and G. Dvali, "The Hierarchy Problem and New Dimensions at a Millimetre," Phys. Lett. B429 (1998) 263, [arXiv:hep-ph/9803315]; I. Antoniadis, N. Arkani-Hamed, S. Dimopoulos and G. R. Dvali, "New dimensions at a millimetre to a Fermi and superstrings at a TeV," Phys. Lett. B436 (1998) 257 [arXiv:hep-ph/9804398].

39. M. Borunda, M. Serone and M. Trapletti, "On the quantum stability of type IIB orbifolds and orientifolds with Scherk-Schwarz SUSY breaking," Nucl. Phys. B 653 (2003) 85 [arXiv:hep-th/0210075]; I. Antoniadis, K. Benakli, A. Laugier and T. Maillard, "Brane to bulk supersymmetry breaking and radion force at micron distances," Nucl. Phys. B662 (2003) 40 [arXiv:hep-ph/0211409].

40. C. Angelantonj and R. Blumenhagen, "Discrete deformations in type I vacua," Phys. Lett. B473 (2000) 86 [arXiv:hep-th/9911190].

41. M. Bianchi, G. Pradisi and A. Sagnotti, "Toroidal compactification and symmetry breaking in open string theories," Nucl. Phys. B376 (1992) 365; E. Witten, "Toroidal compactification without vector structure," JHEP 9802 (1998) 006 [arXiv:hep-th/9712028]; M. Bianchi, "A note on toroidal compactifications of the type I superstring and other superstring vacuum configurations with 16 supercharges," Nucl. Phys. B528 (1998) 73 [arXiv:hep-th/9711201].

42. C. Angelantonj, "Comments on open-string orbifolds with a non-vanishing B_{ab}," Nucl. Phys. B566 (2000) 126 [arXiv:hep-th/9908064].

43. M. Bianchi and A. Sagnotti in [1].

44. A. Sagnotti, "A Note on the Green–Schwarz mechanism in open string theories," Phys. Lett. B294 (1992) 196 [arXiv:hep-th/9210127].

45. C. Angelantonj and A. Sagnotti, "Type-I vacua and brane transmutation," arXiv:hep-th/0010279; R. Blumenhagen, B. Kors and D. Lust, "Type I strings with F- and B-flux," JHEP 0102 (2001) 030 [arXiv:hep-th/0012156].

46. C. Angelantonj and I. Antoniadis, "Suppressing the cosmological constant in non-supersymmetric type-I strings," Nucl. Phys. B676 (2004) 129 [arXiv:hep-th/0307254].

47. M. Bianchi and A. Sagnotti, "Open Strings And The Relative Modular Group," Phys. Lett. B231 (1989) 389; I. Antoniadis and T. R. Taylor, "Topological masses from broken supersymmetry," Nucl. Phys. B695 (2004) 103 [arXiv:hep-th/0403293].

48. C. Angelantonj and M. Cardella, "Vanishing perturbative vacuum energy in non-supersymmetric orientifolds," Phys. Lett. B595 (2004) 505 [arXiv:hep-th/0403107].

Flux and energy modulation of iron emission due to relativistic effects in NGC 3516

Giovanni Miniutti

Institute of Astronomy, University of Cambridge, Cambridge CB3 0HA, UK

Abstract. We report results from X-ray observations of the Seyfert galaxy NGC 3516. We detect an emission feature at 6.1 keV in the time-averaged spectrum, confirming previous results. By applying a new analysis technique, we find that the feature varies systematically in flux at intervals of 25 ks. The peak moves in energy between 5.7 keV and 6.5 keV. The spectral evolution of the feature agrees with Fe K emission arising from a spot on the accretion disc, illuminated by a corotating flare located between 7 and 16 gravitational radii from the central black hole, and modulated by Doppler and gravitational effects as the flare orbits around in the accretion disc. Combining the orbital time and the location of the orbiting flare, the mass of the central hole is estimated to be $(1–5) \times 10^7\ M_\odot$, which is in agreement with values obtained from other techniques.

IRON EMISSION FROM ACCRETING BLACK HOLES

The X-ray spectra of Active Galactic Nuclei (AGN) often exhibit a narrow emission line at 6.4 keV which is unambiguously identified with Fe fluoresent $K\alpha$ emission from fairly neutral material. The Fe line is emitted by reflection from matter illuminated by the primary X-ray continuum (see e.g. Ref. [5]). A narrow Fe $K\alpha$ emission line is expected in all cases in which the matter responsible for the production of the reflection component is located far from the central black hole, so that orbital motion is too slow to significantly affect the line profile. One example is given in the left panel of Fig. 1 (the Seyfert 2 galaxy Mrk 3 [2]) in which the Fe line is believed to be emitted from the molecular torus surrounding the active nucleus at distances of the order of the parsec.

On the other hand, AGN (and black hole binaries) are powered by accretion onto the central black hole, and the accreting matter is thought to form an accretion disc orbiting the black hole. In principle, its inner radius can be as close to the black hole as the marginal stable circular orbit ($\sim 1.23\ r_g$ where $r_g = GM/c^2$ if the black hole is maximally spinning). Matter in the accretion disc can also act as a reflector producing the Fe line: however, in this case, the reflection spectrum is strongly distorted by Doppler and Gravitational redshifts in the immediate vicinity of the central black hole. Each radius of the accretion disc produces a *symmetric* double-horned line profile corresponding to emission from material on the approaching and receding sides (with respect to the observer) of the disc. Near the central black hole, where the orbital velocities are mildly/highly relativistic, special relativistic beaming enhances the high-energy peak of the line with respect to the low-energy one. Finally, transverse Doppler effect and gravitational redshift manage to shift the emission from each radius to lower energy. Summing the contribution from all radii in the accretion disc gives a characteristic *skewed*,

CP751, *General Relativity and Gravitational Physics, 16th SIGRAV Conference*, edited by G. Esposito et al.

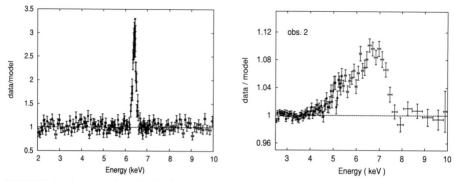

FIGURE 1. **Left panel:** the Fe Kα line observed in the Seyfet 2 galaxy Mrk 3. The line is narrow, as expected if emitted from matter far away from the central black hole. **Right panel:** the Fe line detected in the black hole binary XTE J1650-500. The line is clearly broad and agrees with emission from the accretion disc around a rapidly rotating Kerr black hole.

asymmetric and *broad* line profile.

The profile and variability of broad Fe lines are a powerful tool to probe the geometry of the accretion flow within few r_g from the central black hole. The high-energy (blue) extent of the line is almost completely a function of the observer inclination and provides a quite robust way to observationally measure the inclination of the disc with respect to the observer. On the other hand, the low-energy (red) extent of the line is sensitive to the inner disc radius and, if this is assumed to coincide with the innermost stable circular orbit, provides information on the black hole spin parameter. The most remarkable case of a broad Fe line in an AGN is that of the Seyfert galaxy MCG–6-30-15, where the line is so broad to require emission from as close as $2\ r_g$ from the black hole strongly suggesting the presence of a rapidly rotating Kerr black hole [8, 4]. Another example is that of the black hole binary XTE J1650–500 whose broad Fe line has been observed first with XMM–Newton [12]. In the right panel of Fig. 1 we show the broad line of XTE J1650–500 as observed with the Italian-Dutch X-ray mission BeppoSAX [15]. Moreover, the variability properties of the broad line with respect to the continuum in both objects can be explained by requiring that most of the emission results from within a few gravitational radii from the black hole (see [13, 14, 15] for more details). This opens the exciting prospect of probing the strong field regime of General Relativity via X-ray observations of accreting black holes. Present data are telling us that (in some cases) we are already starting to see relativistic effects on the spectra and variability of such X-ray sources.

Emission lines observed at peculiar energies

In addition to the major line emission around 6.4 keV, transient emission features at energies lower than 6.4 keV are sometimes observed in X-ray spectra of AGN. An early example was found in the ASCA observation of MCG–6-30-15 [9]. More examples followed in recent years with improved sensitivity provided by XMM-Newton and Chandra X-ray Observatory. The most remarkable cases are those of NGC 3516 [20, 1, 10], NGC 7314 [22], ESO 198–G24 [6, 1], Mrk 766 [21], and ESO 113–G10 [18]. In all cases, the energy at which these features are observed lies below 6.4 keV and does not correspond to any atomic transition with large enough probability to occur and therefore to be detected in the X-ray spectra of AGN. The most natural explanation is that these features are Fe Kα emission lines that have been redshifted kinematically and/or gravitationally to the observed energies.

One exciting, though not unique, possibility is that these features are Fe Kα lines emitted in the inner regions of the accretion disc from a spot illuminated by a localized flare above it. If the illuminated spot is close to the central black hole, the line emission is redshifted, depending on the location of the spot on the disc. This simple model has the great advantage of making definite and therefore falsifiable predictions on the short timescale variability of such lines. In particular, since the flare is most likely linked to the disc via magnetic fields, the usual assumption is that the flare corotates with the disc. Therefore, if the flare lasts for more than one orbital period, periodic modulations on both the flux and the energy of the emitted lines should be observed, as the illuminated spot on the disc orbits around the black hole [16, 3].

The Keplerian orbital time around a $10^7 \, M_\odot$ black hole (typical for Seyfert galaxies) is $\sim 10^4$ s at a radius of 10 r_g. Some of the long XMM-Newton observations last for $\sim 10^5$ s without interruptions, unlike low-orbit satellites such as ASCA. Therefore a few cycles of periodic modulations of line emission induced by an orbiting spot can occur within an XMM-Newton long observation. However, it is generally assumed that the sensitivity of currently available X-ray instruments is not appropriate for detecting the relativistic effects induced by the orbital motion close to the black hole. Here, we describe a new analysis technique, devised to search for temporal evolution of the Fe K line, and its application to one of the XMM-Newton datasets in which such a redshifted emission line has been detected.

THE CASE OF NGC 3516

We selected one of the XMM-Newton observations of the bright Seyfert galaxy NGC3516, for which Bianchi et al (2004a) reported excess emission at around 6.1 keV in addition to a stronger 6.4 keV Fe Kα line in the time-averaged spectrum. Our main goal was to study the variability of this 6.1 keV feature to shed light on its origin. The broad-band X-ray spectrum of NGC3516 is very complex, as a result of modification by absorption and reflection. Since our interest is on the behavior of the 6.1 keV feature only, we designed our analysis method as follows to avoid unnecessary complication: i) the energy band is restricted to 5.0–7.1 keV, which is free from absorption effects at energies below and above; ii) the continuum is determined by fitting an absorbed power-law to the data excluding the line band (6.0–6.6 keV), and is subtracted to obtain excess emission which is then corrected for the detector response and recorded (see Ref. [10] for details).

We first investigated the excess emission at resolutions of 5 ks in time and 100 eV

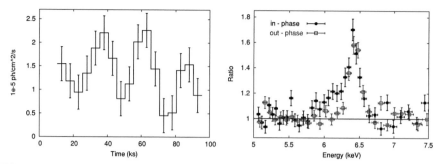

FIGURE 2. **Left panel:** the light curve of the red feature extracted from the filtered image of the excess emission. **Right panel:** the line profiles obtained from the 'on' and 'off' phases. Excess emission around 6.1 keV is detected only in the 'on' phase at the 4σ level.

in energy. An image of the excess emission in the time-energy plane is constructed from individual time intervals. The image suggests the presence of a strong and nearly constant 6.4 keV line which is easily interpreted as Fe Kα emission from matter far from the black hole. However, a feature is also present between 5.8 keV and 6.2 keV which corresponds to the 6.1 keV line detected in the time-averaged spectrum (hereafter this redshifted line will be referred to as 'red feature') and the image suggests short timescale variability. Low pass filtering through weak Gaussian smoothing was applied to the image to suppress random noise between neighbouring pixels. This image filtering brings out the systematic variations of the red feature more clearly. We then extracted light curves from the filtered image in the 5.8–6.2 keV energy band which is the relevant one for the red feature. Error bars were computed by performing extensive simulations as described by Iwasawa, Miniutti & Fabian (2004).

The resulting light curve is shown in the left panel of Fig. 2. The red feature apparently shows a recurrent behavior characterized by 'on' and 'off' phases. The 'on' phases appear at intervals of $\simeq 25$ ks for nearly four cycles. In contrast, the 6.4 keV line core remains largely constant (not shown here). However, less than four cycles cannot secure any periodicity in the red feature with high significance. Whether noise can produce spurious periodic variations as observed at a significant probability has been investigated through extensive simulations (see Ref [10] for details). When compared to measurment error and random noise, the cyclic behavior was found to be highly significant (about 99.8 per cent). Even by assuming a very pessimistic situation in which the red feature light curve has a 'coloured' (e.g., red) noise spectrum with a nearly unplausibly large r.m.s. amplitude of 35 per cent[1], testing our data against simulations revealed a significance of about 98 per cent (see Iwasawa, Miniutti & Fabian 2004 for details).

Using the light curve (Fig. 2, left panel) as a guide, we accumulated two spectra representative of the 'on' and 'off' phases, by combining time intervals chosen in a

---

[1] The r.m.s. variability amplitude of the X-ray continuum in NGC 3516 on a 25 ks timescale is about 2-3 per cent [11].

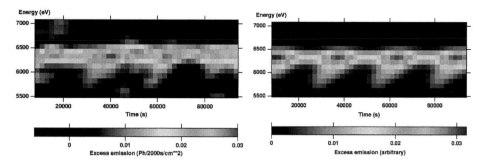

FIGURE 3. **Left panel:** observed image of the excess emission in the time-energy plane. **Right panel:** theoretical model for the image of the excess emission in the time-energy plane (see text for details).

periodic manner. The resulting line profiles are shown in the right panel of Fig. 2 as a ratio with the continuum model. While the 6.4 keV core remains similar between the two, there is a clear difference in the energy range of 5.7–6.2 keV resulting from the presence/absence of the red feature. When modelled with a Gaussian emission line, the 'on' phase spectrum exhibits a red feature at $6.13^{+0.10}_{-0.07}$ keV with a flux of $2.1^{+1.3}_{-0.8} \times 10^{-5}$ ph s^{-1} cm^{-2}. On the other hand, the red feature is not detected in the 'off' phase spectrum with an upper limit on its flux of 0.7×10^{-5} ph s^{-1} cm^{-2}. The variability detected between the two spectra is significant at the 4σ level.

Interpretation and modelling

If the red feature is due to emission from an orbiting spot on the accretion disc, we do expect to see not only a periodic modulation in flux, but also in energy. This is because, as the spot orbits the black hole, it samples different regions of the accretion disc where different Doppler shifts are imprinted. As an example, when the spot is approaching the observer, we expect a brighter and higher energy line, while when it is receding we should observe a fainter and lower energy line, because of the combined effect of relativistic Doppler and beaming effects. The image constructed from individual 5 ks spectra does not show any energy evolution. We therefore decided to investigate the feature evolution at a shorter timescale (2 ks).

The resulting (smoothed) image of the excess emission in the time-energy plane is shown in the left panel of Fig. 3 and exhibits further interesting behavior. Besides the horizontal strip (representing the stable 6.4 keV Fe line) a saw-tooth pattern appears in the red feature energy band suggesting its energy evolution with time. The red feature emerges at about 5.7 keV and then increases its energy with time to merge with the 6.4 keV line. This behavior appears to be repeated in each 'on' phase. If the characteristic variability timescale of 25 ks is interpreted as the orbital period of the emitting orbiting spot, this behavior can be easily understood. Only the bright emission from the approaching side of the disc is seen; when the spot is on the receding side of the disc, line emission is too faint to be detected (because of relativistic beaming away

19

from the observer).

To extract as much information as possible from the observed image, we adopt a simple model in which a flare is located above an accretion disc, corotating with it at a fixed radius. The flare illuminates an underlying region on the disc (or spot) which produces a reflection spectrum, including an Fe $K\alpha$ line. The disc illumination is computed by integrating the photon geodesics in a Kerr spacetime from the flare to the disc, and is converted to local line emissivity. Then, the observed emission line profile is computed through the ray-tracing technique including all special and general relativistic effects (see Miniutti et al 2003; Miniutti & Fabian 2004; Iwasawa, Miniutti & Fabian 2004 for more details). The main parameters of the model are the flare location, specified by its distance r from the black-hole axis and its height above the accretion disc (h), and the accretion-disc inclination i. We have computed the evolution of Fe K emission induced by an orbiting flare and simulated time-energy maps with the same resolution and smoothing as in the excess emission map of Fig. 3 (left panel) exploring the parameter space. A constant, narrow 6.4 keV core representing the Fe $K\alpha$ line from distant material was also added to the model.

The right panel of Fig. 5 shows one of the theoretical time-energy maps we produced. This particular example assumes a flare with $(r, h) = (9, 6)\, r_g$ and a disc inclination of $20°$. By comparing the theoretical images with the data, we estimate that the flare must be located at a radius $r = (7 - 16)\, r_g$. Since the orbital period (that we assume here to be $T = 25 \pm 5$ ks) of the corotating flare is related to its radial position by

$$T = 310 \left[a + (r/r_g)^{3/2} \right] M_7 \quad [\text{seconds}], \qquad (1)$$

where M_7 is the black hole mass in units of $10^7 M_\odot$, our results on T and r can be translated into an estimate of $M_{BH} = (1 - 5) \times 10^7\, M_\odot$ for the black-hole mass in NGC 3516.

Our result on M_{BH} is complementary to those from the reverberation mapping technique which are based on emission from clouds much far away from a central hole and unlikely to see those clouds to complete a whole orbit in a reasonable time. The estimates of M_{BH} in NGC 3516 from reverberation mapping lie in the range between $1 \times 10^7\, M_\odot$ and $4 \times 10^7\, M_\odot$. The most recent result obtained by combining the Hα and Hβ emission lines is $(1.68 \pm 0.33) \times 10^7\, M_\odot$ (Onken et al. 2003), while the previous analysis based on Hβ alone gave $2.3 \times 10^7\, M_\odot$ (Ho 1999; no uncertainty is given). It is remarkable that our estimate of the black-hole mass in NGC 3516 is in excellent agreement with the above results.

CONCLUSIONS

We study the variability of a transient emission feature around 6 keV, detection of which has been reported previously in the time-averaged X-ray spectrum of NGC3516. The feature appears to vary systematically both in flux and energy on a characteristic timescale of 25 ks. The flux and energy evolution of the red feature is consistent with Fe $K\alpha$ emission produced by an illuminated spot on the accretion disc and modulated by Doppler and gravitational effects. Modelling the observed X-ray data with a relativistic model constrains the radial location of the flare to be $r = (7 - 16) \, r_g$. This is combined with the orbital period to provide an estimate of the black-hole mass in NGC 3516 which is $M_{BH} = (1 - 5) \times 10^7 \, M_\odot$, in excellent agreement with estimates obtained with other techniques. Our results indicate that present X-ray missions such as XMM-Newton are close to probing the spacetime geometry in the vicinity of supermassive black holes if their observational capabilities are pushed to the limit. Future observatories such as *XEUS* and *Constellation-X*, which are planned to have much larger collecting area at 6 keV, will be able to exploit this potential and map the strong-field regime of general relativity with great accuracy.

ACKNOWLEDGMENTS

I am grateful to the SIGRAV for inviting me at the 2004 SIGRAV meeting in Vietri sul Mare and for the 2004 SIGRAV Prize award. The results presented here would not have been reached withouth the collaboration of Kazushi Iwasawa and Andy Fabian. I also want to thank my PhD supervisor, Valeria Ferrari: most of the GR expertise I was able to apply later to X-ray data comes from her. I thank Donato Bini, Bob Jantzen, Leonardo Gualtieri, and José Pons for helping me during my Laurea and PhD in Rome. I thank the UK's PPARC for support.

REFERENCES

1. Bianchi S. et al. 2004a, A&A, 422, 65.
2. Bianchi S., Miniutti G., Fabian A.C., Iwasawa K., Matt G., 2004b, in preparation.
3. Dovčiak M., Bianchi S., Guainazzi M., Karas V., Matt G., 2004, MNRAS, 350, 745.
4. Fabian A.C. et al., 2002, MNRAS, 335, L1.
5. George I.M., Fabian A.C., 1991, MNRAS, 249, 352.
6. Guainazzi M., 2003, A&A, 401, 903.
7. Ho L.C., 1999, in Observational Evidence for Black Holes in the Universe, 157, ed. K. Chakrabarti (Kluwer, Dordrecht).
8. Iwasawa K. et al., 1996, MNRAS, 282, 1038.
9. Iwasawa K. et al, 1999, MNRAS, 306, L19.
10. Iwasawa K., Miniutti G., Fabian A.C., 2004, MNRAS in press, astro-ph/0409293.
11. Markowitz A., et al, 2003, ApJ, 593, 96.
12. Miller J.M. et al., 2002, ApJ, 570, L69.
13. Miniutti G., Fabian A.C., Goyder R., Lasenby A.N., 2003, MNRAS, 344, L22.
14. Miniutti G., Fabian A.C., 2004, MNRAS, 349, 1435.
15. Miniutti G., Fabian A.C., Miller J.M., 2004, 351, 466.
16. Nayakshin S., Kazanas D., 2001, ApJ, 553, 885.
17. Onken C.A., Peterson B.M., Dietrich M., Robinson A., Salamanca I.M., 2003, ApJ, 585, 121.
18. Porquet D. et al., 2004, A&A, 413, 913.
19. Reynolds, C.S., *PhD thesis*, 1996, University of Cambridge.
20. Turner T.J. et al., 2002, ApJ, 574, L123.
21. Turner T.J., Kraemer S.B., Reeves J.N., 2004, ApJ, 603, 62.
22. Yaqoob T. et al., 2003, ApJ, 596, 85.

PLENARY SESSION

Gravitational Energies
and Generalized Geometric Entropy

Gianluca Allemandi*, Lorenzo Fatibene*, Marco Ferraris*, Mauro
Francaviglia* and Marco Raiteri*

*Dipartimento di Matematica, Università di Torino
Via C. Alberto 10, 10123 Torino (Italy)

Abstract. A geometrical covariant definition of the variation of conserved quantities is introduced
for Lagrangian field theories, suitable for both metric and affine gravitational theories. When this
formalism is applied to the Hilbert Lagrangian we obtain a covariant definition of the Hamiltonian
(and consequently a definition of the variation of Energy) for a gravitational system. The definition
of the variation of Energy depends on boundary conditions one imposes. Different boundary conditions are introduced to define different energies: the *gravitational heat* (corresponding to Neumann
boundary conditions) and the Brown–York quasilocal energy (corresponding to Dirichlet boundary
conditions) for a gravitational system. An analogy between the behavior of a gravitational system
and a macroscopical thermodynamical system naturally arises and relates control modes for the
thermodynamical system with boundary conditions for the gravitational system. This geometrical
and covariant framework enables one to define entropy of gravitational systems, which turns out
to be a geometric quantity with well-defined cohomological properties arising from the obstruction
to foliate spacetimes into spacelike hypersurfaces. This definition of gravitational entropy is found
to be very general: it can be generalized to causal horizons and multiple-horizon spacetimes and
applied to define entropy for more *exotic* singular solutions of the Einstein field equations. The
same definition is also well-suited in higher dimensions and in the case of alternative gravitational
theories (e.g. Chern–Simons theories, Lovelock Gravity).

BLACK HOLE ENTROPY: AN OVERVIEW

The first law of Classical Thermodynamics was first introduced by Clausius for isolated
macroscopical systems under the form

$$\delta U = T\delta S - \delta W, \qquad (1)$$

where U denotes the internal energy, T the temperature, W the work done by the system
and S the entropy of the system. One can use the first law (1) to define S classically,
provided the other quantities are known. The first (macroscopical) law of Thermodynamics gives therefore the definition of entropy, while the second principle states that
entropy should be a never decreasing quantity. The first law of Thermodynamics can be
applied to general thermodynamical systems once the thermodynamical parameters and
the control mode for the system are chosen [17]. The amount of energy which can be
transformed into work, by means of a reversible transformation, is called the free energy
F of the thermodynamical system, while the degenerate part of energy transforming into
heat is $E - F = TS$ (at least in transformations where the temperature of the initial and
final state are the same); see [17].

CP751, *General Relativity and Gravitational Physics, 16th SIGRAV Conference*, edited by G. Esposito et al.
© 2005 American Institute of Physics 0-7354-0236-1/05/$22.50

However, the definition of entropy has a deeper (microscopical) interpretation, related with information theory and the lack of information about a system. Entropy is related with that part of energy transforming into heat and hence it is related to the degenerate part of energy which cannot be converted into work anymore. The information enclosed in this degenerate part of energy ($E - F = TS$) is hidden to any observer (as it cannot be transformed anymore) and it is consequently related to the lack of information of observers on the thermodynamical system itself. In Information Theory, a fundamental definition of entropy was given by Shannon [33]: entropy of a stream of characters is related with the probability of characters to occur. Shannon's entropy is a *positive definite, additive* quantity, which vanishes when the stream is constant and is maximal for completely random sequences. Despite Shannon's entropy has not a direct application to thermodynamical systems, most of physical entropies have been recognized to be a by-product of this definition.

When we deal with *fundamental theories* for physical systems, such as statistical mechanics and quantum (statistical) mechanics, we obtain information about the microstates of the system. Within this framework, the same macrostate of the system (once the measurable parameters are fixed) can be realized by a set of microscopical states, which is called *ensemble*. Under suitable hypotheses (see for example [27]) the counting of microstates realizing the ensemble gives the probability of a macroscopical state to occur and thus allows to introduce a Shannon-like definition of entropy for a physical system. This entropy is thus related to the lack of information of an observer on the microscopical structure of the system and on its evolution. The more information we have about the microscopical states of a system, the more energy we are able to extract from it: this explains the link with the classical (*á la Clausius*) definition of entropy.

When specializing to statistical mechanics or to the more general case of quantum statistical mechanics, the definition of entropy is thus strictly related to the probability function and to the statistical (probability) operator respectively [27], thus reproducing the famous Boltzmann formula

$$S = -\kappa \ln Z(\beta_i), \tag{2}$$

where Z represents the partition function of the system (κ is the Boltzmann constant), while β_i are the control parameters of the thermodynamical system considered [17], [27].

J.D. Bekenstein was the first to state that a formal first law (of thermodynamics) should hold for stationary black holes [4], noticing an analogy between the geometrical properties of a black hole and the thermodynamical quantities appearing in the first law of thermodynamics. Moreover, as much as in Shannon's approach, the starting point for the definition of entropy of black holes was the lack of information of an external observer on the internal degrees of freedom of the black hole, behaving like an ensemble. It was stated by Bekenstein that the entropy of the (stationary) black hole should be proportional to the area A of its horizon, which is the one-way membrane hiding the information inside it to outer observers.

The variational equation (the first law of black hole thermodynamics) for the macroscopical (thermodynamical) quantities of the black hole is (see [4], [12], [29] for details)

$$\delta M = T\delta S + \Omega \delta J + b^a \delta Q_a, \tag{3}$$

where $T = \frac{\kappa_H}{2\pi}$ is the temperature of the black hole (here κ_H is the surface gravity on the horizon H); Ω is the angular velocity of the horizon; M is the mass of the black hole solution; J represents its angular momentum and $S = \frac{A_H}{4}$ is its entropy (A_H is the area of the horizon); the parameters b^a are the Lagrangian multipliers for possible gauge charges Q_a. At this level, however, this is just a formal analogy and it should be suitably endowed with physical significance by means of the identification of physical processes, corresponding to thermodynamical transformations of the black hole.

As we stated before the *fundamental* definition of entropy passes through a microstates counting of the system and thus (for BH) through a quantum theory of gravity, but our knowledge about this topic is nowadays only preliminary. There are however some efforts and formulations, for example in the framework of quantum geometry and string theory [11]. Many efforts at an intermediate level, i.e. semi-classically, have also been done in preliminary theories of quantum gravity [23]. An interesting derivation of the first principle (3) has also been obtained for spatially confined systems [7], based on a statistical approach to the gravitating system considered as a micro-canonical ensemble. It is worth noticing that the definition of black hole entropy obtained in the framework of quantum statistical theories of gravity (should) reproduce the area law for stationary black holes in the classical limit. These results show that it is worth investigating the relations between black hole dynamics (geometry) and thermodynamics also at a classical level, in a framework as general as possible, which could at least provide the low-energy limit of the *fundamental* definition of entropy. This leads us to the belief that the *geometrical interpretation of black hole entropy* and its generalization to more general singular spacetimes at a classical level is still worth being studied nowadays.

The first law (3) has, first of all, a macroscopical significance as it is related to the macroscopical parameters of the theory as much as in the Clausius formulation of Thermodynamics. A black hole can be considered as a macroscopical (thermodynamical) system: this fact is supported by the *no-hair theorem* stating that the status of the black hole (after collapse) is described by very few parameters. Any state function, such as entropy, should be a function of these parameters. This implies that entropy, mass and angular momentum (and possibly others charges) for the solution can be analyzed in the macroscopical approach which follows from the Noether theorem applied to the gravitational system, without reference to the microscopical behavior. The parameters T and Ω are assumed in this framework to be external parameters of the macroscopical thermodynamical theory, which have to be obtained by means of physical considerations on the geometry and the dynamics of the black hole. Temperature is thus calculated by using the Hawking radiation effect for the black hole and it corresponds to Hawking's temperature [23]. This could appear as a contradiction, by virtue of the quantum nature of Hawking's radiation; however, this is analogue with Classical Thermodynamics, where temperature has to be measured independently.

A preliminary proposal to define the entropy of black holes via the Noether theorem has been first formulated in [29], where it has been applied to solutions admitting a bifurcate Killing horizon and a flat asymptotical behavior. However this (non covariant) definition suffers from the drawback that it cannot be easily generalized to the case of non-stationary black holes and to non-trivial spacetimes.

On extending the approach of [29], the definition of black hole entropy via the Noether theorem was generalized by some of us in [12] in a geometrical and fully covariant framework. Moreover, some significant mathematical properties hold for this new and more general definition. Basically it does not require the solution to admit a bifurcate Killing horizon, nor to have any specified asymptotical behavior [12]. Furthermore, this definition can be applied to a much wider class of black-hole solutions [13] and it can be generalized also to spacetimes admitting (multiple causal) horizons [22] and to more *exotic* spacetimes (e.g. Taub-bolt) [14], [21]. This definition of entropy has been shown to be independent of the specific Lagrangian formulation of the theory and a fully covariant Hamiltonian formalism has been developed to define conserved quantities and entropy, by linking the definitions of the thermodynamical potentials with the boundary conditions for the gravitational system [19], [22]. Even more, this formulation admits applications to higher-dimensional and higher-order theories of gravity [2]. We will try in the following to give an overview of the most striking applications of this important geometrical definition of entropy for gravitational systems.

Geometrical definition of Black Hole entropy

The entropy of a stationary black hole solution is defined as that macroscopical quantity which satisfies the first principle of thermodynamics (3) and it is therefore related to the conserved charges of the black hole solution by means of an integrating factor (temperature).

The problem can thus be divided into two parts: the calculation of conserved quantities (energy, angular momentum...) for the spacetime under analysis and the identification of temperature for the singularities present in that spacetime. As we already explained before, in this macroscopical and classical approach the temperature for the singularities of spacetime has to be provided by means of physically-motivated reasons. Basically it is identified with the surface gravity κ_H for causal horizons $T = \frac{\kappa_H}{2\pi}$ and it corresponds to Hawking's temperature for black hole solutions or for cosmological horizons, where κ_H is the surface gravity. Temperature can be equivalently interpreted as the inverse of the period of Euclidean time (when dealing with spacetimes with single horizon) [7], [23]. When multiple singularities (horizons) are present in the spacetime under analysis, temperature assumes a *local* meaning on each horizon [31].

Provided this definition of temperature has been given, we shall focus on the definition of conserved quantities and their properties. According to our view, each theory is required to be covariant: this means that conserved quantities have to be defined in a covariant framework, so that they can have an immediate geometrical and global interpretation [12], [19].

A field theory on a m-dimensional spacetime M (with local coordinates $\{x^\mu\}$) is constructed on a configuration bundle $\pi : C \to M$ with fibered coordinates $\{x^\mu, y^i\}$. The configuration of the physical system near a point $x \in M$ is determined by a local section $\sigma \in \Gamma_{loc}(U)$, $U \subset M$ (where $x \in U$) of the configuration bundle C and in local coordinates it reads as $\sigma : x^\mu \mapsto (x^\mu, y^i = \sigma^i(x))$. Therefore the space of fields will be identified by all (local) sections of the configuration bundle. Field equations dictate how the configuration of the physical system evolves. A *Lagrangian of order* k is a bundle morphism between the k-jet bundle[1] $J^k C$ of the configuration bundle C and the bundle $A_m(M)$ of m-forms on spacetime [18], [34]:

$$L : J^k C \to A_m(M). \tag{4}$$

In local coordinates

$$L = \mathscr{L}(x^\mu, y^i, y^i_\mu, \ldots, y^i_{\mu_1 \ldots \mu_k}) \, dx^1 \wedge \ldots dx^m. \tag{5}$$

We consider a vertical vector field $X = \delta y^i \frac{\partial}{\partial y^i}$ on C (i.e. a variation of fields); each section of C can be Lie-dragged along the flow of X defining a global bundle morphism δL, which is the *Lagrangian variation* and splits into a morphism called the Euler–Lagrange morphism and the so-called Poincaré–Cartan morphism(s)[2] (see a detailed explanation of this formalism in [15]):

$$< \delta L \mid j^k X > = < \mathbb{E}(L) \mid X > + \mathrm{Div} < \mathbb{F}(L) \mid j^{k-1} X >, \tag{6}$$

where $< \cdot \mid \cdot >$ denotes the natural duality between the vertical bundle $V(J^k C)$ and $V^*(J^k C)$; see [15]. Field equations can be obtained by means of Hamilton's principle, requiring the variation of the action functional $A = \int_D L \circ j^k \sigma$ to vanish (D is assumed to be a compact region in M and $j^k \sigma$ is the k-jet prolongation of σ). On imposing $j^{k-1} X \mid_{\partial D} = 0$, the vanishing of the Euler–Lagrange morphism provides the field equations, while the Poincaré–Cartan morphism vanishes on ∂D by Stokes' theorem and boundary conditions.

For gauge-natural theories[3] (and for natural theories as a particular case, e.g. for General Relativity) it is possible to construct a vector bundle, whose sections Ξ (the infinitesimal generators of symmetries, projecting onto vector fields ξ on M) are in one-to-one correspondence with the symmetries of the theory.

The Noether theorem for geometrical field theories (see e.g. [15], [18]) makes it possible to associate to each (Lagrangian) symmetry on spacetime a conserved current $\mathscr{E}(L, \Xi, \sigma)$, which is covariantly conserved on-shell, i.e.

$$\mathrm{Div}(\mathscr{E}(L, \Xi, \sigma)) = 0. \tag{7}$$

The Noether current naturally splits into a bulk part, which is exactly vanishing on shell, and a boundary part, which is the superpotential of the theory [15], [18]. The superpotential leads to the natural definition of conserved charges, when integrated on

---

[1] For a detaild theory of jet bundles and their application to field theories see [15].

[2] This property is related with the geometrical properties of the definitions of bundle morphisms and in particular with Spencer's cohomology; Poincaré–Cartan morphisms form a family, while the Euler–Lagrange morphism is unique, see [15].

[3] For an introduction to gauge-natural theories, see [15].

29

the boundary of (a part of) a spacelike hypersurface. However conserved quantities, simply defined by means of the superpotential, are usually afflicted by the *anomalous factor problem* (see [18], [20] and references therein). This is related to the fact that in classical (field) theories the concept of vacuum is not naturally defined, which is instead deeply related with the quantum (field) theory [15]. The concept of zero point for conserved quantities, as much as the introduction of a reference background (which are indeed strictly related) has to be introduced *by hand* in classical field theories. We shall therefore better define finite conserved quantities relative to different field configurations and define in this way the conserved quantities of a solution with respect to a suitably chosen reference background (which can be arbitrarily assumed as a zero point for conserved quantities). Within this framework it is possible to proceed by following two different ways: we can introduce the reference background into a redefined covariant Lagrangian (the so-called *first-order Lagrangian* for General Relativity), such that we obtain immediately the corrected relative conserved quantities [20], [21]. Otherwise it is possible to define the variation of conserved quantities, resulting from the Noether theorem [20], [29] or directly from field equations [19]. This latter definition postpones the problem of choosing a background solution and it is much closer in spirit with the symplectic and Hamiltonian formalism for field theories.

Despite this technical difference, the meaning of the variation of conserved quantities is not different from the reference background framework. As a matter of fact, the variation of conserved quantities is interpreted as the infinitesimal generator along a family of solutions. Such a prescription hence represents the relative infinitesimal amount of conserved quantities needed to deform infinitesimally an unperturbed solution $\bar{\sigma}$ (to obtain, roughly speaking, $\sigma = \bar{\sigma} + \varepsilon \delta \sigma$). The unperturbed solution $\bar{\sigma}$ enters the framework exactly as a reference background. This analogy between the two approaches can be shown to be rigorous (see [16]). The variation $\delta_X Q$ of the conserved quantity relative to an infinitesimal symmetry Ξ is

$$\delta_X Q(\Xi) = \int_B [\delta_X U(L, \Xi) - i_\xi < \mathbb{F}(L) \mid j^{k-1} X >], \qquad (8)$$

where ξ is the projection onto M of Ξ and B is assumed to be the outer boundary of a spacelike hypersurface Σ; here X denotes a vertical vector field on C, as stated before. Note that (8) gives $\delta_X Q$ but not Q; the integrability properties will be discussed in Section II. This formula has remarkable properties: it is independent of the addition of divergence terms to the Lagrangian; it is independent of a specific representation of the boundary B, inside the same homology class (i.e. the integrated form in (8) is a closed form; see [12]), under the hypotheses that $\pounds_\xi \sigma = 0$, i.e. ξ is a Killing vector for the solution σ. The above formula is moreover linear in the infinitesimal generator of symmetries Ξ and its covariant derivatives up to a fixed finite order, which is determined by the theory.

In gravitational field theories, described by the Hilbert Lagrangian, the variation of mass and the angular momentum of a stationary black hole are *naively* obtained by means of (8) as the conserved quantities related to the timelike Killing vector field ξ_m and the rotational Killing vector field ξ_J over spacetime, respectively. The definition of mass and angular momentum implies that the boundary B of the integration region can be pushed to a boundary homologically equivalent to spatial infinity in order to calculate the conserved quantities of the whole spacetime [12], [29], [32].

On substituting in (3) the expression for mass, angular momentum and charge following from (8), and recalling that U is linear in Ξ, we obtain the definition of the variation of entropy for a stationary black hole:

$$
\begin{aligned}
\delta_X S(L_H, \Xi, \sigma) &= \tfrac{1}{T}(\delta M - \Omega \delta J - b_a \xi^a) \\
&= \tfrac{1}{T} \int_B [\delta_X U(L_H, \Xi, g) - i_\xi < \mathbb{F}(L_H) \circ j^{2k-1} g | j^{k-1} X >],
\end{aligned}
\tag{9}
$$

where the vector field $\Xi = \xi_m + \Omega \xi_J + b^a \xi_a$ projects over the vector $\xi = \xi_m + \Omega \xi_J$ on spacetime and B is homologically equivalent to spacelike infinity. No assumption is required on the asymptotical behavior of the solutions involved; the boundary B is a generic $(m-2)$-surface embedded into spacetime, which is just required to be cohomological to spacelike infinity (in the case of stationary black holes); formula (9) is independent of boundary terms added to the Lagrangian. The definition of entropy obtained in this covariant framework has been successfully applied to a wide number of solutions (besides the *trivial* cases): such as the BTZ solution is 3-dimensional spacetimes [13] and in the framework of Chern–Simons theories [2], the Taub-bolt solution (we remark that in this case the topology of spacetime in highly non-trivial) [14]. We shall analyze in the following the integrability conditions for the above formula (9).

Energy and Boundary conditions

The famous paper [3] written by Arnowitt, Deser and Misner was the cornerstone of a wide literature developed during the last decades about the Hamiltonian approach to conserved quantities; see e.g. [5], [6], [7]. The Hamiltonian structure for General Relativity is identified by projecting the dynamical fields onto a spacelike hypersurface Σ with an induced metric h defined on it and writing field equations over this surface. This formulation is related to the initial-value problem (the surface Σ has to be chosen to be a spacelike Cauchy surface) and with boundary conditions over Σ. Different boundary conditions (which correspond to different choices of the control mode of the physical system [30]) lead to the definition of different Hamiltonians and consequently to the definition of different energies for the gravitational system (defined as the on-shell value for the Hamiltonians). The bulk terms in the Hamiltonian vanish by virtue of the constraint equations, and the energy, or quasi-local energy, is a pure boundary term evaluated on a $(m-2)$-dimensional surface $\partial\Sigma$; see [7], [10], [24]. This result can be seen as a generalization of the Gauss theorem, in strict analogy with what we found before in a Lagrangian framework. We stress that boundary terms in the Hamiltonian are suitably chosen to match the variational principle with the *a priori* assigned boundary conditions. Therefore the energy of the system depends on the choice of boundary conditions.

From a physical viewpoint it was noticed that a gravitational system in thermal equilibrium must feature a finite spatial extent: a system of infinite spatial extent at fixed temperature is thermodynamically unstable (see e.g. [8]). A large number of definitions of energy has been consequently given for spatially bounded gravitational systems, where boundary conditions are imposed on the worldtube boundary [7]. We follow a Noether-like approach to conserved quantities, where the Hamiltonian is simply defined as the integral of the Noether current relative to a transversal (timelike) vector field over (portion of) a Cauchy surface. We perform a Regge–Teitelboim-like analysis of the variation of conserved quantities obtained by means of the Noether theorem, which allows to handle boundary terms in such a way that quasilocal Noether charges can be defined avoiding the anomalous factor problem. Once the variation δQ of the corrected conserved quantity is defined, it remains to be seen whether the variation δQ can be integrated to give (apart from a constant of integration) the conserved quantity Q in relation with boundary conditions. When δQ is integrable we shall obtain the conserved quantity $Q - Q_0$ up to an integration constant Q_0. The latter can be fixed as a zero-level for the conserved quantity or, in other words, as a background reference. Moreover, it is important to notice that for thermodynamical systems there exist different kinds of energy (such as the internal energy, the free energy, ...), each one corresponding to different choices of boundary conditions and/or of control variables.

We start by considering a region D in spacetime, foliated by spacelike hypersurfaces Σ_t, with $(m-2)$-dimensional boundary B_t and we set $\mathscr{B} = \cup_t B_t$; see [7], [25]. We denote by u^μ the future-directed unit normal to Σ_t and we denote by n_μ the outward pointing unit normal of B_t in Σ_t. The *time evolution field* ξ in D is defined through the (local) rule $\xi^\mu \nabla_\mu t = 1$ and, on the boundary $\mathscr{B}$, it is tangent to the boundary itself and can be decomposed as

$$\xi^\mu = N u^\mu + N^\mu, \tag{10}$$

where N is the *lapse* and the *shift* vector N^μ is tangent to the hypersurfaces Σ_t. Let us define $h_{\mu\nu}$, $\gamma_{\mu\nu}$ and $\sigma_{\mu\nu}$ to be the metrics induced on Σ_t, $\mathscr{B}$ and B_t by the metric $g_{\mu\nu}$, and $P^{\mu\nu}$ and $\Pi^{\mu\nu}$ the momenta (conjugate to $h_{\mu\nu}$ and $\gamma_{\mu\nu}$) of the hypersurface Σ_t and $\mathscr{B}$; see [7], [25]. The extrinsic curvatures of $\mathscr{B}$ in M and of B_t in Σ_t are respectively defined as $\Theta_{\mu\nu} = -\gamma_\mu^\alpha \nabla_\alpha n_\nu$ and $\mathscr{K}_{\mu\nu} = -\sigma_\mu^\alpha D_\alpha n_\nu$, where D denotes the covariant derivative on Σ_t compatible with $h_{\mu\nu}$. We can define the (variation of the) Hamiltonian of the theory as the (variation of the) conserved Noether current (8) evaluated on Σ_t and relative to the evolution vector field ξ (see, e.g. [5], [6]):

$$\delta_X H(\xi, \Sigma_t) = \delta_X \int_{\Sigma_t} \{N\mathscr{H} + \mathscr{N}^\alpha \mathscr{H}_\alpha\} d^3x \tag{11}$$

$$+ \int_{B_t} d^2x \left\{ N\delta(\sqrt{\sigma}\,\varepsilon) - N^\alpha \delta(\sqrt{\sigma}\, j_\alpha) + \frac{N\sqrt{\sigma}}{2} s^{\alpha\beta} \delta\sigma_{\alpha\beta} \right\}.$$

The bulk term is here related with the standard Hamiltonian constraints ($\mathscr{H}$ and $\mathscr{H}_\alpha$), which vanish on-shell. The boundary term is otherwise related with the energy of the system, defined as the on-shell value of the Hamiltonian. In equation (11): $\varepsilon = \frac{1}{\kappa}\mathscr{K}$ is the *surface quasilocal energy* ($\kappa = 8\pi$ in geometric units); $j_\alpha = -\frac{2}{\sqrt{\gamma}}\sigma_{\alpha\mu}\Pi^{\mu\nu}u_\nu$ is the *surface momentum* and $s^{\alpha\beta} = \frac{1}{\kappa}\left[(n^\mu a_\mu)\sigma^{\alpha\beta} - \mathscr{K}\,\sigma^{\alpha\beta} + \mathscr{K}^{\alpha\beta}\right]$ (with $a_\mu = u^\nu \nabla_\nu u_\mu$) is the *surface stress tensor* which describes the stress energy–momentum content of the gravitational field inside B_t (see [5], [6], [7]). The physical interpretation of this formula is very important and it has been analyzed in detail in [5]; the surface quasilocal energy is related to the internal energy of the localized system, the surface momentum is related with the rotation of the boundary in its evolution, while the surface stress tensor is related with the deformations of the boundary itself. The above formula, evaluated on-shell, can be conveniently rewritten in an explicit covariant form as follows:

$$\delta_X E(\xi, B_t) = \int_{B_t} \delta\left[\frac{\sqrt{g}}{2\kappa} \nabla^{[\beta} \xi^{\alpha]}\right] ds_{\alpha\beta} + \int_{B_t} \frac{\sqrt{g}}{2\kappa} g^{\mu\nu} \delta u_{\mu\nu}^{[\beta} \xi^{\alpha]} ds_{\alpha\beta} \tag{12}$$

$$= \frac{1}{2\kappa} \int_{B_t} d^2x\, \delta\left[2\sqrt{\sigma} u^{\mu} \Theta_{\mu}^{\alpha} \xi_{\alpha}\right] - \int_{B_t} d^2x\, \gamma_{\mu\nu}\, \delta\Pi^{\mu\nu}$$

$$= \delta_X \int_{B_t} U_{Kom}(\xi) + \int_{B_t} U_{CADM}(\xi, X),$$

with $u_{\mu\nu}^{\beta} = \Gamma_{\mu\nu}^{\beta} - \delta_{(\mu}^{\beta} \Gamma_{v)\rho}^{\rho}$, which comes naturally into play when attempting to describe the gravitational system in terms of thermodynamical variables. We remark that the first term resembles the variation of the Komar superpotential, while the second term is the ADM covariant correction; see [22]. To obtain the energy of the system it is necessary to impose the boundary conditions on the system and integrate (if possible) the above variational equation.

If we impose Dirichlet boundary conditions on the boundary $\delta\gamma_{\mu\nu}|_{\mathscr{B}} = 0$, the above expression can be integrated to obtain ([21], [22])

$$E_D(\xi, B_t) = \int_{B_t} d^2x\sqrt{\sigma}\left\{N(\varepsilon - \varepsilon_0) - N^{\alpha}(j_{\alpha} - j_{0\alpha})\right\}, \tag{13}$$

where the subscript 0 refers to a background solution g_0 with the same boundary metric $\gamma_{\mu\nu}$ (see [21], [22]) reproducing the Brown and York quasi-local energy.

If we instead impose (weak)-Neumann boundary conditions $\gamma_{\mu\nu}\, \delta\Pi^{\mu\nu}|_{\mathscr{B}} = 0$, the variational formula (12) can be integrated to obtain the Neumann energy, which is just related to the Komar superpotential $\delta E_{[N]}(\xi, B_t) = \delta \int_{B_t} U_{Kom}(\xi)$ by

$$E_N(\xi, B_t) = \int_{B_t} U_{Kom}(\xi, g) - \int_{B_t} U_{Kom}(\xi, g_0) \tag{14}$$

$$= \frac{1}{\kappa} \int_{B_t} \sqrt{\sigma} d^2x \left\{N n^{\mu} a_{\mu} - N^{\alpha} K_{\alpha\beta} n^{\beta}\right\} - Q_0,$$

where Q_0 is the Neumann energy of a background solution which satisfies the (weak) Neumann boundary conditions. We stress that E_N and E_D are related by means of a Legendre transformation on the control parameters of the boundary [22], [30].

For causal horizons $\mathscr{H}$, defined as the boundary of the past of timelike curves representing the observers' worldlines, see [22]. The *horizon entropy* is defined, for any cross-section H of $\mathscr{H}$ in Σ_t, as the quantity satisfying the first principle of thermodynamics (for closed systems):

$$\delta E_N(\xi, H) = T\delta S. \tag{15}$$

Unlike the variation (9), this formula is surface adapted, in that it is computed directly on the horizon $\mathscr{H}$. However, the two definitions (9) an (15) coincide provided that the boundary B in (9) is homologically equivalent to H (and ξ is a Killing vector for the solution). In spacetimes with multiple horizons each horizon gives its contribution to the variation of conserved quantities, B being homological to the the union of all horizon surfaces $\sum_i H_i$. This property resembles the additive property for entropy which is required from information theory, and the definition of entropy splits as

$$\delta E(\xi, B) = \sum_i \delta E(\xi, H_i) = \sum_i T_i \delta S_i, \qquad (16)$$

where T_i and S_i are the *local expression for entropy and temperature* defined on each horizon cross-section. This means that the geometrical definition of entropy is thus deeply related with the topological obstructions to globally foliate spacetime into space-like hypersurfaces. Each obstruction has to be taken into account into the computation of the total entropy (16); see [12], [22].

On horizon cross sections (where ξ is chosen to be the null Killing vector) the weak Neumann conditions ensure that the surface gravity (and hence the temperature on H) are constant [22]. This implies that from (14) and (15) we can immediately integrate the variational equation (once weak Neumann boundary conditions are imposed) and we see that $E_N(\xi, H) = TS$. This is a striking result, stating that the Neumann energy is exactly the gravitational heat TS. From (14) it thus follows that gravitational heat corresponds exactly to the Komar superpotential. This provides a further interpretation for the anomalous factor problem: the Komar superpotential simply provides a definition of energy reproducing the gravitational heat (and not the total mass-energy[4]). We can furthermore identify the second term in (12) with the free energy of the system in strict analogy with the Gibbs–Duhem formula $\delta E = \delta(TS) + \delta(F)$. This implies that

$$\delta_X F(\xi, H_i) = - \int_{H_i} \gamma_{\mu\nu} \delta \Pi^{\mu\nu} d^2x = \int_{H_i} U_{CADM}(\xi, X). \qquad (17)$$

This result completes the identification between thermodynamical quantities of a self-gravitating system and the geometrical characteristic of the system itself. Although these formulae have a classical origin and a classical definition, they provide results in agreement (at least in the low-energy limit) with the semi-classical statistical approach, where S is instead related to the microcanonical action functional and the free energy of the system is proportional to the partition function $F = -T \ln(Z(\beta_i))$; see e.g. [7].

The definition of energy obtained by means of this geometrical and covariant method has relaxed the hypotheses first imposed in [29]. The formalism developed here is suitable for application to the more general cases of spacetimes encompassing multiple causal horizons [22], isolated horizons [1] and more exotic solutions (e.g. Taub-bolt; see [14] and [22]). We remark that in the simple case of black hole solutions, the standard Bekenstein–Hawking *one quarter area law* is reproduced, as expected. We also remark that, on the contrary, in the Taub-bolt case the area law is not respected (as it was first pointed out in [28]).

---

[4] For example, for the Schwarzshild solution Energy $= M$, Komar $= TS = \frac{M}{2}$, $T = \frac{1}{8\pi M}$ and $S = 4\pi M^2$; see [22].

Acknowledgements

This work is partially supported by GNFM-INdAM research project "*Metodi geometrici in meccanica classica, teoria dei campi e termodinamica*" and by MIUR: PRIN 2003 on "*Conservation laws and thermodynamics in continuum mechanics and field theories*". G.A. is also supported by the I.N.d.A.M. grant: "Assegno di collaborazione ad attività di ricerca a.a. 2002-2003".

REFERENCES

1. G. Allemandi, M. Francaviglia, M. Raiteri, Class. Quantum Grav. **19**, 2633 (2002).
2. G. Allemandi, M. Francaviglia, M. Raiteri, Class. Quantum Grav. **20**, 483 (2003).
3. R. Arnowitt, S. Deser, C. W. Misner in: *Gravitation: An Introduction to Current Research*, L. Witten Ed., Wiley, 227, (New York, 1962).
4. J. D. Bekenstein, Nuovo Cim. Lett.. **4**, 737 (1972), J. D. Bekenstein, Phys. Rev. D **7**, 2333 (1973), J. D. Bekenstein, Phys. Rev. D **9**, 3292 (1974).
5. I. Booth, gr-qc/0008030; I. Booth, R.B. Mann, Phys. Rev. D **59**, 064021; I. Booth, R.B. Mann, Phys. Rev. D, 124009 (1999).
6. J. D. Brown, S. R. Lau, J. W. York, Ann. Phys. (N.Y.) 297, 175 (2002).
7. J. D. Brown and J. W. York, Phys. Rev. D **47**, 1407 (1993); J. D. Brown and J. W. York, Phys. Rev. D **47**, 1420 (1993).
8. J.D. Brown, J. Creighton, R.B. Mann, Phys. Rev. D **50**, 6394 (1994).
9. C. Caratheodory, *Untersuchungen über die Grundlagen der Thermodynamik*, in Math. Ann. (Berlin) **67**, 335 (1909).
10. C.-M. Chen, J. M. Nester, Gravitation & Cosmology **6**, 257 (2000).
11. M. Domagala, J. Lewandowski, (gr-qc/0407051).
12. L. Fatibene, M. Ferraris, M. Francaviglia, M. Raiteri, Ann. Phys. (N.Y.) **275**, 27 (1999).
13. L. Fatibene, M. Ferraris, M. Francaviglia, M. Raiteri, Phys. Rev. D **60**, 124012 (1999); L. Fatibene, M. Ferraris, M. Francaviglia, M. Raiteri, Phys. Rev. D **60**, 124013 (1999).
14. L. Fatibene, M. Ferraris, M. Francaviglia, M. Raiteri, Ann. Phys. (N.Y.) **284**, 197 (2000).
15. L. Fatibene and M. Francaviglia, *Natural and gauge natural formalism for classical field theories: a geometric perspective including spinors and gauge fields,* Kluwer Academic Publishers, Dordrecht (2003).
16. L. Fatibene, M. Ferraris, M. Francaviglia, S. Mercadante, *In preparation.*
17. E. Fermi, *Thermodynamics*, Dover Publications (1937).
18. M. Ferraris, M. Francaviglia, in: *Mechanics, Analysis and Geometry: 200 Years after Lagrange*, Editor: M. Francaviglia, Elsevier Science Publishers B.V., (Amsterdam, 1991) 451.
19. M. Ferraris, M. Francaviglia, M. Raiteri: Class. Quantum Grav. **20**, 4043 (2003).
20. M. Ferraris and M. Francaviglia, Atti Sem. Mat. Univ. Modena, **37**, 61 (1989); M. Ferraris and M. Francaviglia, Gen. Rel. Grav., **22**, 965 (1990); M. Ferraris, M. Francaviglia, I. Sinicco, Nuovo Cim. B **107**, 1303 (1992).
21. M. Francaviglia, M. Raiteri, Class. Quantum Grav. **19**, 237 (2002).
22. M. Francaviglia, M. Raiteri, Class. Quantum Grav. **21**, 3459 (2004).
23. G. W. Gibbons, S. W. Hawking, Phys. Rev. D **15**, 2752 (1977); G. W. Gibbons, S. W. Hawking, Phys. Rev. D **15**, 2738 (1977); S. W. Hawking, Phys. Rev. Lett. **26**, 1344 (1971); S. W. Hawking, Commun. Math. Phys. **43**, 199 (1975).
24. G. Hayward, Phys. Rev. D **47**, 3275 (1993).
25. S. W. Hawking, G.F.R. Ellis, *The Large Scale Structure of Space–Time* (Cambridge University Press, Cambridge, 1973).
26. S. W. Hawking, C. J. Hunter, Class. Quantum Grav. **13**, 2735 (1996).
27. K. Huang, *Statistical Mechanics*, J. Wiley and Sons Inc. (1987), 2nd edition.
28. S. W. Hawking, C. J. Hunter, D. N. Page, Phys. Rev. D **59**, 044033 (1999); S. W. Hawking, C. J. Hunter, Phys. Rev. D **59**, 044025 (1999).
29. V. Iyer and R. Wald, Phys. Rev. D **50**, 846 (1994); R.M. Wald, J. Math. Phys., **31**, 2378 (1993).

30. J. Kijowski, Gen. Rel. Grav. **29**, 307 (1997).
31. T. Padmanabhan, gr-qc/0311036; T. Padmanabhan, gr-qc/0309053; T. Padmanabhan, gr-qc/0308070.
32. T. Regge, C. Teitelboim, Ann. Phys. (N.Y.) **88**, 286 (1974).
33. C.E. Shannon and W. Weaver, *The Mathematical Theory of Communications*, University of Illinois Press, Urbana (1949).
34. A. Trautman, in: *Gravitation: An Introduction to Current Research*, L. Witten ed. (Wiley, New York, 1962) 168; A. Trautman, Commun. Math. Phys., **6**, 248 (1967).

Inertial effects of an accelerating black hole

D. Bini*, C. Cherubini† and B. Mashhoon**

*Istituto per le Applicazioni del Calcolo "M. Picone", CNR, I-00161 Rome, Italy and
International Center for Relativistic Astrophysics - I.C.R.A.
University of Rome "La Sapienza", I-00185 Rome, Italy
†Faculty of Engineering, University Campus Bio-Medico of Rome, via E. Longoni 47, 00155 Rome,
Italy,
International Center for Relativistic Astrophysics - I.C.R.A.
University of Rome "La Sapienza", I-00185 Rome, Italy
**Department of Physics and Astronomy, University of Missouri-Columbia, Columbia, Missouri
65211, USA

Abstract. We consider the static vacuum C metric that represents the gravitational field of a black hole of mass m undergoing uniform translational acceleration A such that $mA < 1/(3\sqrt{3})$. The influence of the inertial acceleration on the exterior perturbations of this background are investigated. In particular, we find no evidence for a direct spin-acceleration coupling.

INTRODUCTION

We study the motion of test particles and the propagation of wave fields on the exterior vacuum C metric background, which can be thought of as a nonlinear superposition of Schwarzschild and Rindler spacetimes. We find geodesic orbits that are circles about the direction of acceleration. Moreover, we consider the massless field perturbations of the C metric in search of a *direct coupling* between the spin of the perturbing field and the acceleration of the background, in complete analogy with the well-known spin-rotation coupling [1]. The results indicate that such a coupling does not exist. Furthermore, in the linear approximation to the C metric, we show that the propagation of the scalar field on this background entails a "gravitational Stark effect" that is analogous to the motion of an electron in the Stark effect.

VACUUM C METRIC

The vacuum C metric was first discovered by Levi-Civita [2] in 1918 within a class of Petrov type D (degenerate) static vacuum metrics. However, over the years it has been rediscovered many times: by Newman and Tamburino [3] in 1961, by Robinson and Trautman [4] in 1961 and again by Ehlers and Kundt [5] —who called it the C metric— in 1962. The charged C metric has been studied in detail by Kinnersley and Walker [6, 7]. In general the spacetime represented by the C metric contains one or, via an extension, two uniformly accelerated particles as explained in [7, 8]. A description of the geometric properties as well as the various extensions of the C metric is contained in [9], which

CP751, *General Relativity and Gravitational Physics, 16th SIGRAV Conference*, edited by G. Esposito et al.

should be consulted for a more complete list of references. The main property of the C metric is the existence of two hypersurface-orthogonal Killing vectors, one of which is timelike (showing the static property of the metric) in the spacetime region of interest in this work. The most familiar form of the C metric is [6, 7]

$$ds^2 = \frac{-1}{A^2(\tilde{x}+\tilde{y})^2}[(\tilde{F}dt^2 - \tilde{F}^{-1}d\tilde{y}^2) - (\tilde{G}^{-1}d\tilde{x}^2 + \tilde{G}d\tilde{z}^2)], \tag{1}$$

where

$$\tilde{F}(\tilde{y}) = -1+\tilde{y}^2 - 2mA\tilde{y}^3, \qquad \tilde{G}(\tilde{x}) = 1-\tilde{x}^2 - 2mA\tilde{x}^3, \qquad \tilde{G}(\tilde{x}) = -\tilde{F}(-\tilde{x}). \tag{2}$$

These coordinates are adapted to the hypersurface-orthogonal Killing vector $\kappa = \partial_t$, the spacelike Killing vector $\partial_{\tilde{z}}$ and $\partial_{\tilde{x}}$, which is aligned along the non-degenerate eigenvector of the hypersurface Ricci tensor. The constants $m \geq 0$ and $A \geq 0$ denote the mass and acceleration of the source, respectively. Unless specified otherwise, we choose units such that the gravitational constant and the speed of light in vacuum are unity. Moreover, we assume that the C metric has signature +2; to preserve this signature, we must have $\tilde{G} > 0$. We assume further that $\tilde{F} > 0$; it turns out that the physical region of interest in this case corresponds to $mA < 1/(3\sqrt{3})$ [10, 11, 12] .

Working with the metric in the form (1), the Schwarzschild limit ($A = 0$) is not immediate. Therefore, it is useful to introduce the retarded time coordinate u, the radial coordinate r and the azimuthal coordinate ϕ:

$$u = \frac{1}{A}\left[t + \int^{\tilde{y}} \tilde{F}^{-1}(\xi)d\xi\right], \qquad r = \frac{1}{A(\tilde{x}+\tilde{y})}, \qquad \phi = \tilde{z}, \tag{3}$$

so that the metric can be cast in the form

$$ds^2 = -\tilde{H}du^2 - 2dudr - 2Ar^2dud\tilde{x} + \frac{r^2}{\tilde{G}}d\tilde{x}^2 + r^2\tilde{G}d\phi^2, \tag{4}$$

where

$$\tilde{H}(r,\tilde{x}) = 1 - \frac{2m}{r} - A^2r^2(1-\tilde{x}^2 - 2mA\tilde{x}^3) - Ar(2\tilde{x}+6mA\tilde{x}^2) + 6mA\tilde{x}. \tag{5}$$

The norm of the hypersurface-orthogonal Killing vector κ is determined by $\tilde{H}$, $\kappa_\alpha \kappa^\alpha = -r^2\tilde{F} = -\tilde{H}/A^2$, so that this Killing vector is timelike for $\tilde{H} > 0$. We find it convenient to work with the $\{u, r, \theta, \phi\}$ coordinate system, where (r, θ, ϕ) are spherical polar coordinates with $\tilde{x} = \cos\theta$. Thus, the C metric takes the form

$$ds^2 = -Hdu^2 - 2dudr + 2Ar^2 \sin\theta dud\theta + \frac{r^2 \sin^2\theta}{G}d\theta^2 + r^2Gd\phi^2, \tag{6}$$

where G and H are given by

$$
\begin{aligned}
G(\theta) &= \sin^2\theta - 2mA\cos^3\theta, \\
H(r,\theta) &= 1 - \frac{2m}{r} - A^2r^2(\sin^2\theta - 2mA\cos^3\theta) - 2Ar\cos\theta(1+3mA\cos\theta) \\
&+ 6mA\cos\theta.
\end{aligned}
\tag{7}
$$

To study the location of horizons it is useful to introduce an acceleration length scale based on $A > 0$ given by $L_A = 1/(3\sqrt{3}A)$. It turns out that the modification of the horizons is related to the ratio of m and L_A. The event horizons of the vacuum C metric are Killing horizons given by $H = 0$ [7]. The solution of $H = 0$ can be written as $r^{-1} = A(\cos\theta + W^{-1})$, where W is a solution of $W^3 - W + 2mA = 0$. There are three cases depending on whether m is less than, equal to or greater than L_A. We have assumed at the outset that $m < L_A$; therefore, we expect that the two horizons of the Schwarzschild ($r = 2m$) and the Rindler ($r = [A(1 + \cos\theta)]^{-1}$) metrics will be somewhat modified. In fact let

$$\frac{1}{\sqrt{3}}\left(-\frac{m}{L_A} + i\sqrt{1 - \frac{m^2}{L_A^2}}\right)^{1/3} = \hat{U} + i\hat{V}, \tag{8}$$

then there are three real solutions for W given by $W = 2\hat{U}$, which results in $r = 2m$ for $A \to 0$, $W = -\hat{U} + \sqrt{3}\hat{V}$, which results in $r^{-1} = A(1 + \cos\theta)$ for $m \to 0$, and $W = -\hat{U} - \sqrt{3}\hat{V}$, which results in $r^{-1} = A(\cos\theta - 1)$ for $m \to 0$ and is therefore unacceptable. In the next two sections we will discuss the motion of test particles and the propagation of wave fields in the exterior spacetime region.

TEST PARTICLE MOTION: CIRCULAR ORBITS

Imagine the exterior of a spherically symmetric gravitational source that is uniformly accelerated along the $\theta = \pi$ direction with acceleration A. In the rest frame of the source, it is possible to find circular orbits about the direction of acceleration. In fact, in the Newtonian limit, a test particle can follow such an orbit of radius $r\sin\theta$ for fixed r and θ in the natural spherical polar coordinate system (r, θ, ϕ). In this case, $(m/r^2)\cos\theta = A$ and the speed of circular motion v is given by $v^2 = (m/r)\sin^2\theta$. It follows that circular orbits are possible for $0 < \theta < \pi/2$. The situation in general relativity is very similar, but somewhat more complicated. Indeed, timelike circular orbits exist for $\theta_0 < \theta < \pi/2$, where $G(\theta_0) = 0$. Moreover, for $\theta = \pi/2$, the circular orbit is null and is given by $r = 3m$ for all A such that $p \equiv mA < 1/(3\sqrt{3})$. Finally, there are spacelike circular orbits for $\pi/2 < \theta < \theta_c$, where $\theta_c(p)$ is a critical polar angle; for details, see [13].

WAVE MOTION: PERTURBATIONS

A master equation, analogous to the one derived in the Kerr spacetime [14, 15, 16, 17, 18, 19, 20] and describing massless field perturbations of any spin, has been studied by Prestidge [21] on the C-metric background. However, the physical content of this equation is not yet completely understood, because the master equation cannot be integrated exactly but only separated in $\{t, \tilde{x}, \tilde{y}, \tilde{z}\}$ coordinates.

We present the master equation for the C metric in a slightly different form compared with the one obtained by Prestidge [21]. In fact, we use here a principal NP frame which is also Kinnersley-like, i.e. it has the NP spin coefficient $\varepsilon = 0$. This allows some further simplification and puts this development in a form very close to the black hole case, where the master equation formalism has been successfully developed. Details for the derivation of the master equation in this case can be found in [22, 23].

With the C metric in the form (1) and switching the signature to -2 to agree with the standard Newman-Penrose formalism, a Kinnersley-like NP principal null tetrad can be easily constructed with

$$\mathbf{l} = A(\tilde{x}+\tilde{y})^2\left(\frac{1}{\tilde{F}}\partial_t + \partial_{\tilde{y}}\right), \quad \mathbf{n} = \frac{A}{2}\left(\partial_t - \tilde{F}\partial_{\tilde{y}}\right), \quad \mathbf{m} = \frac{\tilde{G}^{1/2}A(\tilde{x}+\tilde{y})}{\sqrt{2}}\left(\partial_{\tilde{x}} + \frac{i}{\tilde{G}}\partial_{\tilde{z}}\right). \quad (9)$$

The nonvanishing spin coefficients are

$$\mu = \frac{A^2\tilde{F}}{2\rho}, \quad \tau = \frac{A}{\sqrt{2}}\tilde{G}^{1/2} = -\pi, \quad \rho = A(\tilde{x}+\tilde{y}), \quad \beta = \frac{\rho}{4\sqrt{2}}\frac{\tilde{G}'}{\tilde{G}^{1/2}},$$

$$\alpha = \frac{A\tilde{G}^{-1/2}}{4\sqrt{2}}[\tilde{G}-\tilde{y}\tilde{G}'+3-\tilde{x}^2], \quad \gamma = \frac{A}{4(\tilde{x}+\tilde{y})}[\tilde{F}+\tilde{x}\dot{\tilde{F}}+3-\tilde{y}^2], \quad (10)$$

while the only surviving Weyl scalar is $\psi_2 = -mA^3(\tilde{x}+\tilde{y})^3$; here a prime and a dot denote differentiation with respect to $\tilde{x}$ and $\tilde{y}$, respectively. Following the approach of Prestidge [21], rescaling the unknown ψ_s of the master equation (for the various ψ_s satisfying the master equation, see e.g. [24]) we find that

$$\psi_s = (\tilde{x}+\tilde{y})^{(2s+1)}e^{-i\omega t}e^{ik_3\tilde{z}}X_s(\tilde{x})Y_s(\tilde{y}) \quad (11)$$

gives separated equations for X_s and Y_s, i.e.

$$X_s'' + \frac{\tilde{G}'}{\tilde{G}}X_s' + \left[\frac{-4S-s^2+2p\tilde{x}(s^2-4)}{4\tilde{G}}\right.$$
$$\left. -\frac{(-24pk_3+s)s\tilde{x}^2+2s(9ps-4k_3)\tilde{x}+3s^2+4k_3^2}{4\tilde{G}^2}\right]X_s = 0,$$

$$\ddot{Y}_s + \frac{\dot{\tilde{F}}(s+1)}{\tilde{F}}\dot{Y}_s + \left[\frac{S+s(s+1)-2p\tilde{y}(s+1)(2s+1)}{\tilde{F}} + \frac{\omega(\omega-is\dot{\tilde{F}})}{\tilde{F}^2}\right]Y_s = 0, (12)$$

where S is a separation constant. Because of the symmetry of the metric under the exchange of $\tilde{x}$ and $\tilde{y}$, one would expect a similar property to hold for these two equations. It can be shown that this is exactly the case (modulo further replacement of $\tilde{y} \to -\tilde{x}$, $\omega \to ik_3$, $s \to -s$) when one uses the following rescaling for $X_s(\tilde{x})$ and $Y_s(\tilde{y})$: $X_s(\tilde{x}) \to X_s(\tilde{x})/\tilde{G}^{1/2}$, $Y_s(\tilde{y}) \to Y_s(\tilde{y})/\tilde{F}^{(s+1)/2}$. Thus, without any loss of generality one can consider the equation for X_s only. This equation, in turn, cannot be solved exactly, unless $p = 0$. In this limit, with $\tilde{x} = \cos\theta$, one gets

$$\frac{d^2X_s}{d\theta^2} + \cot\theta\frac{dX_s}{d\theta} - \left[S + \frac{s^2-2k_3s\cos\theta+k_3^2}{\sin^2\theta}\right]X_s = 0, \quad (13)$$

so that with $S = -l(l+1)$ and $\tilde{z} = \phi$, it is easy to show that $X_s(\tilde{x})e^{ik_3\tilde{z}}$ reduces to the standard spin-weighted spherical harmonics.

Let us then consider the equation for X_s in (12), where we set $\tilde{x} = \cos\theta$ and use the rescaling $X_s(\theta) = \sin\theta\,\mathscr{T}_s(\theta)/\tilde{G}^{1/2}$. The equation for $\mathscr{T}_s$ is then

$$\frac{\mathrm{d}^2\mathscr{T}_s}{\mathrm{d}\theta^2} + \cot\theta\,\frac{\mathrm{d}\mathscr{T}_s}{\mathrm{d}\theta} - \mathscr{V}\mathscr{T}_s = 0, \tag{14}$$

where $\mathscr{V}$ is given by

$$\mathscr{V} = \frac{1}{(1 - 2p\cos\theta\cot^2\theta)^2}\left[p^2\mathscr{V}_{(2)}(\theta) + p\mathscr{V}_{(1)}(\theta) + \mathscr{V}_{(0)}(\theta)\right], \tag{15}$$

and the coefficients

$$
\begin{aligned}
\mathscr{V}_{(2)}(\theta) &= (1-s^2)\cos^2\theta - (1+s^2)\cot^2\theta + 4\cot^6\theta, \\
\mathscr{V}_{(1)}(\theta) &= 2\cos\theta\left[2s^2(1+\cot^2\theta) - S\cot^2\theta - 2(1+\cot^2\theta)^2\right] - 6k_3s\cot^2\theta, \\
\mathscr{V}_{(0)}(\theta) &= S + \frac{s^2 - 2k_3s\cos\theta + k_3^2}{\sin^2\theta},
\end{aligned} \tag{16}
$$

which do not depend on p. We recall that in the case under consideration here $p < 1/(3\sqrt{3})$. For $p \ll 1$, it is straightforward to develop a perturbation series solution to equation (14) in powers of p. In this way, terms of the form $ps = msA$ and higher order appear in $\mathscr{V}$, but a *direct* spin-acceleration coupling term sA that would be independent of mass m does not exist in X_s and hence ψ_s; therefore, we may conclude that this coupling does not exist. To see this in a more transparent way we will consider in the next section a linearization of the C metric.

LINEARIZED C METRIC

In the metric (6) let us consider the coordinate transformation $\{u, r, \theta, \phi\} \to \{T, X, Y, Z\}$, where

$$T = -u - \left[r + 2m\ln\left(\frac{r}{2m} - 1\right)\right] - Ar^2\cos\theta,$$

$$X = r\sin\theta\cos\phi, \quad Y = r\sin\theta\sin\phi, \quad Z = r\cos\theta + \frac{1}{2}Ar^2. \tag{17}$$

The transformed metric becomes

$$\mathrm{d}s^2 = \left(1 - \frac{2m}{R} - 2AZ\right)\mathrm{d}T^2 - \frac{2m}{R^3}(X\mathrm{d}X + Y\mathrm{d}Y + Z\mathrm{d}Z)^2 - \mathrm{d}X^2 - \mathrm{d}Y^2 - \mathrm{d}Z^2, \tag{18}$$

where $R = \sqrt{X^2 + Y^2 + Z^2}$ and we have neglected m^2, mA, A^2 and higher-order terms. Next, introduce polar coordinates Θ and Φ such that

$$X = R\sin\Theta\cos\Phi, \quad Y = R\sin\Theta\sin\Phi, \quad Z = R\cos\Theta. \tag{19}$$

With respect to these, the metric (18) becomes

$$ds^2 = \left(1 - \frac{2m}{R} - 2AR\cos\Theta\right)dT^2 - \left(1 + \frac{2m}{R}\right)dR^2 - R^2(d\Theta^2 + \sin^2\Theta\,d\Phi^2). \quad (20)$$

Finally, by introducing the isotropic radial coordinate ρ,

$$R = \left(1 + \frac{m}{2\rho}\right)^2\rho = \rho + m + \frac{m^2}{4\rho} \simeq \rho + m, \quad (21)$$

we get the linear metric in standard form

$$ds^2 = \left(1 - \frac{2m}{\rho} - 2A\hat{Z}\right)dT^2 - \left(1 + \frac{2m}{\rho}\right)(d\hat{X}^2 + d\hat{Y}^2 + d\hat{Z}^2), \quad (22)$$

where

$$\hat{X} = \rho\sin\Theta\cos\Phi, \quad \hat{Y} = \rho\sin\Theta\sin\Phi, \quad \hat{Z} = \rho\cos\Theta. \quad (23)$$

Gravitational Stark effect

Consider the massless scalar field equation $\nabla^\mu\nabla_\mu\chi = 0$ on the background spacetime given by the metric (22). To first order in m and A, χ can be separated by introducing parabolic coordinates in analogy with the Stark effect, which is the shift in the energy levels of an atom in an external electric field corresponding to the eigenvalues of a Schrödinger equation with a Coulomb potential $-k/r$ plus the potential due to a constant electric field $\mathbf{E} = E\,\hat{\mathbf{z}}$, i.e. $-k/r + eEz$, where $-e$ is the charge of the electron. In this gravitoelectromagnetic counterpart of the Stark effect, we set

$$\hat{X} = \sqrt{\xi\eta}\cos\psi, \quad \hat{Y} = \sqrt{\xi\eta}\sin\psi, \quad \hat{Z} = \frac{1}{2}(\xi - \eta), \quad (24)$$

and assume that

$$\chi(T,\xi,\eta,\psi) = e^{-i\omega T}\,e^{i\nu\psi}U(\xi)V(\eta), \quad (25)$$

where $\xi \geq 0$, $\eta \geq 0$, ψ takes values from 0 to 2π, ω is a constant and ν is an integer.

It follows from the scalar wave equation that

$$U_{\xi\xi} + \frac{1}{\xi}\left(1 - \frac{1}{2}A\xi\right)U_\xi + \left[\frac{\omega^2}{4}(1 + \xi A) + \frac{1}{\xi}(m\omega^2 - C) - \frac{\nu^2}{4\xi^2}\right]U = 0,$$

$$V_{\eta\eta} + \frac{1}{\eta}\left(1 + \frac{1}{2}A\eta\right)V_\eta + \left[\frac{\omega^2}{4}(1 - \eta A) + \frac{1}{\eta}(m\omega^2 + C) - \frac{\nu^2}{4\eta^2}\right]V = 0, \quad (26)$$

where C is the separation constant and $U_\xi = dU/d\xi$, etc. Note that the second equation for $V(\eta)$ can be obtained from the first one for $U(\xi)$ by replacing $A \to -A$ and $C \to -C$. On introducing a new constant β by

$$C = \frac{1}{2}\left(\beta - \frac{A}{2}\right), \quad (27)$$

and rescaling U and V,

$$U(\xi) = \left(1 + \frac{A\xi}{4}\right) a(\xi), \qquad V(\eta) = \left(1 - \frac{A\eta}{4}\right) b(\eta), \tag{28}$$

eqs. (26) become

$$\frac{d}{d\xi}\left(\xi \frac{da}{d\xi}\right) + \left[\frac{\omega^2\xi}{4} - \frac{v^2}{4\xi} + \frac{A\omega^2}{4}\xi^2 + \left(m\omega^2 + \frac{\beta}{2}\right)\right] a = 0,$$

$$\frac{d}{d\eta}\left(\eta \frac{db}{d\eta}\right) + \left[\frac{\omega^2\eta}{4} - \frac{v^2}{4\eta} - \frac{A\omega^2}{4}\eta^2 + \left(m\omega^2 - \frac{\beta}{2}\right)\right] b = 0. \tag{29}$$

These equations can be put in exact correspondence with the Schrödinger equation for the hydrogen atom in a constant electric field that results in the Stark effect [25]. For details see [22].

Let us note here again the close formal correspondence between the quantum theory of the Stark effect in hydrogen and the theory of a classical massless scalar field on the linearized C-metric background. Extension of this result to massless fields with nonzero spin present difficulties, as we have already seen in section IV.

Finally, for many laboratory applications, the potential associated with the gravitational Stark effect can be written as $m/\rho + A\rho\cos\Theta$ with $\rho = \rho_\oplus + \zeta$, where $\rho_\oplus$ is the average radius of the Earth and ζ is the local vertical coordinate in the laboratory. On using the local acceleration of gravity, $g = m/\rho_\oplus^2$, the effective *Newtonian* gravitational potential is then $-m/\rho_\oplus + g\zeta - A(\rho_\oplus + \zeta)\cos\Theta$; some of the applications of this potential are discussed in the next subsection.

Acceleration-induced phase shift

From the gravitational Stark effect we have learned that wave phenomena in the exterior spacetime represented by (22) are affected by the acceleration A. Consider then wave fields in a laboratory fixed on the Earth, assumed to undergo a small uniform nongravitational acceleration (e.g. solar radiation pressure or Mathisson–Papapetrou coupling of the curvature of the solar gravitational field with the angular momentum of the Earth). Estimates suggest that such accelerations are very small and at a level below $\sim 10^{-10}$ cm/s^2. In this sense, the total field of the Earth (nonrotating, spherical and endowed with a very small acceleration) is taken into account by the linearized C metric and we expect that the Earth's acceleration will introduce a very small shift in the phase of a wave propagating in the gravitational field of the Earth. Consider, for instance, the gravitationally induced quantum interference of neutrons as in the COW experiment [26, 27]. Let us imagine for the sake of simplicity that the $\hat{Z}$-axis of the system $\{T, \hat{X}, \hat{Y}, \hat{Z}\}$ of the metric (22) makes an angle Θ with the vertical direction in our local laboratory, and hence an otherwise free particle in the laboratory is subject to the effective Newtonian gravitational acceleration $g - A\cos\Theta$. The corresponding neutron phase shift in the COW experiment would then be given by

$$\Delta\varphi = (g - A\cos\Theta)\frac{\mathscr{A}\omega}{v}\sin\alpha, \tag{30}$$

where ω is the de Broglie frequency of the neutron, $\mathscr{A}$ is the area of the interferometer, α is the inclination angle of the interferometer plane with respect to the horizontal plane in the laboratory and v is the neutron speed. When $A = 0$ or $\Theta = \pi/2$, this formula reduces to the standard formula of the COW experiment [27]. A complete discussion of the neutron phase shift for nonzero A is beyond the scope of this work.

Pioneer anomaly

Imagine an inertial reference frame and a star of mass m such that its center of mass accelerates with a constant acceleration $\mathbf{A} = A\hat{\mathbf{z}}$, with $A > 0$. Thus, the motion of a planet or a satellite about the star in terms of a noninertial coordinate system $\{t, x, y, z\}$ in which the star is at rest with its center of mass at the origin of the spatial coordinates, is given to lowest order by

$$\frac{d^2\mathbf{r}}{dt^2} + \frac{m\mathbf{r}}{r^3} = -\mathbf{A}, \tag{31}$$

in agreement with Newtonian physics. Within the context of general relativity, the equation of motion of the test planet or satellite is given by the geodesic equation in the vacuum C metric.

Let us now apply these ideas to the anomalous acceleration of Pioneer 10 and Pioneer 11 [28, 29, 30], launched over thirty years ago to explore the outer solar system. The analysis of Doppler tracking data from Pioneer 10/11 spacecraft (moving away from the solar system in almost opposite directions) is consistent with the existence of a small anomalous acceleration of about 10^{-7} cm/s^2 toward the Sun.

Let $\hat{\mathbf{P}}$ and $\hat{\mathbf{P}}'$ be unit vectors that indicate the radial directions of motion with respect to the Sun of Pioneer 10 and Pioneer 11, respectively. Suppose that the smaller angle between these directions is given by $\pi - 2\beta$, where $\beta \simeq 7°$. Then, $\mathbf{A}$ can be expressed as

$$\mathbf{A} = \frac{A_0}{2\sin\beta}(\hat{\mathbf{P}} + \hat{\mathbf{P}}'), \tag{32}$$

where $A_0 \simeq 10^{-6}$ cm/s^2 is the magnitude of the vector $\mathbf{A}$ and is such that, with $\sin 7° \simeq 0.12$, $A_0\sin\beta$ is the magnitude of the anomalous acceleration. It follows that $\mathbf{A}\cdot\hat{\mathbf{P}} = \mathbf{A}\cdot\hat{\mathbf{P}}' = A_0\sin\beta$. It is therefore possible to find a vector $-\mathbf{A}$ that generates the Pioneer anomaly; however, the problem is then shifted to explaining the origin of such an acceleration of the center of mass of the Sun.

One possibility could be a recoil acceleration resulting from the anisotropic emission of solar radiation. But estimates for this effect give $A < 10^{-10}$ cm/s^2, so that it does not seem possible to account for the Pioneer anomaly in this way.

44

REFERENCES

1. B. Mashhoon, Phys. Rev. Lett. **61**, 2639 (1988).
2. T. Levi-Civita, Rend. Accad. Naz. Lincei **27**, 343 (1918).
3. E. Newman and L. Tamburino, J. Math. Phys. **2**, 667 (1961).
4. I. Robinson and A. Trautman, Proc. Roy. Soc. (London) **A265**, 463 (1962).
5. J. Ehlers and W. Kundt, in *Gravitation: An Introduction to Current Research*, ed L. Witten, Wiley, New York (1962).
6. W. Kinnersley, Phys. Rev. **186**, 1335 (1969).
7. W. Kinnersley and M. Walker, Phys. Rev. D **2**, 1359 (1970).
8. W.B. Bonnor, Gen. Rel. Grav. **15**, 535 (1983).
9. H. Stephani, D. Kramer, M.A.H. MacCallum, C. Hoenselaers and E. Herlt, *Exact Solutions of Einstein's Theory*, Cambridge Univ. Press, Cambridge, second edition (2003).
10. H. Farhoosh and L. Zimmerman, Phys. Rev. D **21**, 317 (1980).
11. V. Pravda and A. Pravdová, Czech. J. Phys. **50**, 333 (2000).
12. J. Podolský and J.B. Griffiths, Gen. Rel. Grav. **33**, 59 (2001).
13. D. Bini, C. Cherubini, A. Geralico, R.T. Jantzen, in preparation (2004).
14. R. Güven, Phys. Rev. D **22**, 2327 (1980).
15. S.A. Teukolsky, Phys. Rev. Lett. **29**, 1114 (1973).
16. S.A. Teukolsky, Astrophys. J. **185**, 635 (1973).
17. S.L. Detweiler and J.R. Ipser, Astrophys. J. **185**, 675 (1973).
18. J. Wainwright J. Math. Phys. **12**, 828 (1971).
19. G.F.T. Del Castillo and G. Silva-Ortigoza, Phys. Rev. D **42**, 4082 (1990).
20. A.L. Dudley and J.D. Finley III, Phys. Rev. Lett. **38**, 1505 (1977); Errata: Phys. Rev. Lett. **39**, 367 (1977).
21. T. Prestidge, Phys. Rev. D **58**, 124022 (1998).
22. D. Bini, C. Cherubini and B. Mashhoon, Phys. Rev. D, **70**, 044020 (2004).
23. D. Bini, C. Cherubini and B. Mashhoon, Class. Quantum Grav. **21**, 3893 (2004).
24. D. Bini, C. Cherubini, R.T. Jantzen and B. Mashhoon, Phys. Rev. D **67**, 084013 (2003).
25. L.D. Landau and E.M. Lifshitz, *Quantum Mechanics*, Pergamon Press, Oxford (1965).
26. R. Colella, A.W. Overhauser and S.A. Werner, Phys. Rev. Lett. **34**, 1472 (1975).
27. H. Rauch and S.A. Werner, *Neutron Interferometry*, Clarendon Press, Oxford (2000).
28. J.D. Anderson et al., Phys. Rev. Lett. **81**, 2858 (1998).
29. J.D. Anderson et al., Phys. Rev. D **65**, 082004 (2002).
30. J.D. Anderson and B. Mashhoon, Phys. Lett. A **315**, 199 (2003).

Particle Dark Matter

A. Bottino

Università di Torino, Dipartimento di Fisica Teorica
and Istituto Nazionale di Fisica Nucleare, Sezione di Torino
Via P. Giuria 1, 10125, Torino, Italia, E-mail: bottino@to.infn.it

Abstract. Extensions of the standard model of particle physics offer a number of interesting candidates for dark matter. We discuss some of these candidates, focussing in particular on the ones predicted by supersymmetric theories. We discuss their cosmological properties and strategies for detection.

DARK CONSTITUENTS IN THE UNIVERSE

A consistent picture for the matter/energy contents in our Universe is provided by an impressive number of independent cosmological observations. These may be summarised as follows: i) a host of observational data on galactic halos, clusters of galaxies, and large scale structures point to the following range for the matter density: $0.2 \lesssim \Omega_m \lesssim 0.4$ (notice that for any constituent i we define, as usual, $\Omega_i \equiv \rho_i/\rho_{crit}$, where ρ_i is the density of that constituent and $\rho_{crit} \equiv 3H_0^2/(8 \pi G) = 1.88 \times 10^{-29} \, h^2 \, g \cdot cm^{-3}$; h is the Hubble parameter, defined as $h = H_0/(100 \, km \cdot s^{-1} \cdot Mpc^{-1})$); ii) measurements of the cosmological microwave background (CMB)(see, for instance, Refs. [1, 2, 3]) show that the total density in the Universe is close to the critical one: $\Omega \simeq 1$; iii) high-redshift SNIa measurements (High-z SN Search [4], SN Cosmology Project [5]) give $0.8 \, \Omega_m - 0.6 \, \Omega_\Lambda \simeq -0.2 \pm 0.1$. Using $\Omega = \Omega_m + \Omega_\Lambda$ and any pair of the previous three points, one obtains $\Omega_m \sim 0.3$, $\Omega_\Lambda \sim 0.7$.

More recently, combined analyses of large scale structure (LSS) properties and CMB data provided new, accurate determinations of various cosmological parameters (though dependent on the assumption of some priors). By combining WMAP results with the 2dF Galaxy Survey and Lyman α forest data, in Ref. [6] one derives the following range for the CDM relic abundance: $\Omega_{CDM}h^2 = 0.1126 \pm 0.009$. Based on these results, in the following we will assume that the cold dark matter is bounded at 2σ level by the values: $(\Omega_{CDM}h^2)_{min} = 0.095$ and $(\Omega_{CDM}h^2)_{max} = 0.131$. An independent determination (Ω_{CDM} is provided by the Sloan Digital Sky Survey Collaboration [7]; this new data agrees with the results of Ref. [6].

Do we have some other indications about the nature of the matter density? As far as visible matter is concerned, observationally we have $\Omega_{vis} \lesssim 0.01$. Furthermore, from primordial nucleosynthesis it turns out that the baryonic abundance is $0.019 \lesssim \Omega_b h^2 \lesssim 0.021$ or, equivalently, $0.03 \lesssim \Omega_b \lesssim 0.05$ (in very good agreement with measurements of CMB acoustic peaks). Thus, combining these data with the previously quoted range $0.2 \lesssim \Omega_m \lesssim 0.4$, one concludes that: i) some dark matter is baryonic, but ii) most of it is

CP751, *General Relativity and Gravitational Physics, 16th SIGRAV Conference*, edited by G. Esposito et al.
© 2005 American Institute of Physics 0-7354-0236-1/05/$22.50

non-baryonic.

This last property implies that particles suitable to constitute the big bulk of dark matter can only be found in extensions of the Standard Model. One trivial extension of the SM is provided by light neutrinos ($m_v < 1\ MeV$). However, these particles fall into the category of hot relics (that is, particles which decouple from the primordial plasma when they are relativistic).

The theory of formation of cosmological structures implies that small-scale structures are erased by hot relics, because of free-streaming. The suppression of power spectrum on small scales induced by light neutrinos can be quantified as $\frac{\Delta P_m}{P_m} \simeq -8\frac{\Omega_v}{\Omega_m}$ [8]. Using the data of the 2dF Galaxy Redshift Survey [9] one puts a bound on the density contribution due to light neutrinos as compared to the total matter contribution: $\frac{\Omega_v}{\Omega_m} < 0.13$.

This means that most of dark matter has to be made up of cold relics, that is, particles which decouple from the primordial plasma when they are non-relativistic. These particles have to be massive, stable (or their lifetime must be at least of the order of the age of the Universe), and have to be only weakly interacting. Hence their generic name: WIMPs.

Many various extensions of the Standard Model provide good candidates for WIMPs. No doubt that one of the most interesting extensions of the SM is represented by supersymmetric theories, for a number of basic reasons related to particle physics. As an extra bonus, supersymmetric theories can naturally offer very appealing candidates for cold dark matter, if R-parity is conserved. Indeed, under this assumption, the Lightest Supersymmetric Particle (LSP) is stable. If uncharged and uncoloured, the LSP is a nice realization of a WIMP. Another interesting candidate, which is receiving careful consideration, is the Lightest Kaluza–Klein Particle (LKP) in theories with compactified extra dimensions. In the present report we discuss some properties connected to candidates within these two theoretical schemes.

SUPERSYMMETRIC CANDIDATES

Supersymmetric models

Though theoretically well motivated, supersymmetric theories still miss experimental validation. Thus, there exist a large variety of schemes; this situation prevent the theory from being really predictive.

One of the major unknowns is due to the supersymmetry-breaking mechanism. Three main schemes are usually investigated: the gravity-mediated mechanism, the gauge-mediated, and the anomaly-mediated one. The nature of the LSP depends on the susy-breaking mechanism and on the region of the parameter space. In what follows we discuss some phenomenological implications of the Minimal Supersymmetric extension of the Standard Model (MSSM) in the gravity-mediated scheme and in regions of the parameter space where the LSP is the neutralino [10].

The model we employ here is an effective MSSM scheme at the electroweak scale, defined in terms of a minimal number of parameters, only those necessary to shape the essentials of the theoretical structure of MSSM and of its particle content. In our model no gaugino-mass unification at a Gut scale is assumed. The assumptions that we impose at the electroweak scale are: a) all squark soft–mass parameters are degenerate: $m_{\tilde{q}_i} \equiv m_{\tilde{q}}$; b) all slepton soft–mass parameters are degenerate: $m_{\tilde{l}_i} \equiv m_{\tilde{l}}$; c) all trilinear parameters vanish except those of the third family, which are defined in terms of a common dimensionless parameter A: $A_{\tilde{b}} = A_{\tilde{t}} \equiv A m_{\tilde{q}}$ and $A_{\tilde{\tau}} \equiv A m_{\tilde{l}}$. As a consequence, the supersymmetric parameter space consists of the following independent parameters: $M_2, \mu, \tan\beta, m_A, m_{\tilde{q}}, m_{\tilde{l}}, A$ and R. In the previous list of parameters we have denoted by μ the Higgs mixing mass parameter, by $\tan\beta$ the ratio of the two Higgs v.e.v.'s, by m_A the mass of the CP-odd neutral Higgs boson and by R the ratio of the U(1) gaugino mass to the SU(2) one, i.e.: $R \equiv M_1/M_2$.

This model has been recently discussed in Refs. [11, 12, 13, 14]. In this series of papers it has been derived that, in models where gaugino-mass unification is not assumed, the present lower limit on neutralino mass is provided by the cosmological upper bound on the cold dark matter abundance (one finds $m_\chi \gtrsim 7$ GeV). In Refs. [12, 13, 14] theoretical expectations for direct and indirect searches of relic neutralinos are discussed, with particular emphasis on the light ones (i.e. those with $m_\chi \lesssim 50$ GeV). Note that, in models with gaugino mass unification, the lower bound $m_\chi \gtrsim 50$ GeV follows from the LEP lower bound on the chargino mass.

Here we report some results derived in [12, 13] for the expected rates for direct detection. In the numerical random scanning of the supersymmetric parameter space the following ranges are used: $1 \le \tan\beta \le 50$, 100 GeV $\le |\mu|$, $M_2 \le 1000$ GeV, 100 GeV $\le m_{\tilde{q}}, m_{\tilde{l}} \le 1000$ GeV, $\mathrm{sign}(\mu) = -1, 1$, 90 GeV $\le m_A \le 1000$ GeV, $-3 \le A \le 3$, $0.01 \le R \le 0.5$. The following experimental constraints are imposed: accelerators data on supersymmetric and Higgs boson searches, measurements of the $b \to s + \gamma$ decay and of the muon anomalous magnetic moment $a_\mu \equiv (g_\mu - 2)/2$. The range used here for the $b \to s + \gamma$ branching ratio is $2.18 \times 10^{-4} \le BR(b \to s + \gamma) \le 4.28 \times 10^{-4}$. For the deviation of the current experimental world average of a_μ from the theoretical evaluation within the Standard Model we use the 2σ range: $-142 \le \Delta a_\mu \cdot 10^{11} \le 474$; this interval takes into account the recent evaluations of Refs. [15, 16]. Also the current upper limit of the branching ratio of $B_s \to \mu^+ + \mu^-$ is included [14].

Neutralino relic abundance

The neutralino relic abundance is given by

$$\Omega_\chi h^2 = \frac{x_f}{g_\star(x_f)^{1/2}} \frac{3.3 \cdot 10^{-38} \text{ cm}^2}{<\widetilde{\sigma_{ann}v}>}, \tag{1}$$

where $<\widetilde{\sigma_{ann}v}> \equiv x_f \langle \sigma_{ann} v \rangle_{int}$, $\langle \sigma_{ann} v \rangle_{int}$ being the integral from the present temperature up to the freeze-out temperature T_f of the thermally averaged product of the annihilation cross-section times the relative velocity of a pair of neutralinos, x_f is defined as $x_f \equiv \frac{m_\chi}{T_f}$ and $g_\star(x_f)$ denotes the relativistic degrees of freedom of the thermodynamic bath at x_f. Detailed calculation of the neutralino relic abundance are found in the literature. Of particular interest is the behavior of this abundance for low neutralino masses

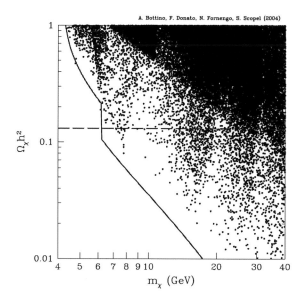

A. Bottino, F. Donato, N. Fornengo, S. Scopel (2004)

FIGURE 1. Neutralino relic abundance $\Omega_\chi h^2$ as a function of the mass m_χ. The solid curve denotes the minimal value of $\Omega_\chi h^2$ [12] for $T_{QCD} = 300$ MeV. Dashed and dot-dashed curves refer to the representative values $T_{QCD} = 100$ MeV, $T_{QCD} = 500$ MeV, respectively. The two horizontal lines denote two representative values of $\Omega_{CDM}h^2$: $\Omega_{CDM}h^2 = 0.3$ (short-dashed line) and $\Omega_{CDM}h^2 = 0.131$ (long-dashed line). The scatter plot is obtained by a full scanning of the supersymmetric parameter space.

(*i.e.* $m_\chi \lesssim 50$ GeV). This is analyzed in Refs. [11, 12], where a lower bound on m_χ is derived by requiring that $\Omega_\chi h^2 \leq (\Omega_{CDM}h^2)_{max}$. Using the most recent determinations of the restrictions on supersymmetric constraints mentioned above, this lower bound turns out to be: $m_\chi \lesssim 7$ Gev. The behavior of $\Omega_\chi h^2$ at low neutralino masses is shown in Fig. 1. The solid, short-dashed and dot-dashed lines are provided by an analytical analysis with different representative values for the hadron-quark transition temperature T_{QCD} [12]. The scatter plot is obtained by a full scanning of the supersymmetric parameter space.

Neutralino direct detection

The most natural way of detecting relic neutralinos (as other CDM candidates) is to search for effects resulting from the energy released by the neutralino, when it hits the nucleus of an appropriate detector. In this case, the detection rate is proportional to the neutralino-nucleus cross-section. For neutralino-nucleus interactions, coherent effects

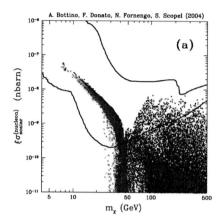
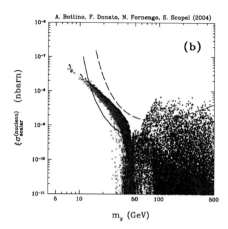

FIGURE 2. Scatter plot of $\xi\sigma_{scalar}^{(nucleon)}$ versus m_χ. Crosses (red) and dots (blue) denote neutralino configurations with $\Omega_\chi h^2 \geq (\Omega_{CDM}h^2)_{min}$ and $\Omega_\chi h^2 < (\Omega_{CDM}h^2)_{min}$, respectively ($(\Omega_{CDM}h^2)_{min} = 0.095$) (a) The curves delimit the DAMA region where the likelihood-function values are distant more than 4σ from the null (absence of modulation) hypothesis [18]; this region is the union of the regions obtained by varying the WIMP DF over the set considered in Ref. [21]. (b) Solid and dashed lines are the experimental upper bounds provided by the CDMS [19] and the EDELWEISS [20] Collaborations, respectively, under the hypothesis that the WIMP DF is given by an isothermal distribution with a standard set of astrophysical parameters.

systematically dominate over spin-dependent ones, hence the rates for direct detection are conveniently expressed in terms of the quantity $\xi\sigma_{scalar}^{(nucleon)}$ [17], where $\sigma_{scalar}^{(nucleon)}$ is the neutralino–nucleon scalar cross–section and ξ is a rescaling factor between the neutralino local matter density ρ_χ and the total local dark matter density ρ_0: $\xi \equiv \rho_\chi/\rho_0$. Following a standard assumption, ξ is taken as $\xi = \min(1, \Omega_\chi h^2/(\Omega_{CDM}h^2)_{min})$.

In Fig. 2 we display the scatter plot of the quantity $\xi\sigma_{scalar}^{(nucleon)}$ as a function of m_χ. This scatter plot shows that, in the mass range 6 GeV $\lesssim m_\chi \lesssim$ 25 GeV, the quantity $\xi\sigma_{scalar}^{(nucleon)}$ falls in a narrow funnel; this funnel is delimited from below by configurations with $\Omega_\chi h^2 \sim (\Omega_{CDM}h^2)_{max} = 0.131$, and delimited from above by supersymmetric configurations with a very light Higgs boson (close to its lower experimental bound of 90 GeV) and with an $\Omega_\chi h^2$ below $(\Omega_{CDM}h^2)_{min}$. For $m_\chi \lesssim$ 10 GeV only values of $30 \lesssim \tan\beta \leq 50$ and 100 GeV $\leq |\mu| \lesssim$ 300 GeV contribute, while in the interval 10 GeV $\lesssim m_\chi \lesssim$ 25 GeV $\tan\beta$ extends also to lower values around 8 and $|\mu|$ is not significantly constrained. Moreover, for $m_\chi \lesssim$ 20 GeV, m_A is strongly bounded from above by $(\Omega_{CDM}h^2)_{max}$. Note that the dip at $\simeq$ 45 GeV results from the Z–pole in the annihilation cross–section.

It is also remarkable that, within the funnel, the size of $\xi \sigma_{\text{scalar}}^{\text{(nucleon)}}$ is large enough to make light relic neutralinos explorable by WIMP direct experiments with the current sensitivities. To illustrate this point, in Fig. 2 the theoretical predictions of Refs. [12, 13] are compared with the experimental data of Refs. [18, 19, 20]. In panel (a) the contour line of the annual modulation of Ref. [18] is shown, in panel (b) the upper bounds of Refs. [19, 20] are displayed.

In deriving its contour line, the DAMA Collaboration has taken into account a rather large class of possible phase–space distribution functions (DF) for WIMPs in the galactic halo. The categories of DFs considered in Ref. [18] are those analyzed in Ref. [21]; the annual–modulation region displayed in Fig. 2, panel (a), is the union of the regions obtained by varying over the set of the DFs considered in Ref. [21]. From Fig. 2(a) it is derived that the entire population of relic neutralinos with $m_\chi \lesssim 25$ GeV as well as a significant portion of those with a mass up to about 50 GeV are within the annual–modulation region of the DAMA Collaboration. Thus, this yearly effect could be due to relic neutralinos of light masses, in alternative to the other possibility already discussed in the papers of Ref. [17] for neutralinos with masses above 50 GeV.

The experimental upper bounds of Refs. [19, 20], displayed in panel (b) of Fig. 2, are derived under the assumption of an isothermal distribution and for a single set of the astrophysical parameters: $\rho_0 = 0.3$ GeV $\cdot$ cm^{-3}, $v_0 = 220$ km $\cdot$ s^{-1} (v_0 is the local rotational velocity). For the case of light neutralinos the EDELWEISS bound (dashed line) is marginal, the one from CDMS (solid line) can potentially put constraints on neutralino masses in the 10 GeV $\lesssim m_\chi \lesssim 20$ GeV . However, to set a solid constraint on the theoretical predictions, it is necessary to derive from the experimental data the upper bounds on $\xi \sigma_{\text{scalar}}^{\text{(nucleon)}}$ for a large variety of DFs and of the corresponding astrophysical parameters (with their own uncertainties); the intersection of these bounds would provide an absolute limit to be used to possibly exclude a subset of supersymmetric population. An investigation by the CDMS Collaboration along these lines would be very interesting. It is worth noticing that a more effective comparison of theoretical results with experimental data will only be feasible when the analysis of different experimental results in terms of $m_\chi - \xi \sigma_{\text{scalar}}^{\text{(nucleon)}}$ is presented for each analytic form of the DF, separately. This is also the unique way of comparing results of different experiments among themselves.

Searches for relic neutralinos can also be performed by looking at possible indirect signals, based on products of the pair-annihilation of neutralinos as free particles in the galactic halo or bound inside macroscopic celestial objects (Earth and Sun). Evaluations of the relevant detection rates can be found, for instance, in Ref. [14].

LIGHTEST KALUZA–KLEIN PARTICLE

In theories with extra dimensions [22], models where all fields of the Standard Model propagate into compactified extra dimensions (Universal Extra Dimensions (UED) models [23]) are of particular phenomenological interest, one of the characteristic feature being that (at tree level) momentum is conserved.

Indeed, the mass-shell relation in the $(4+D)$-dimensional space

$$P^2 = p_0^2 - p_1^2 - p_2^2 - p_3^2 - \sum_i p_i^2 = m_0^2, \tag{2}$$

when viewed in four-dimensional space, makes it possible to define a Kaluza–Klein mass M_{KK} as

$$M_{KK}^2 \equiv p_0^2 - p_1^2 - p_2^2 - p_3^2 = m_0^2 + \sum_i p_i^2. \tag{3}$$

If one introduces a condition of periodicity of the wave function along any compact dimension, i.e. $p_i = \frac{n_i}{R_i}$ (where n_i = mode number, R_i = size of the compact dimension), one obtains

$$M_{KK}^2 = m_0^2 + \sum_i \frac{n_i^2}{R_i^2}. \tag{4}$$

Thus, above each SM particle there exists a tower of states. In turn, conservation of Kaluza–Klein parity makes the Lightest Kaluza–Klein Particle (LKP) stable: the LKP becomes a good candidate for CDM [24]. In this UED scenario the most interesting candidates for CDM are: the first level KK modes of neutral gauge bosons ($B^{(1)}$ and $W_3^{(1)}$) and the first level KK mode of the neutrino.

Relic abundances for DM KK-candidates have been studied in detail in Ref. [24]. Using these results and the present observational determinations of Ω_{CDM}, one finds for instance that the LKP related to $B(1)$ can give a sizeable contribution to CDM provided its mass is of order of a few hundreds of GeV (with an upper bound of about 800 GeV).

Sizes of direct/indirect detection rates for LKP candidates have been evaluated, for instance, in Refs. [25, 26, 27, 28].

CONCLUSIONS

Many attractive candidates for dark matter exist in various extensions of the Standard Model of particle physics. In the present report we have discussed some aspects of two of the most noticeable scenarios: supersymmetric theories and theories with compact extra dimensions. Both provide interesting candidates, with sizeable relic densities and direct and/or indirect detection rates at the level of experimental sensitivities, either already currently available or reachable in a near future. Parallel searches at the future particle accelerators will hopefully clarify which particle-physics scenario is actually realized in Nature.

REFERENCES

1. R. Stompor et al., Astrophys. J. **561**, L7 (2001).
2. J. E. Ruhl et al., Astrophys. J. **599**, 786 (2003).
3. D.N. Spergel et al., Astrophys. J. Suppl. **148**, 175 (2003).
4. A. Riess et al., Astron. J. **116**, 1009 (1998).
5. S. Perlmutter et al., Astrophys. J. **517**, 565 (1999).

6. D.N. Spergel *et al.*, Astrophys. J. Suppl. **148**, 175 (2003).
7. M. Tegmark et al., in press in Phys. Rev. D, astro-ph/0310723.
8. W. Hu, D.J. Eisenstein and M. Tegmark, Phys. Rev. Lett. **80**, 5255 (1998).
9. O. Elgaroy et al., Phys. Rev. Lett. **89**, 061301 (2002).
10. The literature concerning theoretical analyses of the neutralino as the LSP, in various supersymmetric models, is extremely vast. Among these papers: L. Bergström and P. Gondolo, Astroparticle Phys. **5**, 263 (1996); E. Gabrielli, S. Khalil, C. Munoz, E. Torrente-Lujan, Phys. Rev. D **63**, 025008 (2001); V.A. Bednyakov and H.V. Klapdor-Kleingrothaus, Phys. Rev. D **63**, 095005 (2001); J.L. Feng, K.T. Matchev and F. Wilczek, Phys. Lett. B **482**, 388 (2000); A. Corsetti and P. Nath, Phys. Rev. D **64**, 125010 (2001); J.R. Ellis, Keith A. Olive, Y.Santoso, V.C. Spanos, Phys. Lett. B **565**, 176 (2003); R. Arnowitt and B. Dutta, hep-ph/0204187; A.B. Lahanas, D.V. Nanopoulos and V.C. Spanos, Nucl. Phys. Proc. Suppl. **124**, 159 (2003).
11. A. Bottino, N. Fornengo and S. Scopel, Phys. Rev. D **67**, 063519 (2003).
12. A. Bottino, F. Donato, N. Fornengo and S. Scopel, Phys. Rev. D **68**, 043506 (2003).
13. A. Bottino, F. Donato, N. Fornengo and S. Scopel, Phys. Rev. D **69**, 0307303 (2004).
14. A. Bottino, F. Donato, N. Fornengo and S. Scopel, Phys. Rev. D **70**, 015005 (2004).
15. M. Davier et al., Eur. Phys. J. C **31**, 503 (2003).
16. K. Hagiwara et al., hep-ph/0312250.
17. A. Bottino, F. Donato, N. Fornengo, S. Scopel, Phys. Lett. B **423**, 109 (1998); Phys. Rev. D **62**, 056006 (2000); Phys. Rev. D **63**, 125003 (2001).
18. R. Bernabei *et al.*, Riv. N. Cim. **26** n. 1, 1 (2003).
19. D.S. Akerib *et al.*, hep-ex/0405033.
20. A. Benoit *et al.*, Phys. Lett. B **545**, 43 (2002).
21. P. Belli, R. Cerulli, N. Fornengo and S. Scopel, Phys. Rev. D **66**, 043503 (2002).
22. See, for instance, I. Antoniadis, Nucl. Phys. Proc. Suppl. **127**, 8 (2004).
23. T. Appelquist, H.-C. Cheng and B. A. Dobrescu, Phys. Rev. D **64**, 035002 (2001).
24. G. Servant and T.M.P. Tait, Nucl. Phys. B **650**, 391 (2003).
25. H.-C. Cheng, J.L. Feng and K.T. Matchev, Phys. Rev. Lett. **89**, 211301 (2002).
26. G. Servant and T.M.P. Tait, New J. Phys. **4**, 99 (2002).
27. G. Bertone, G. Servant and G. Sigl, Phys. Rev. D **68**, 044008 (2003).
28. D. Hooper and G.D. Kribs, hep-ph/0406026.

Higher Order Curvature Theories of Gravity Matched with Observations: a Bridge Between Dark Energy and Dark Matter Problems

S. Capozziello*, V.F. Cardone*, S. Carloni† and A. Troisi*

*Dipartimento di Fisica "E.R. Caianiello" and INFN, Università di Salerno, Italy
†Department of Mathematics and Applied Mathematics, University of Cape Town, South Africa.

Abstract. Higher order curvature gravity has recently received a lot of attention since it gives rise to cosmological models which seem capable of solving dark energy and quintessence issues without using "ad hoc" scalar fields. Such an approach is naturally related to fundamental theories of quantum gravity which predict higher order terms for loop expansions of quantum fields in curved spacetimes. In this framework, we obtain a class of cosmological solutions which are fitted against cosmological data. We reproduce reliable models able to fit high redshift supernovae and WMAP observations. The age of the universe and other cosmological parameters are recovered in this context. Furthermore, in the weak field limit, we obtain gravitational potentials which differ from the Newtonian one because of repulsive corrections increasing with distance. We evaluate the rotation curve of our Galaxy and compare it with the observed data in order to test the viability of these theories and to estimate the scale-length of the correction. It is remarkable that the Milky Way rotation curve is well fitted without the need of any dark matter halo, and similar results hold also for other galaxies.

Introduction

The Hubble diagram of type Ia supernovae (hereafter SNeIa) [1], the anisotropy spectrum of the cosmic microwave background radiation (hereafter CMBR) [2], the matter power spectrum determined by the large scale distribution of galaxies [3] and by the data on the Lyα clouds [4] are evidences in favor of a new picture of the universe, which is spatially flat and undergoing an accelerated expansion driven by a negative pressure fluid nearly homogeneously distributed and constituting up to $\sim 70\%$ of the energy content. This is called *dark energy*, while the model is usually referred to as the *concordance model*. Even if supported by the available astrophysical data, this new picture is not free of problems. Actually, while it is clear how dark energy works, its nature remains an unsolved problem. The simplest explanation claims for the cosmological constant Λ thus leading to the so called ΛCDM model [5]. Although being the best fit to most of the available astrophysical data [2], the ΛCDM model is also plagued by many problems on different scales. If interpreted as vacuum energy, Λ is up to 120 orders of magnitudes smaller than the predicted value. Furthermore, one should also solve the *coincidence problem*, i.e. the nearly equivalence, in orders of magnitude, of matter and Λ contributions to the total energy density. In order to address these issues, much interest has been devoted to models with dynamical vacuum energy, the so called *quintessence*. These models typically involve a scalar field rolling down its self-interaction potential thus allowing the vacuum energy to become dominant at present epoch. Although quintessence by a scalar field is the most studied candidate for dark energy, it generally does not avoid *ad hoc* fine tuning to solve the coincidence problem. Moreover, it is not clear where is this scalar field from and how to choose the self-interaction potential. Actually, there is a different way to face the problem of cosmic acceleration. It is possible that the observed acceleration is not the manifestation of another ingredient in the cosmic pie, but rather the first signal of a breakdown of our understanding of the laws of gravitation. From this point of view, it is thus tempting to modify the Friedmann equations to see whether it is possible to fit the astrophysical data with a model comprising only standard matter. In this framework, there is the attractive possibility to consider the Einstein gravity as a particular case of a more general theory. This is the underlying philosophy of what are referred to as $f(R)$ theories [6, 7, 8]. In this case, the Friedmann equations have to be given away in favor of a modified set of cosmological equations that are obtained by varying a generalized gravity Lagrangian where the scalar curvature R has been replaced by a generic function $f(R)$. The standard general relativity is recovered in the limit $f(R) = R$, while different results may be obtained for other choices of $f(R)$. With this paradigm in mind, the problems of dark energy and dark matter could be geometrically interpreted, giving rise to a completely new picture of gravitational interactions. From a cosmological point of view, the key point of $f(R)$ theories is the presence of modified Friedmann equations, obtained by varying the generalized Lagrangian. However, here lies also the main problem of this approach since it is not clear how the variation should be performed. Actually, once the Friedmann–Robertson–Walker (FRW) metric has been assumed, the equations governing the dynamics of the universe are different depending on whether one varies with respect to the metric only or with respect to the metric components and the connections. It is usual to refer to these two possibilities as the *metric approach* and the *Palatini approach* respectively. The two methods give the same results only in the case $f(R) = R$, while they lead to significantly different dynamical equations for every other choice of $f(R)$ (see [8] and references therein). The debate on what is the true physical approach is still open [9], nevertheless several positive results have been achieved in both of them. In [6] and then in [7, 8], it has been showed that it is possible to obtain the observed accelerating dynamics of the universe expansion by taking into account higher order curvature terms into the gravitational Lagrangian. Furthermore, in [6], a successful test with SNeIa data has been performed. Having tested such a scheme on cosmological scales, it is straightforward to try to complement the approach by analyzing the low energy limit of these theories in order to see whether this approach is consistent with the *local* physics, i.e. on galactic scale. In [10], it has been found that, in the weak-field limit, the Newtonian potential is modified by an additive term which scales with the distance r as a power law. Having obtained the corrected gravitational potential, the theoretical rotation curve of our Galaxy has been evaluated and compared with the observational data. This test shows that the correction term allows to well fit the Milky Way rotation curve without the need of dark matter. These results suggest that considering $f(R)$ theories of gravity can provide both an explanation to dark energy and dark matter issues. In this lecture, we outline the basic features of the $f(R)$-theories in the metric approach regarding the dark energy and the dark matter problems, stressing, in particular, the matching with astrophysical and cosmological data. Far from being exhaustive on the whole argument, we want to point out that these families of extended theories of gravity have to be seriously taken into account since they give rise to viable and reliable pictures of the observed universe.

Curvature Quintessence

A generic fourth–order theory of gravity, in four dimensions, is given by the action [6],

$$\mathscr{A} = \int d^4x\sqrt{-g}\left[f(R)+\mathscr{L}_{(matter)}\right], \tag{1}$$

where $f(R)$ is a function of Ricci scalar R and $\mathscr{L}_{(matter)}$ is the standard matter Lagrangian density. We are using physical units $8\pi G_N = c = \hbar = 1$. The field equations are

$$G_{\alpha\beta} = R_{\alpha\beta} - \frac{1}{2}g_{\alpha\beta}R = T_{\alpha\beta}^{(curv)} + T_{\alpha\beta}^{(matter)}, \tag{2}$$

where the stress-energy tensor has been defined for the curvature contributions

$$T_{\alpha\beta}^{(curv)} = \frac{1}{f'(R)}\left\{\frac{1}{2}g_{\alpha\beta}\left[f(R)-Rf'(R)\right]+f'(R)^{;\mu\nu}(g_{\alpha\mu}g_{\beta\nu}-g_{\alpha\beta}g_{\mu\nu})\right\} \tag{3}$$

and the matter contributions

$$T_{\alpha\beta}^{(matter)} = \frac{1}{f'(R)}\tilde{T}_{\alpha\beta}^{(matter)}. \tag{4}$$

We have taken into account the nontrivial coupling to geometry; prime means the derivative with respect to R. If $f(R) = R+2\Lambda$, we recover the standard second–order Einstein gravity (plus a cosmological constant term). In a FRW metric, the action (1) reduces to the point-like one:

$$\mathscr{A}_{(curv)} = \int dt\left[\mathscr{L}(a,\dot{a};R,\dot{R})+\mathscr{L}_{(matter)}\right], \tag{5}$$

where the dot means the derivative with respect to the cosmic time. In this case the scale factor a and the Ricci scalar R are the canonical variables. It has to be stressed that the definition of R in terms of $a,\dot{a},\ddot{a}$ introduces a constraint in the action (5) [6], by which we obtain

$$\mathscr{L} = a^3\left[f(R)-Rf'(R)\right]+6a\dot{a}^2f'(R)+6a^2\dot{a}\dot{R}f''(R)-6kaf'(R)+a^3p_{(matter)}, \tag{6}$$

where the standard fluid matter contribution acts essentially as a pressure term. The Euler–Lagrange equations resulting from (6) are the system

$$2\left(\frac{\ddot{a}}{a}\right)+\left(\frac{\dot{a}}{a}\right)^2+\frac{k}{a^2} = -p_{(tot)}, \tag{7}$$

and

$$f''(R)\left\{R+6\left[\frac{\ddot{a}}{a}+\left(\frac{\dot{a}}{a}\right)^2+\frac{k}{a^2}\right]\right\} = 0, \tag{8}$$

constrained by the energy condition

$$\left(\frac{\dot{a}}{a}\right)^2 + \frac{k}{a^2} = \frac{1}{3}\rho_{(tot)}.$$ (9)

On using Eq. (9), it is possible to write Eq. (7) in the form

$$\left(\frac{\ddot{a}}{a}\right) = -\frac{1}{6}\left[\rho_{(tot)} + 3p_{(tot)}\right].$$ (10)

The accelerated behavior of the scale factor is achieved for

$$\rho_{(tot)} + 3p_{(tot)} < 0.$$ (11)

To understand the actual effect of these terms, we can distinguish between matter and geometrical contributions, i.e.

$$p_{(tot)} = p_{(curv)} + p_{(matter)}, \quad \rho_{(tot)} = \rho_{(curv)} + \rho_{(matter)}.$$ (12)

Assuming that all matter components have non-negative pressure, Eq. (11) becomes

$$\rho_{(curv)} > \frac{1}{3}\rho_{(tot)}.$$ (13)

The curvature contributions result from the stress-energy tensor (3) and then the *curvature pressure* is

$$p_{(curv)} = \frac{1}{f'(R)}\left\{2\left(\frac{\dot{a}}{a}\right)\dot{R}f''(R) + \ddot{R}f''(R) + \dot{R}^2 f'''(R) - \frac{1}{2}\left[f(R) - Rf'(R)\right]\right\}, \quad (14)$$

and the *curvature energy-density* is

$$\rho_{(curv)} = \frac{1}{f'(R)}\left\{\frac{1}{2}\left[f(R) - Rf'(R)\right] - 3\left(\frac{\dot{a}}{a}\right)\dot{R}f''(R)\right\}, \quad (15)$$

which account for the geometrical contributions to the thermodynamical variables. It is clear that the form of $f(R)$ plays an essential role for this model. For the sake of simplicity, we choose the $f(R)$ function as a generic power law of the scalar curvature and we ask also for power law solutions of the scale factor, i.e.

$$f(R) = f_0 R^n, \quad a(t) = a_0\left(\frac{t}{t_0}\right)^\alpha.$$ (16)

The interesting cases are for $\alpha \geq 1$ which give rise to accelerated expansion. For $\rho_{(matter)} = 0$ and for spatially flat space-time ($k = 0$), we get algebraic equations for n and α reading as

$$\alpha[\alpha(n-2) + 2n^2 - 3n + 1] = 0, \quad \alpha[n^2 + \alpha(n-2-n-1)] = n(n-1)(2n-1), \quad (17)$$

from which the allowed solutions are

$$\alpha = 0 \rightarrow n = 0, 1/2, 1, \quad \alpha = \frac{2n^2 - 3n + 1}{2 - n}, \forall n \neq 2.$$ (18)

57

The solutions for $\alpha = 0$ are not interesting since they provide static cosmologies with a non-evolving scale factor. On the other hand, the cases with generic α and n provide an entire family of significant cosmological models. We see that such a family of solutions admits negative and positive values of α which give rise to accelerated behaviors (see also [11] for a detailed discussion). The curvature equation of state is given by

$$w_{(curv)} = - \left(\frac{6n^2 - 7n - 1}{6n^2 - 9n + 3} \right) , \tag{19}$$

which clearly is $w_{(curv)} \to -1$ for $n \to \infty$. This fact shows that the approach is compatible with the recovery of a cosmological constant. The accelerated behavior is allowed only for $w_{(curv)} < 0$, as requested for a cosmological fluid with negative pressure. From these straightforward considerations, the accelerated phase of the universe expansion can be described as an effect of higher order curvature terms which provide an effective negative pressure contribution. In order to see if such a behavior is possible for today epoch, we have to match the model with observational data. The presence of standard fluid matter $(\rho_{(matter)} \neq 0)$ does not affect greatly this overall behavior as widely discussed in [11].

Matching with dark energy observations

To verify if the curvature quintessence approach is an interesting perspective, we have to match the model with the observational data. In this way, we can constrain the parameters of the theory to significant values. First we compare our theoretical setting with the SNeIa results. As a further analysis, we check also the capability of the model of predicting the age of the universe. It is worth noticing that the SNeIa observations have represented a cornerstone in the recent cosmology, pointing out that we live in an expanding accelerating universe. This result has been possible in relation to the main feature of supernovae, which can be considered reliable standard candles thanks to the *Phillips amplitude-luminosity relation*. To test our cosmological model, we have taken into account the supernovae observations reported in [1] and compiled a combined sample of these data. Starting from these data, it is possible to perform a comparison between the theoretical expression of the distance modulus and its experimental value for SNeIa. The best fit is performed by minimizing the χ^2 calculated between the theoretical and the observational value of distance modulus. In our case, the luminosity distance is

$$d_L(z, H_0, n) = \frac{c}{H_0} \left(\frac{\alpha}{\alpha - 1} \right) (1 + z) \left[(1 + z)^{\frac{\alpha}{\alpha - 1}} - 1 \right] , \tag{20}$$

where c is the speed of light and z is the red-shift. The range of n can be divided into intervals, taking into account the existence of singularities in (20). In order to define a limit for H_0, we have to note that the Hubble parameter, being a function of n, has the same trend of α. We find that for n lower than -100, the trend is strictly increasing while for n positive, greater than 100, it is strictly decreasing. The results of the fit are showed in Table 1.

TABLE 1. Results obtained by fitting the curvature quintessence models against SNeIa data. The first column indicates the range of n, column 2 gives the relative best fit value of H_0, column 3 n^{best}, column 4 the χ^2 index.

Range	$H_0^{best}(km\,s^{-1}Mpc^{-1})$	n^{best}	χ^2
$-100 < n < 1/2(1-\sqrt{3})$	65	-0.73	1.003
$1/2(1-\sqrt{3}) < n < 1/2$	63	-0.36	1.160
$1/2 < n < 1$	100	0.78	348.97
$1 < n < 1/2(1+\sqrt{3})$	62	1.36	1.182
$1/2(1+\sqrt{3}) < n < 3$	65	1.45	1.003
$3 < n < 100$	70	100	1.418

TABLE 2. The results of the age test. The first column presents the tested range. Column 2 shows the 3σ-range for H_0 obtained by SNeIa test, while in the third we give the n intervals, i.e. the values of n which allow to obtain ages of the universe ranging between $10\,Gyr$ and $18\;Gyr$. In the last column, the best fit age values of each interval are reported.

Range	$\Delta H(km\,s^{-1}Mpc^{-1})$	Δn	$t(n^{best})(Gyr)$
$-100 < n < 1/2(1-\sqrt{3})$	$50-80$	$-0.67 \leq n < -0.37$	23.4
$1/2(1-\sqrt{3}) < n < 1/2$	$57-69$	$-0.37 < n \leq -0.07$	15.6
$1 < n < 1/2(1+\sqrt{3})$	$56-70$	$1.28 \leq n < 1.36$	15.3
$1/2(1+\sqrt{3}) < n < 2$	$54-78$	$1.37 < n \leq 1.43$	24.6

The age of the universe can be obtained, from a theoretical point of view, if one knows the current value of the Hubble parameter. In our case, it is

$$t = \left(\frac{2n^2 - 3n + 1}{2 - n}\right) H_0^{-1}. \tag{21}$$

We evaluate the age by taking into account the intervals of n and the 3σ-range of variability of the Hubble parameter deduced from the SNeIa fit. We have considered, as good predictions, age estimates included between $10\,Gyr$ and $18\,Gyr$. By this test, we are able of refining the allowed values of n. The results are shown in Table 2. First of all, we discard the intervals of n which give negative values of t. Conversely, the other ranges, tested by SNIa fit (Tab.1), become narrower, strongly constraining n.

Another check for the allowed values of n is to verify if the interesting ranges of n provide also accelerated expansion rates. This test can be easily performed by considering the definition of the deceleration parameter $q_0 = -(\ddot{a}a)/(\dot{a}^2)_0$, using the relation (16) and the definition of α in terms of n. To obtain an accelerated expanding behavior, the scale factor $a(t) = a_0 t^\alpha$ has to get negative (pole-like) or positive values of α greater than one. We obtain that only the intervals $-0.67 \leq n \leq 0.37$ and $1.37 \leq n \leq 1.43$ provide a negative deceleration parameter with $\alpha > 1$. Conversely, the other two

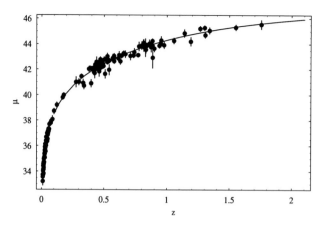

FIGURE 1.   Best fit curve to the SNeIa Hubble diagram for the power-law Lagrangian model.

intervals of Tab. 2 do not give interesting cosmological dynamics, because $q_0 > 0$ and $0 < \alpha < 1$ (standard Friedmann behavior).

A further test of the model can be performed by the age estimate obtained by the WMAP campaign [2]. Using these data, we can improve the constraints on n in relation to the very low error (1%) of WMAP age estimator which ranges between $13.5 Gyr$ and $13.9 Gyr$ [6].

Matching with dark matter: The Milky way rotation curve

Besides cosmology, the consistency of $f(R)$ gravity may be verified also at shorter astrophysical scales, e.g. at galactic scales, in order to check the full viability of the theory. In the low-energy limit, assuming as above $f(R) = f_0 R^n$, we obtain the gravitational potential [10]

$$\Psi(r) = -\frac{c^2}{2}\left[\left(\frac{r}{\xi_1}\right)^{-1} - \left(\frac{r}{\xi_2}\right)^{\beta(n)}\right].$$ (22)

where c is the speed of light,

$$\beta(n) = \sqrt{\frac{4n-1}{2(n-1)}} \times [\mathscr{P}(n) + \mathscr{Q}(n)],$$ (23)

and $\xi_{1,2}$ are scale-lengths. A first estimate of ξ_1 may be obtained by pointing out that, for $r \ll \xi_2$, Eq. (22) reduces to

$$\Psi(r) \sim -\frac{c^2}{2}\left(\frac{r}{\xi_1}\right)^{-1}.$$

Since we have to recover the Newtonian potential at these scales, we have to fix

$$\xi_1 = \frac{2GM}{c^2} \simeq 9.6 \times \frac{M}{M_\odot} \times 10^{-17} \text{ kpc} ,$$

with $M_\odot$ the mass of the Sun. The value of ξ_2 is a free parameter of the theory. So far, we can only say that ξ_2 should be much larger than the Solar System scale to avoid violating the constraints obtained from local gravity experiments. Eq. (22) gives the gravitational potential of a pointlike source. Since real galaxies are not pointlike, we have to generalize Eq. (22) to an extended source. For this purpose, we may suppose to divide the Milky Way in infinitesimal mass elements, to evaluate the contribution to the potential of each mass element and then to sum up these terms to get the final potential. In order to test whether the theory is in agreement with observations and to determine the parameter ξ_2, we have computed the Milky Way rotation curve by modelling our Galaxy as a two-component system, a spheroidal bulge and a thin disk. In particular, we assume

$$\rho_{bulge} = \rho_0 \left(\frac{m}{r_0}\right)^{-1.8} \exp\left(-\frac{m^2}{r_t^2}\right) , \quad \rho_{disk} = \frac{\Sigma_0}{2z_d} \exp\left(-\frac{R}{R_d} - \left|\frac{z}{z_d}\right|\right) , \quad (24)$$

where $m^2 = R^2 + z^2/q^2$, R is the radial coordinate and z is the height coordinate. The central densities ρ_0 and Σ_0 are conveniently related to the bulge total mass M_{bulge} and the local surface density $\Sigma_\odot$ by the following two relations:

$$\rho_0 = \frac{M_{bulge}}{4\pi q \times 1.60851} , \quad \Sigma_0 = \Sigma_\odot \exp\left(\frac{R_0}{R_d}\right) ,$$

$R_0 = 8.5$ kpc being the distance of the Sun to the Galactic Centre. We fix the Galactic parameters as follows:

$$M_{bulge} = 1.3 \times 10^{10} \, M_\odot , \; r_0 = 1.0 \text{ kpc} , \; r_t = 1.9 \text{ kpc} ,$$

$$\Sigma_\odot = 48 \, M_\odot \text{ pc}^{-2} , \; R_d = 0.3R_0 , \; z_d = 0.18 \text{ kpc} .$$

The Milky Way rotation curve $v_c(R)$ can be reconstructed starting from the data on the observed radial velocities v_r of test particles. We have used the data coming from the H II regions, molecular clouds and those coming from classical Cepheids in the outer disc obtained by Pont et al. [13].

For a given n, we perform a χ^2 test to see whether the modified gravitational potential is able to fit the observed rotation curve and to constrain the value of ξ_2. Since a priori we do not know what is the range for ξ_2, we get a first estimate of ξ_2 by a simple approach. For a given R, we compute ξ_2 by imposing that the theoretical rotation curve should be equal to the observed one. Then, we study the distribution of the ξ_2 values thus obtained and evaluate both the median ξ_2^{med} and the median deviation $\delta\xi_2$. The usual χ^2 test is then performed with the prior that ξ_2 lies in the range $(\xi_2^{med} - 5 \, \delta\xi_2, \xi_2^{med} + 5 \, \delta\xi_2)$. As a first test, we arbitrarily fix $n = 0.35$. We get $\xi_2 = 14.88$ kpc, $\chi^2 = 0.96$. In Fig. 2, we show both the theoretical rotation curve for $(n, \xi_2) = (0.35, 14.88)$ and the observed data. The agreement is quite good even if we have not added any dark matter component to the

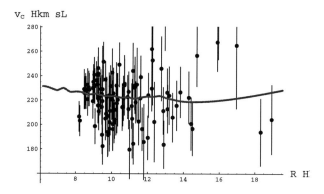

v_c Hkm sL

R H

FIGURE 2. Observed data and theoretical Milky Way rotation curve computed using the modified gravitational potential with $n = 0.35$ and $\xi_2 = 14.88$ kpc. Note that the points with R between 15.5 and 17.5 kpc are likely affected by systematic errors as discussed in [13].

Milky Way model. This result seems to suggest that our modified theory of gravitation is able to fit galaxy rotation curves without the need of dark matter. As a final remark, we note that $\xi_2^{med} = 14.37$ kpc is quite similar to the best fit value. Actually, a quite good estimate is also obtained on considering the value of ξ_2 evaluated by using the observed rotational velocity at R_0. This suggests that a quick estimate of ξ_2 for other values of n may be directly obtained by imposing $v_{c,theor}(R_0; n, \xi_2) = v_{c,obs}(R_0)$.

Conclusions

In this lecture, we have considered $f(R)$ theories of gravity to address the problems of dark energy and dark matter. Such an approach has a natural background in several attempts to quantize gravity, because higher-order curvature invariants occur in the renormalization process of quantum field theories in curved space-time. We have obtained a family of cosmological solutions [6] which we have fitted against several classes of observational data. A straightforward test is a comparison with SNIa observations [1]. The model fits these data and provides a constraint on the family of possible cosmological solutions. To improve this result, we have performed a test with the age of the universe, giving encouraging results in the range between $10Gyr$ and $18Gyr$. In order to better refine these ranges, we have then considered a test based on WMAP age evaluation. In this case, the age ranges between $13.5Gyr$ and $13.9Gyr$. In conclusion, we can say that a fourth-order theory of gravity of the form $f(R) = f_0 R^{1+\varepsilon}$ with $\varepsilon \simeq -0.6$ or $\varepsilon \simeq 0.4$ can give rise to reliable cosmological models which well fit SNeIa and WMAP data. In this sense, we need only "small" corrections to the Einstein gravity in order to achieve quintessence issues. Indications in this sense can be found also in a detailed analysis of $f(R)$ cosmological models performed against CMBR constraints, as shown in [12].

Furthermore, we have analyzed the low-energy limit of $f(R) = f_0 R^n$ theories of gravity by considering stationary solutions. An exact solution of the field equations has been obtained. The resulting gravitational potential for a point-like source is the sum of a Newtonian term and a contribution whose rate depends on a function of the exponent n of the Ricci scalar. The potential agrees with experimental data if n ranges into the interval $(0.25, 1)$, so that the correction term scales as r^β with $\beta > 0$. The following step is the generalization of this result to an extended source as a galaxy. For this purpose, the experimental data and the theoretical prediction for the rotation curve of Milky Way have been compared. The final result has been that the modified potential is able to provide a rotation curve which fits data *without adding any dark matter component*. This result has to be tested further before drawing a definitive conclusion against the need for galactic dark matter. Indeed, one has to show that a potential like that predicted by our model is able to fit rotation curves of a homogeneous sample of external galaxies with both well measured rotation curves and detailed surface photometry. In particular, the exponent n obtained from the fit must be the same for all galaxies, while ξ_2 might be different, being related to the scale where deviations from the Newtonian potential sets in.

In conclusion, we have given indications that it is possible to reduce the dark energy and dark matter issues under the same standard of $f(R)$ theories of gravity, which could give rise to realistic models working at very large scales (cosmology) and astrophysical scales (galaxies).

REFERENCES

1. S. Perlmutter et al., ApJ, 517 (1999) 565; A.G. Riess et al., ApJ, 607 (2004) 665.
2. P. de Bernardis et al., Nature, 404 (2000) 955; D.N. Spergel et al. ApJS, 148 (2003) 175.
3. A.C. Pope et al., ApJ, 607 (2004) 655.
4. R.A.C. Croft et al., ApJ, 495 (1998) 44.
5. V. Sahni, A. Starobinski, Int. J. Mod. Phys. D, 9 (2000) 373.
6. S. Capozziello, Int. J. Mod. Phys. D, 11 (2002) 483; S. Capozziello, V.F. Cardone, S. Carloni, A. Trosi, Int. J. Mod. Phys. D 12 (2003) 1969.
7. S.M. Carroll, V. Duvvuri, M. Trodden, M. Turner, astro-ph/0306438 (2003); S. Nojiri and S.D. Odintsov, Phys. Rev. D, 68 (2003) 123512.
8. D.N. Vollick, Phys. Rev. 68 (2003) 063510; E.E. Flanagan, Phys. Rev. Lett. 92 (2004) 071101; G. Allemandi, A. Borowiec, M. Francaviglia, hep-th/0403264 (2004).
9. G. Allemandi, M. Capone, S. Capozziello, M. Francaviglia, hep-th/0409198 (2004).
10. S. Capozziello, V.F. Cardone, S. Carloni, A. Troisi, *Phys. Lett.* A 326 (2004) 292.
11. S. Carloni, P.K.S. Dunsby, S. Capozziello, A. Troisi, gr-qc/0410046 (2004).
12. J. Hwang, H. Noh, *Phys. Lett.* B 506 (2001) 13.
13. Brand J., Blitz L., A&A, 275 (1993) 67; Pont F. et al., A&A, 318 (1997) 416.

Gravity and Yang–Mills Fields: Geometrical Approaches

Roberto Cianci and Stefano Vignolo

Dipartimento di Ingegneria della Produzione e Modelli Matematici
Università di Genova
Piazzale Kennedy, Pad. D - 16129 Genova (Italia)
E-mail: cianci@dipem.unige.it, vignolo@dipem.unige.it

Abstract. A new geometrical framework for a tetrad-affine formulation of gravity coupled with Yang–Mills fields is proposed.

THE NEW GEOMETRICAL FRAMEWORK

The mathematical arena suitable for formulating gravity in the tetrad formalism is the gauge bundle framework (see, for example, [1] and references therein), where dynamical fields are metric connections (not a priori torsion-free) and pseudo-orthonormal tetrads.

Metric connections turn out to be principal connections on the structure bundle $(P, M, \pi, SO(1,3))$ over spacetime M, while tetrads are seen to be sections of a suitable associated bundle with P, locally diffeomorphic to the co-frame bundle.

More precisely, denoting by $L(M)$ the frame bundle over M, a tetrad field is seen to be a section of a $GL(4, \Re)$ bundle $\mathscr{E}$, which is the bundle associated with $P \times L(M)$ through the left action

$$\lambda : (SO(1,3) \times GL(4,\Re)) \times GL(4,\Re) \to GL(4,\Re), \quad \lambda(\Lambda, J; X) := \Lambda \cdot X \cdot J^{-1}.$$

We can refer $\mathscr{E}$ to local fibered coordinates of the kind x^i, e_i^μ $(i, \mu = 1, \dots, 4)$ undergoing the transformation laws

$$\bar{x}^j = \bar{x}^j(x^i), \qquad \bar{e}_j^\mu = e_i^\sigma \Lambda^\mu{}_\sigma(x) \frac{\partial x^i}{\partial \bar{x}^j}, \tag{1}$$

where $\Lambda^\mu{}_\sigma(x) \in SO(1,3), \forall x \in M$.

The assignment of a tetrad field $x \to e^\mu(x) = e_i^\mu(x) dx^i$ gives rise to a metric $g := \eta_{\mu\nu} e^\mu \otimes e^\nu$ over M, with $\eta := \mathrm{diag}(-1,1,1,1)$. We may then identify the structure bundle $(P, M, \pi, SO(1,3))$ with the bundle of orthonormal frames associated with the metric g: for each tetrad e there is an isomorphism between P and $SO(M,g)$.

In view of this, the quotient bundle $\mathscr{C} := J_1 P / SO(1,3)$ of principal connections over the structure bundle P is in turn naturally identified with the bundle of g-metric compatible linear connections over M.

CP751, *General Relativity and Gravitational Physics, 16th SIGRAV Conference*, edited by G. Esposito et al.
© 2005 American Institute of Physics 0-7354-0236-1/05/$22.50

The manifold $\mathscr{C}$ can be referred to local coordinates $x^i, \omega_i{}^{\mu\nu}(:= \omega_i{}^\mu{}_\sigma \eta^{\sigma\nu})$ with $\mu < \nu$, subject to the transformation laws

$$\bar{\omega}_i{}^{\mu\nu} = \Lambda^\mu{}_\sigma(x)\Lambda^\nu{}_\gamma(x)\frac{\partial x^j}{\partial \bar{x}^i}\omega_j{}^{\sigma\gamma} - \Lambda_\sigma{}^\eta(x)\frac{\partial \Lambda^\mu{}_\eta(x)}{\partial x^h}\frac{\partial x^h}{\partial \bar{x}^i}\eta^{\sigma\nu}, \tag{2}$$

where $\omega_j{}^{\sigma\gamma} := -\omega_j{}^{\gamma\sigma}$ whenever $\sigma > \gamma$.

This formalism is able to embody automatically the Lorentz invariance of the theory in the geometry; here, in fact, tetrads are truly gauge natural objects, sensitive to the transformations of the structure bundle P. Moreover, the adoption of the gauge natural bundle framework is motivated also by the requirement to not be forced to assume the existence of a globally defined tetrad field.

If we want to describe gravity coupled with Yang–Mills fields, we have to introduce a principal fiber bundle over space–time $Q \to M$, with structure group G.

Principal connections of $Q \to M$ are additional dynamical fields; the latter may be represented as sections of the affine bundle $J_1 Q/G$ (the space of principal connections), referred to local coordinates $x^i, a_i^A, A = 1, \ldots, r = \dim G$.

Local sections $x \to \gamma^A(x)$ of $Q \to M$, trivializing Q, induce corresponding changes of coordinates in $J_1 Q/G$ of the form

$$\bar{a}_i^A = \left[Ad(\gamma^{-1})_B^A a_j^B + W_B^A(\gamma^{-1}, \gamma)\frac{\partial \gamma^B}{\partial x^j} \right]\frac{\partial x^j}{\partial \bar{x}^i}, \tag{3}$$

where Ad_B^A and W_B^A denote respectively the adjoint representation of G and the differential of the left multiplication in G.

In the combined theory of gravitational and Yang–Mills fields, the configuration space is then the fibered product $\mathscr{E} \times_M \mathscr{C} \times_M J_1 Q/G$ ($\mathscr{E} \times \mathscr{C} \times J_1 Q/G$ for short) over M.

Following [2, 3, 4], we construct a new affine bundle over the configuration space of the theory, by changing the standard definition of jet–equivalence.

To start with, let $J_1(\mathscr{E} \times \mathscr{C} \times J_1 Q/G, M)$ be the first jet–bundle associated with $\mathscr{E} \times \mathscr{C} \times J_1 Q/G \to M$, referred to jet–coordinates $x^i, e_i^\mu, \omega_i{}^{\mu\nu}, a_i^A, e_{ij}^\mu(\simeq \frac{\partial e_i^\mu}{\partial x^j}), \omega_{ij}{}^{\mu\nu}(\simeq \frac{\partial \omega_i{}^{\mu\nu}}{\partial x^j}), a_{ij}^A(\simeq \frac{\partial a_i^A}{\partial x^j})$.

We introduce the following equivalence relation on the bundle $J_1(\mathscr{E} \times \mathscr{C} \times J_1 Q/G, M)$: given two points $z = (x^i, e_i^\mu, \omega_i{}^{\mu\nu}, a_i^A, e_{ij}^\mu, \omega_{ij}{}^{\mu\nu}, a_{ij}^A)$ and $\hat{z} = (\hat{x}^i, \hat{e}_i^\mu, \hat{\omega}_i{}^{\mu\nu}, \hat{a}_i^A, \hat{e}_{ij}^\mu, \hat{\omega}_{ij}{}^{\mu\nu}, \hat{a}_{ij}^A) \in J_1(\mathscr{E} \times \mathscr{C} \times J_1 Q/G, M)$, we set $z \sim \hat{z} \Leftrightarrow x^i = \hat{x}^i, e_i^\mu = \hat{e}_i^\mu, \omega_i{}^{\mu\nu} = \hat{\omega}_i{}^{\mu\nu}, a_i^A = \hat{a}_i^A$ and $(e_{ij}^\mu - e_{ji}^\mu) = (\hat{e}_{ij}^\mu - \hat{e}_{ji}^\mu), (\omega_{ij}{}^{\mu\nu} - \omega_{ji}{}^{\mu\nu}) = (\hat{\omega}_{ij}{}^{\mu\nu} - \hat{\omega}_{ji}{}^{\mu\nu}), (a_{ij}^A - a_{ji}^A) = (\hat{a}_{ij}^A - \hat{a}_{ji}^A)$.

The transformation laws of $J_1(\mathscr{E} \times \mathscr{C} \times J_1 Q/G, M)$ ensure that the above equivalence relation is independent of the choice of local coordinates and therefore geometrically significant.

We denote by $\mathscr{J}(\mathscr{E} \times \mathscr{C} \times J_1 Q/G)$ the quotient space $J_1(\mathscr{E} \times \mathscr{C} \times J_1 Q/G, M)/\sim$ and by $\rho : J_1(\mathscr{E} \times \mathscr{C} \times J_1 Q/G, M) \to \mathscr{J}(\mathscr{E} \times \mathscr{C} \times J_1 Q/G)$ the corresponding canonical projection.

The manifold $\mathscr{J}(\mathscr{E} \times \mathscr{C} \times J_1Q/G, M)$ is an affine bundle; we refer it to local $\mathscr{J}$-coordinates $x^i, e^\mu_i, \omega_i^{\ \mu\nu}, a^A_i, E^\mu_{ij} := \frac{1}{2}(e^\mu_{ij} - e^\mu_{ji}), \Omega_{ij}^{\ \ \mu\nu} := \frac{1}{2}(\omega_{ij}^{\ \ \mu\nu} - \omega_{ji}^{\ \ \mu\nu}), A^A_{ij} := \frac{1}{2}(a^A_{ij} - a^A_{ji})$ $(i < j)$.

Changes of trivialization of the bundles P and Q induce coordinate transformations in $\mathscr{J}(\mathscr{E} \times \mathscr{C} \times J_1Q/G, M)$ of the form (1), (2), (3) together with

$$\bar{e}^\mu_{jk} = e^\sigma_{ih} \frac{\partial x^h}{\partial \bar{x}^k} \Lambda^\mu_{\ \sigma} \frac{\partial x^i}{\partial \bar{x}^j} + e^\sigma_i \frac{\partial \Lambda^\mu_{\ \sigma}}{\partial x^h} \frac{\partial x^h}{\partial \bar{x}^k} \frac{\partial x^i}{\partial \bar{x}^j} + e^\sigma_i \Lambda^\mu_{\ \sigma} \frac{\partial^2 x^i}{\bar{x}^k \bar{x}^j}, \tag{4a}$$

$$\begin{aligned}
\bar{\omega}_{ik}^{\ \ \mu\nu} &= \Lambda^\mu_{\ \sigma} \Lambda^\nu_{\ \gamma} \frac{\partial x^j}{\partial \bar{x}^i} \frac{\partial x^h}{\partial \bar{x}^k} \omega_{jh}^{\ \ \sigma\gamma} + \frac{\partial \Lambda^\mu_{\ \sigma}}{\partial x^h} \frac{\partial x^h}{\partial \bar{x}^k} \Lambda^\nu_{\ \gamma} \frac{\partial x^j}{\partial \bar{x}^i} \omega_j^{\ \sigma\gamma} \\
&\quad + \Lambda^\mu_{\ \sigma} \frac{\partial \Lambda^\nu_{\ \gamma}}{\partial x^h} \frac{\partial x^h}{\partial \bar{x}^k} \frac{\partial x^j}{\partial \bar{x}^i} \omega_j^{\ \sigma\gamma} + \Lambda^\mu_{\ \sigma} \Lambda^\nu_{\ \gamma} \frac{\partial^2 x^j}{\partial \bar{x}^k \partial \bar{x}^i} \omega_j^{\ \sigma\gamma} \\
&\quad - \frac{\partial \Lambda_\sigma^{\ \eta}}{\partial x^s} \frac{\partial x^s}{\partial \bar{x}^k} \frac{\partial \Lambda^\mu_{\ \eta}}{\partial x^h} \frac{\partial x^h}{\partial \bar{x}^i} \eta^{\sigma\nu} - \Lambda_\sigma^{\ \eta} \frac{\partial^2 \Lambda^\mu_{\ \eta}}{\partial x^s \partial x^h} \frac{\partial x^s}{\partial \bar{x}^k} \frac{\partial x^h}{\partial \bar{x}^i} \eta^{\sigma\nu} - \Lambda_\sigma^{\ \eta} \frac{\partial \Lambda^\mu_{\ \eta}}{\partial x^h} \frac{\partial^2 x^h}{\partial \bar{x}^k \partial \bar{x}^i} \eta^{\sigma\nu},
\end{aligned} \tag{4b}$$

and

$$\begin{aligned}
\bar{A}^A_{ik} &= \frac{\partial x^j}{\partial \bar{x}^i} \frac{\partial x^h}{\partial \bar{x}^k} \left[\mathrm{Ad}(\gamma^{-1})^A_B A^B_{jh} + \frac{1}{2} \left(\frac{\partial \mathrm{Ad}(\gamma^{-1})^A_B}{\partial x^h} a^B_j - \frac{\partial \mathrm{Ad}(\gamma^{-1})^A_B}{\partial x^j} a^B_h \right) \right. \\
&\quad + \left. \frac{1}{2} \left(\frac{\partial \eta^A_j}{\partial x^h} - \frac{\partial \eta^A_h}{\partial x^j} \right) \right], \tag{4c}
\end{aligned}$$

with $\eta^A_j(x) := W^A_B(\gamma^{-1}(x), \gamma(x)) \frac{\partial \gamma^B(x)}{\partial x^j}$, and the identifications $E^\mu_{ij} = -E^\mu_{ji}$, $\Omega_{ij}^{\ \ \mu\nu} = -\Omega_{ji}^{\ \ \mu\nu}$ and $A^A_{ij} = -A^A_{ji}$ implicitly understood whenever $i > j$.

We may endow the affine bundle $\mathscr{J}(\mathscr{E} \times \mathscr{C} \times J_1Q/G) \to \mathscr{E} \times \mathscr{C} \times J_1Q/G$ with the following geometrical structures, reproducing some standard results of jet-bundle geometry in $\mathscr{J}(\mathscr{E} \times \mathscr{C} \times J_1Q/G)$:

- $\mathscr{J}$-extension of sections. Given a (local) section $\sigma : M \to \mathscr{E} \times \mathscr{C} \times J_1Q/G$, we define its $\mathscr{J}$-extension $\mathscr{J}\sigma : M \to \mathscr{J}(\mathscr{E} \times \mathscr{C} \times J_1Q/G)$ as

$$\mathscr{J}\sigma := \rho \circ j_1\sigma,$$

$j_1\sigma : M \to j_1(\mathscr{E} \times \mathscr{C} \times J_1Q/G, M)$ denoting the standard first jet-extension of σ. Any section $\gamma : M \to \mathscr{J}(\mathscr{E} \times \mathscr{C} \times J_1Q/G)$ is said *holonomic* if there exists a section $\sigma : M \to \mathscr{E} \times \mathscr{C} \times J_1Q/G$ such that $\gamma = \mathscr{J}\sigma$.

Every holonomic section $\gamma : x \to (x^i, e^\mu_i(x), \omega_i^{\ \mu\nu}(x), a^A_i(x), E^\mu_{ij}(x), \Omega_{ij}^{\ \ \mu\nu}(x), A^A_{ij}(x))$ satisfies

$$E^\mu_{ij}(x) = \frac{1}{2} \left(\frac{\partial e^\mu_i(x)}{\partial x^j} - \frac{\partial e^\mu_j(x)}{\partial x^i} \right),$$

$$\Omega_{ij}^{\ \ \mu\nu}(x) = \frac{1}{2} \left(\frac{\partial \omega_i^{\ \mu\nu}(x)}{\partial x^j} - \frac{\partial \omega_j^{\ \mu\nu}(x)}{\partial x^i} \right),$$

$$A_{ij}^A(x) = \frac{1}{2}\left(\frac{\partial a_i^A(x)}{\partial x^j} - \frac{\partial a_j^A(x)}{\partial x^i}\right).$$

• *Contact forms.* We may introduce in $\mathscr{J}(\mathscr{E} \times \mathscr{C} \times J_1 Q/G)$ the concept of contact form by defining the following 2-forms:

$$\theta^\mu := de_j^\mu \wedge dx^j + E_{ij}^\mu \, dx^i \wedge dx^j, \tag{5a}$$

$$\theta^{\mu\nu} := d\omega_j^{\ \mu\nu} \wedge dx^j + \Omega_{ij}^{\ \ \mu\nu} \, dx^i \wedge dx^j, \tag{5b}$$

$$\theta^A := da_j^A \wedge dx^j + A_{ij}^A \, dx^i \wedge dx^j. \tag{5c}$$

Their behavior under coordinate changes is given by the equations

$$\bar\theta^\mu := d\bar e_j^\mu \wedge d\bar x^j + \bar E_{ij}^\mu \, d\bar x^i \wedge d\bar x^j = \Lambda^\mu_{\ \sigma}(x)\theta^\sigma,$$

$$\bar\theta^{\mu\nu} := d\bar\omega_j^{\ \mu\nu} \wedge d\bar x^j + \bar\Omega_{ij}^{\ \ \mu\nu} \, d\bar x^i \wedge d\bar x^j = \Lambda^\mu_{\ \sigma}(x)\Lambda^\nu_{\ \gamma}(x)\theta^{\sigma\gamma},$$

$$\bar\theta^A := d\bar a_j^A \wedge d\bar x^j + \bar A_{ij}^A \, d\bar x^i \wedge d\bar x^j = Ad(\gamma^{-1}(x))_B^A \theta^B.$$

The bundle spanned locally by the forms $\theta^\mu, \theta^{\mu\nu}, \theta^A$ is then invariant; it is called *contact bundle* and it is denoted by $C(\mathscr{J}(\mathscr{E} \times \mathscr{C} \times J_1 Q/G))$; every section $\eta : \mathscr{J}(\mathscr{E} \times \mathscr{C} \times J_1 Q/G) \to C(\mathscr{J}(\mathscr{E} \times \mathscr{C} \times J_1 Q/G))$ is called a *contact 2-form* on $\mathscr{J}(\mathscr{E} \times \mathscr{C} \times J_1 Q/G)$.

As it happens for standard jet bundles, the contact forms characterize the holonomic sections of $\mathscr{J}(\mathscr{E} \times \mathscr{C} \times J_1 Q/G) \to M$, i.e. we have

Proposition 1 *A section* $\gamma : M \to \mathscr{J}(\mathscr{E} \times \mathscr{C} \times J_1 Q/G)$ *is holonomic if and only if* $\gamma^*(\theta^\mu) = 0$, $\gamma^*(\theta^{\mu\nu}) = 0$ *and* $\gamma^*(\theta^A) = 0$.

• $\mathscr{J}$-*prolongation of morphisms.* It is possible to construct a sort of $\mathscr{J}$-prolongation for a particular family of bundle morphisms Φ of $\mathscr{E} \times \mathscr{C} \times J_1 Q/G$ projecting to diffeomorphisms χ of M.

To see this point, we start by singling out those bundle morphisms (Φ, χ) whose ordinary jet–prolongations $j_1\Phi$ on $J_1(\mathscr{E} \times \mathscr{C} \times J_1 Q/G, M)$ satisfy the requirement

$$\rho \circ j_1\Phi(w_1) = \rho \circ j_1\Phi(w_2), \qquad \forall w_1, w_2 \in \rho^{-1}(z), \tag{6}$$

for any $z \in \mathscr{J}(\mathscr{E} \times \mathscr{C} \times J_1 Q/G)$.

Indeed, for every bundle morphism (Φ, χ) obeying the ansatz (6), it is well defined an associated map $\mathscr{J}\Phi : \mathscr{J}(\mathscr{E} \times \mathscr{C} \times J_1 Q/G) \to \mathscr{J}(\mathscr{E} \times \mathscr{C} \times J_1 Q/G)$, expressed as

$$\mathscr{J}\Phi(z) := \rho \circ j_1\Phi(w), \qquad \forall w \in \rho^{-1}(z), z \in \mathscr{J}(\mathscr{E} \times \mathscr{C} \times J_1 Q/G), \tag{7}$$

henceforth referred to as the $\mathscr{J}$-*prolongation* of (Φ, χ).

It is seen [2, 4] that the most general bundle morphism (Φ, χ) satisfying the ansatz (6) has necessarily the local form

$$
(\Phi, \chi) : \begin{cases}
y^i = \chi^i(x), \\[2ex]
d_i^\nu = \Phi_i^\nu(x, e, \omega, a) = \Gamma_\mu^\nu(x)\dfrac{\partial x^r}{\partial y^i}e_r^\mu + \Gamma_{\gamma\eta}^\nu(x)\dfrac{\partial x^r}{\partial y^i}\omega_r^{\gamma\eta} + \Gamma_B^\nu(x)\dfrac{\partial x^r}{\partial y^i}a_r^B + c_i^\nu(x), \\[2ex]
\eta_i^{\nu\gamma} = \Phi_i^{\nu\gamma}(x, e, \omega, a) = \Gamma_\mu^{\nu\gamma}(x)\dfrac{\partial x^r}{\partial y^i}e_r^\mu + \Gamma_{\mu\sigma}^{\nu\gamma}(x)\dfrac{\partial x^r}{\partial y^i}\omega_r^{\mu\sigma} + \Gamma_B^{\nu\gamma}(x)\dfrac{\partial x^r}{\partial y^i}a_r^B + c_i^{\nu\gamma}(x), \\[2ex]
b_i^A = \Phi_i^B(x, e, \omega, a) = \Gamma_\mu^A(x)\dfrac{\partial x^r}{\partial y^i}e_r^\mu + \Gamma_{\gamma\eta}^A(x)\dfrac{\partial x^r}{\partial y^i}\omega_r^{\gamma\eta} + \Gamma_B^A(x)\dfrac{\partial x^r}{\partial y^i}a_r^B + c_i^A(x),
\end{cases}
$$
$$(8)$$

(8)

where the coefficients $\Gamma(x)$ and $c(x)$ are arbitrary local functions on M (with $\Gamma_{\gamma\eta}^\nu = -\Gamma_{\eta\gamma}^\nu$, $\Gamma_{\mu\sigma}^{\nu\gamma} = -\Gamma_{\sigma\mu}^{\nu\gamma}$ and $\Gamma_{\gamma\eta}^A = -\Gamma_{\eta\gamma}^A$).

The characterization (8) is invariant under coordinate changes (see [4] for the explicit transformation laws of the coefficients Γ). In local coordinates, the explicit expression of (the significant part of) $\mathscr{J}\Phi$ is given by

$$
\mathscr{J}\Phi : \begin{cases}
\begin{aligned}
D_{ij}^{\nu} &= \left(\Gamma_{\mu}^{\nu}E_{ks}^{\mu} + \Gamma_{\sigma\gamma}^{\nu}\Omega_{ks}{}^{\sigma\gamma} + \Gamma_{B}^{\nu}A_{ks}^{B}\right)\frac{\partial x^k}{\partial y^i}\frac{\partial x^s}{\partial y^j} \\
&+ \frac{1}{2}\left[\frac{\partial \Gamma_{\mu}^{\nu}}{\partial x^k}\left(\frac{\partial x^k}{\partial y^j}\frac{\partial x^r}{\partial y^i} - \frac{\partial x^k}{\partial y^i}\frac{\partial x^r}{\partial y^j}\right)e_r^{\mu}\right. \\
&+ \frac{\partial \Gamma_{\sigma\gamma}^{\nu}}{\partial x^k}\left(\frac{\partial x^k}{\partial y^j}\frac{\partial x^r}{\partial y^i} - \frac{\partial x^k}{\partial y^i}\frac{\partial x^r}{\partial y^j}\right)\omega_r{}^{\sigma\gamma} + \frac{\partial \Gamma_{B}^{\nu}}{\partial x^k}\left(\frac{\partial x^k}{\partial y^j}\frac{\partial x^r}{\partial y^i} - \frac{\partial x^k}{\partial y^i}\frac{\partial x^r}{\partial y^j}\right)a_r^{B} \\
&+ \left.\frac{\partial c_i^{\nu}}{\partial x^k}\frac{\partial x^k}{\partial y^j} - \frac{\partial c_j^{\nu}}{\partial x^k}\frac{\partial x^k}{\partial y^i}\right],
\end{aligned} \\[2em]
\begin{aligned}
\Delta_{ij}{}^{\nu\gamma} &= \left(\Gamma_{\mu}^{\nu\gamma}E_{ks}^{\mu} + \Gamma_{\mu\sigma}^{\nu\gamma}\Omega_{ks}{}^{\mu\sigma} + \Gamma_{B}^{\nu\gamma}A_{ks}^{B}\right)\frac{\partial x^k}{\partial y^i}\frac{\partial x^s}{\partial y^j} \\
&+ \frac{1}{2}\left[\frac{\partial \Gamma_{\mu}^{\nu\gamma}}{\partial x^k}\left(\frac{\partial x^k}{\partial y^j}\frac{\partial x^r}{\partial y^i} - \frac{\partial x^k}{\partial y^i}\frac{\partial x^r}{\partial y^j}\right)e_r^{\mu}\right. \\
&+ \frac{\partial \Gamma_{\mu\sigma}^{\nu\gamma}}{\partial x^k}\left(\frac{\partial x^k}{\partial y^j}\frac{\partial x^r}{\partial y^i} - \frac{\partial x^k}{\partial y^i}\frac{\partial x^r}{\partial y^j}\right)\omega_r{}^{\mu\sigma} + \frac{\partial \Gamma_{B}^{\nu\gamma}}{\partial x^k}\left(\frac{\partial x^k}{\partial y^j}\frac{\partial x^r}{\partial y^i} - \frac{\partial x^k}{\partial y^i}\frac{\partial x^r}{\partial y^j}\right)a_r^{B} \\
&+ \left.\frac{\partial c_i{}^{\nu\gamma}}{\partial x^k}\frac{\partial x^k}{\partial y^j} - \frac{\partial c_j{}^{\nu\gamma}}{\partial x^k}\frac{\partial x^k}{\partial y^i}\right],
\end{aligned} \\[2em]
\begin{aligned}
B_{ij}^{A} &= \left(\Gamma_{\mu}^{A}E_{ks}^{\mu} + \Gamma_{\sigma\gamma}^{A}\Omega_{ks}{}^{\sigma\gamma} + \Gamma_{B}^{A}A_{ks}^{B}\right)\frac{\partial x^k}{\partial y^i}\frac{\partial x^s}{\partial y^j} \\
&+ \frac{1}{2}\left[\frac{\partial \Gamma_{\mu}^{A}}{\partial x^k}\left(\frac{\partial x^k}{\partial y^j}\frac{\partial x^r}{\partial y^i} - \frac{\partial x^k}{\partial y^i}\frac{\partial x^r}{\partial y^j}\right)e_r^{\mu}\right. \\
&+ \frac{\partial \Gamma_{\mu\sigma}^{A}}{\partial x^k}\left(\frac{\partial x^k}{\partial y^j}\frac{\partial x^r}{\partial y^i} - \frac{\partial x^k}{\partial y^i}\frac{\partial x^r}{\partial y^j}\right)\omega_r{}^{\mu\sigma} + \frac{\partial \Gamma_{B}^{A}}{\partial x^k}\left(\frac{\partial x^k}{\partial y^j}\frac{\partial x^r}{\partial y^i} - \frac{\partial x^k}{\partial y^i}\frac{\partial x^r}{\partial y^j}\right)a_r^{B} \\
&+ \left.\frac{\partial c_i^{A}}{\partial x^k}\frac{\partial x^k}{\partial y^j} - \frac{\partial c_j^{A}}{\partial x^k}\frac{\partial x^k}{\partial y^i}\right].
\end{aligned}
\end{cases}
\tag{9}
$$

Proposition 2 *A bundle automorphism* (Ψ,χ) *of* $\mathscr{J}(\mathscr{E}\times\mathscr{C}\times J_1Q/G)\to M$ *satisfies* $\Psi^*(\eta)\in Span\{\theta^{\sigma},\theta^{\sigma\nu},\theta^{A}\}\,\forall\,\eta\in Span\{\theta^{\sigma},\theta^{\sigma\nu},\theta^{A}\} \Leftrightarrow \Psi = \mathscr{J}\Phi$ *for some bundle morphism* (Φ,χ) *of* $\mathscr{E}\times\mathscr{C}\times J_1Q/G\to M$.

Proposition 3 *Given a bundle automorphism* (Ψ,χ) *of* $\mathscr{J}(\mathscr{E}\times\mathscr{C}\times J_1Q/G)\to M$, *one has* $\Psi\circ\mathscr{J}\sigma\circ\chi^{-1}$ *is a* $\mathscr{J}$-*extension for every section* $\sigma : M\to\mathscr{E}\times\mathscr{C}\times J_1Q/G \Leftrightarrow \Psi = \mathscr{J}\Phi$ *for some bundle morphism* (Φ,χ) *of* $\mathscr{E}\times\mathscr{C}\times J_1Q/G\to M$.

- $\mathscr{J}$-*prolongation of vector fields.* As was done for bundle morphisms, it is possible to define a $\mathscr{J}$-prolongation for a suitable family of vector fields X on $\mathscr{E}\times\mathscr{C}\times J_1Q/G$, projecting to M.

This aim is achieved by characterizing those vector fields X on $\mathscr{E} \times \mathscr{C} \times J_1Q/G$, projecting to M, whose first jet-prolongations $J_1(X)$ on $J_1(\mathscr{E} \times \mathscr{C} \times J_1Q/G, M)$ pass to the quotient $\mathscr{J}(\mathscr{E} \times \mathscr{C} \times J_1Q/G)$.

Indeed, if X is any such vector field, it is well defined its $\mathscr{J}$-prolongation $\mathscr{J}(X)$: $\mathscr{J}(\mathscr{E} \times \mathscr{C} \times J_1Q/G) \to T\mathscr{J}(\mathscr{E} \times \mathscr{C} \times J_1Q/G)$ as

$$\mathscr{J}(X)(z) := \rho_{*\rho^{-1}(z)}(j_1(X)), \qquad \forall z \in \mathscr{J}(\mathscr{E} \times \mathscr{C} \times J_1Q/G), \tag{10}$$

amounting to taking the standard first jet–prolongation $J_1(X)$ and to projecting it on $\mathscr{J}(\mathscr{E} \times \mathscr{C} \times J_1Q/G)$.

It can be seen [2, 4] that the most general vector field X satisfying the required ansatz is of the form

$$X = \varepsilon^i(x)\frac{\partial}{\partial x^i} + \left(-\frac{\partial \varepsilon^k}{\partial x^q}e_k^\mu + D_v^\mu(x)e_q^v + D_{\gamma\sigma}^\mu(x)\omega_q{}^{\gamma\sigma} + +D_B^\mu(x)a_q^B + G_q^\mu(x)\right)\frac{\partial}{\partial e_q^\mu}$$

$$+ \sum_{\mu<v}\left(-\frac{\partial \varepsilon^k}{\partial x^q}\omega_k{}^{\mu v} + D_\sigma^{\mu v}(x)e_q^\sigma + D_{\gamma\sigma}^{\mu v}(x)\omega_q{}^{\gamma\sigma} + D_B^{\mu v}(x)a_q^B + G_q^{\mu v}(x)\right)\frac{\partial}{\partial \omega_q{}^{\mu v}}$$

$$+ \left(-\frac{\partial \varepsilon^k}{\partial x^q}a_k^A + D_v^A(x)e_q^v + D_{\gamma\sigma}^A(x)\omega_q{}^{\gamma\sigma} + D_B^A(x)a_q^B + G_q^A(x)\right)\frac{\partial}{\partial a_q^A}, \tag{11}$$

where the coefficients $\varepsilon(x)$, $D(x)$ and $G(x)$ are arbitrary local functions on M (with $D_{\gamma\sigma}^\mu = -D_{\sigma\gamma}^\mu$ and $D_{\gamma\sigma}^{\mu v} = -D_{\sigma\gamma}^{\mu v}$).

Again, the representations (11) are invariant under coordinate changes [3, 4].

It is seen [2] that the local expression of $\mathscr{J}$-prolongations is given by

$$\mathscr{J}(X) = \varepsilon^i(x)\frac{\partial}{\partial x^i} + \left(-\frac{\partial \varepsilon^k}{\partial x^q}e_k^\mu + D_v^\mu(x)e_q^v + D_{\gamma\sigma}^\mu(x)\omega_q{}^{\gamma\sigma} + D_B^\mu(x)a_q^B + G_q^\mu(x)\right)\frac{\partial}{\partial e_q^\mu}$$

$$+ \sum_{\mu<v}\left(-\frac{\partial \varepsilon^k}{\partial x^q}\omega_k{}^{\mu v} + D_\sigma^{\mu v}(x)e_q^\sigma + D_{\gamma\sigma}^{\mu v}(x)\omega_q{}^{\gamma\sigma} + D_B^{\mu v}(x)a_q^B + G_q^{\mu v}(x)\right)\frac{\partial}{\partial \omega_q{}^{\mu v}}$$

$$+ \left(-\frac{\partial \varepsilon^k}{\partial x^q}a_k^A + D_v^A(x)e_q^v + D_{\gamma\sigma}^A(x)\omega_q{}^{\gamma\sigma} + D_B^A(x)a_q^B + G_q^A(x)\right)\frac{\partial}{\partial a_q^A}$$

$$+ \sum_{i<j}h_{ij}^\mu\frac{\partial}{\partial E_{ij}^\mu} + \sum_{\mu<v}\sum_{i<j}h_{ij}{}^{\mu v}\frac{\partial}{\partial \Omega_{ij}{}^{\mu v}} + \sum_{i<j}h_{ij}^A\frac{\partial}{\partial A_{ij}^A}, \tag{12}$$

where

$$h_{ij}^{\mu} := \frac{1}{2}\left(\frac{\partial D_{\nu}^{\mu}}{\partial x^{j}}e_{i}^{\nu} - \frac{\partial D_{\nu}^{\mu}}{\partial x^{i}}e_{j}^{\nu} + \frac{\partial D_{\sigma\gamma}^{\mu}}{\partial x^{j}}\omega_{i}{}^{\sigma\gamma} - \frac{\partial D_{\sigma\gamma}^{\mu}}{\partial x^{i}}\omega_{j}{}^{\sigma\gamma} + \frac{\partial D_{B}^{\mu}}{\partial x^{j}}a_{i}^{B} - \frac{\partial D_{B}^{\mu}}{\partial x^{i}}a_{j}^{B}\right.$$
$$\left. + \frac{\partial G_{i}^{\mu}}{\partial x^{j}} - \frac{\partial G_{j}^{\mu}}{\partial x^{i}}\right) + D_{\nu}^{\mu}E_{ij} + D_{\sigma\gamma}^{\mu}\Omega_{ij}{}^{\sigma\gamma} + D_{B}^{\mu}A_{ij}^{B} + \left(E_{ki}^{\mu}\frac{\partial \varepsilon^{k}}{\partial x^{j}} - E_{kj}^{\mu}\frac{\partial \varepsilon^{k}}{\partial x^{i}}\right),$$

$$h_{ij}{}^{\mu\nu} := \frac{1}{2}\left(\frac{\partial D_{\sigma}^{\mu\nu}}{\partial x^{j}}e_{i}^{\sigma} - \frac{\partial D_{\sigma}^{\mu\nu}}{\partial x^{i}}e_{j}^{\sigma} + \frac{\partial D_{B}^{\mu\nu}}{\partial x^{j}}a_{i}^{B} - \frac{\partial D_{B}^{\mu\nu}}{\partial x^{i}}a_{j}^{B} + \frac{\partial D_{\gamma\sigma}^{\mu\nu}}{\partial x^{j}}\omega_{i}{}^{\gamma\sigma} - \frac{\partial D_{\gamma\sigma}^{\mu\nu}}{\partial x^{i}}\omega_{j}{}^{\gamma\sigma}\right.$$
$$\left. + \frac{\partial G_{i}{}^{\mu\nu}}{\partial x^{j}} - \frac{\partial G_{j}{}^{\mu\nu}}{\partial x^{i}}\right) + D_{\sigma}^{\mu\nu}E_{ij}^{\sigma} + D_{\sigma\gamma}^{\mu\nu}\Omega_{ij}{}^{\sigma\gamma} + D_{B}^{\mu\nu}A_{ij}^{B} + \left(\Omega_{ki}{}^{\mu\nu}\frac{\partial \varepsilon^{k}}{\partial x^{j}} - \Omega_{kj}{}^{\mu\nu}\frac{\partial \varepsilon^{k}}{\partial x^{i}}\right),$$

$$h_{ij}^{A} := \frac{1}{2}\left(\frac{\partial D_{\nu}^{A}}{\partial x^{j}}e_{i}^{\nu} - \frac{\partial D_{\nu}^{A}}{\partial x^{i}}e_{j}^{\nu} + \frac{\partial D_{\sigma\gamma}^{A}}{\partial x^{j}}\omega_{i}{}^{\sigma\gamma} - \frac{\partial D_{\sigma\gamma}^{A}}{\partial x^{i}}\omega_{j}{}^{\sigma\gamma} + \frac{\partial D_{B}^{A}}{\partial x^{j}}a_{i}^{B} - \frac{\partial D_{B}^{A}}{\partial x^{i}}a_{j}^{B}\right.$$
$$\left. + \frac{\partial G_{i}^{A}}{\partial x^{j}} - \frac{\partial G_{j}^{A}}{\partial x^{i}}\right) + + D_{\nu}^{A}E_{ij}^{\nu} + D_{\sigma\gamma}^{A}\Omega_{ij}{}^{\sigma\gamma} + D_{B}^{A}A_{ij}^{B} + \left(A_{ki}^{B}\frac{\partial \varepsilon^{k}}{\partial x^{j}} - A_{kj}^{B}\frac{\partial \varepsilon^{k}}{\partial x^{i}}\right).$$

Proposition 4 *Let $\pi : \mathscr{J}(\mathscr{E} \times \mathscr{C} \times J_{1}Q/G) \to \mathscr{E} \times \mathscr{C} \times J_{1}Q/G$ denote the natural projection. Given a vector field Y on $\mathscr{J}(\mathscr{E} \times \mathscr{C} \times J_{1}Q/G)$, projectable on $\mathscr{E} \times \mathscr{C} \times J_{1}Q/G$, such that its projection $X(z) := \pi_{*\pi^{-1}(z)}(Y)$ ($\forall z \in \mathscr{E} \times \mathscr{C} \times J_{1}Q/G$) defines a vector field on $\mathscr{E} \times \mathscr{C} \times J_{1}Q/G$ of the form (11), then*

$$Y = \mathscr{J}(X) \quad \Leftrightarrow \quad L_{Y}\theta^{\mu}, L_{Y}\theta^{\mu\nu}, L_{Y}\theta^{A} \in Span\{\theta^{\sigma}, \theta^{\sigma\mu}, \theta^{A}\}.$$

Corollary 1 *The $\mathscr{J}$-prolongations form a Lie algebra.*

We conclude this section by introducing new fiber coordinates on $\mathscr{J}(\mathscr{E} \times \mathscr{C} \times J_{1}Q/G)$ of the form ($x^{i} = x^{i}$, $e_{i}^{\mu} = e_{i}^{\mu}$, $\omega_{i}{}^{\mu\nu} = \omega_{i}{}^{\mu\nu}$, $a_{i}^{A} = a_{i}^{A}$)

$$T_{ij}^{\mu} := 2E_{ji}^{\mu} + \omega_{i}{}^{\mu}{}_{\nu}e_{j}^{\nu} - \omega_{j}{}^{\mu}{}_{\nu}e_{i}^{\nu},$$

$$R_{ij}{}^{\mu\nu} := 2\Omega_{ji}{}^{\mu\nu} + \frac{1}{2}\left(\omega_{i}{}^{\mu}{}_{\lambda}\omega_{j}{}^{\lambda\nu} - \omega_{j}{}^{\mu}{}_{\lambda}\omega_{i}{}^{\lambda\nu} - \omega_{i}{}^{\nu}{}_{\lambda}\omega_{j}{}^{\lambda\mu} + \omega_{j}{}^{\nu}{}_{\lambda}\omega_{i}{}^{\lambda\mu}\right),$$

and

$$F_{ij}^{A} := 2A_{ji}^{A} + a_{j}^{B}a_{i}^{C}C_{CB}^{A}.$$

The idea is to take the components of the torsion tensor $T^{\mu} = de^{\mu} + \omega^{\mu}{}_{\nu} \wedge e^{\nu}$ and the curvature tensors $R^{\mu\nu} = d\omega^{\mu\nu} + \omega^{\mu}{}_{\lambda} \wedge \omega^{\lambda\nu}$ and $F^{A} = da^{A} + \frac{1}{2}a^{C} \wedge a^{B}C_{CB}^{A}$ as $\mathscr{J}$-coordinates. The latter undergo the well known transformation laws

$$\bar{T}_{ij}^{\mu} = T_{hk}^{\sigma}\Lambda^{\mu}{}_{\sigma}\frac{\partial x^{h}}{\partial \bar{x}^{i}}\frac{\partial x^{k}}{\partial \bar{x}^{j}},$$

$$\bar{R}_{ij}{}^{\mu\nu} = R_{hk}{}^{\sigma\gamma}\Lambda^{\mu}{}_{\sigma}\Lambda^{\nu}{}_{\gamma}\frac{\partial x^h}{\partial \bar{x}^i}\frac{\partial x^k}{\partial \bar{x}^j},$$

$$\bar{F}_{ij}^A = F_{hk}^B Ad(\gamma^{-1})_B^A \frac{\partial x^h}{\partial \bar{x}^i}\frac{\partial x^k}{\partial \bar{x}^j}.$$

THE FIELD EQUATIONS

A Lagrangian on $\mathscr{J}(\mathscr{E}\times\mathscr{C}\times J_1Q/G)$ is a horizontal 4-form, locally expressed as

$$L = \mathscr{L}(x,e,\omega,a,T,R,F)\,ds,$$

where $ds := dx^1\wedge dx^2\wedge dx^3\wedge dx^4$ and $\mathscr{L}$ is a scalar density.

We associate with any Lagrangian L a corresponding Poincaré–Cartan 4-form on $\mathscr{J}(\mathscr{E}\times\mathscr{C}\times J_1Q/G)$, locally described as

$$\Theta_L := \mathscr{L}\,ds - \frac{1}{2}\theta^\mu\wedge P_\mu - \frac{1}{4}\theta^{\mu\nu}\wedge P_{\mu\nu} - \frac{1}{2}\theta^A\wedge P_A,$$

where $P_\mu := \frac{\partial\mathscr{L}}{\partial T_{ij}^\mu}\,ds_{ij}$, $P_{\mu\nu} := \frac{\partial\mathscr{L}}{\partial R_{ij}{}^{\mu\nu}}\,ds_{ij}$ and $P_A := \frac{\partial\mathscr{L}}{\partial F_{ij}^A}\,ds_{ij}$, with

$$ds_{ij} := \frac{\partial}{\partial x^i}\lrcorner\,\frac{\partial}{\partial x^j}\lrcorner\,ds.$$

Through any Lagrangian L we can define the action functional

$$A_L(\sigma) := \int_D \mathscr{J}\sigma^*(L) = \int_D \mathscr{J}\sigma^*(\Theta_L),$$

$\forall$ section $\sigma : D\subset M\to\mathscr{E}\times\mathscr{C}\times J_1Q/G$, D compact domain.

Now, let Φ_ξ be a 1-parameter group of $\mathscr{J}$-prolongable diffeomorphisms on $\mathscr{E}\times\mathscr{C}\times J_1Q/G$, projecting (for simplicity) on the identity map of M. Let X denote the infinitesimal generator of Φ_ξ; X is a $\mathscr{J}$-prolongable vector field, vertical with respect to the fibration $\mathscr{E}\times\mathscr{C}\times J_1Q/G\to M$.

Given a section $\sigma : M\to\mathscr{E}\times\mathscr{C}\times J_1Q/G$, we can deform it along X by setting $\sigma_\xi := \Phi_\xi\circ\sigma$; then we have $\mathscr{J}\sigma_\xi = \mathscr{J}(\Phi_\xi\circ\sigma) = \mathscr{J}\Phi_\xi\circ\mathscr{J}\sigma$.

Following the usual terminology, we call the expression

$$\frac{\delta A_L}{\delta X}(\sigma) := \frac{d}{d\xi}\int_D \mathscr{J}\sigma_\xi^*\Theta_L\bigg|_{\xi=0} = \int_D \mathscr{J}\sigma^*(\mathscr{J}(X)\lrcorner\,d\Theta_L) + \int_{\partial D}\mathscr{J}\sigma^*(\mathscr{J}(X)\lrcorner\,\Theta_L)$$

the *first variation* of A_L at σ in the direction X.

In connection with this, a section σ is said *critical* if $\frac{\delta A_L}{\delta X}(\sigma) = 0$ for all compact domains D and all deformations $\mathscr{J}\sigma_\xi$ constant on the boundary ∂D. By virtue of this last condition at the boundary, a section σ is critical if and only if the equation

$$\mathscr{J}\sigma^*(\mathscr{J}(X)\lrcorner\,d\Theta_L) = 0 \tag{13}$$

holds, for all $\mathscr{J}$-prolongable vector fields X.

In particular, if we consider the Lagrangian

$$L = \frac{1}{4}(e_i^\mu e_j^\nu R_{kl}{}^{\lambda\rho}\varepsilon^{ijkl}\varepsilon_{\mu\nu\lambda\rho} - F_{ij}^A F_{pq}^B \gamma_{AB}\eta^{\mu\nu}e_\mu^p e_\nu^i \eta^{\lambda\sigma}e_\lambda^q e_\sigma^j e)\,ds,$$

where $e := \det\left(e_i^\mu\right)$, e_μ^i denotes the inverse matrix of e_i^μ and γ_{AB} indicates the Cartan–Killing metric of the Lie algebra of the semisimple group G, expressing gravity coupled with a Yang–Mills field, eq. (13) is seen to yield the well known field equations

$$\frac{1}{2}e_j^\nu R_{kl}{}^{\sigma\lambda}\varepsilon^{ijkl}\varepsilon_{\mu\nu\sigma\lambda} = -\left(F_j^{Ai}F_{Ak}^j e_\mu^k - \frac{1}{4}F_{jk}^A F_A^{jk}e_\mu^i\right)e,$$

$$-2D_k(e_p^\sigma)e_q^\lambda\varepsilon^{pqki}\varepsilon_{\sigma\lambda\mu\nu} = 0,$$

$$D_k\left(F_A^{ik}e\right) = 0,$$

where $F_A^{ij} := F_{kh}^B\gamma_{BA}\eta^{\mu\nu}e_\mu^k e_\nu^i \eta^{\sigma\lambda}e_\sigma^h e_\lambda^j$, $D_k(e_p^\sigma) = \frac{\partial e_p^\sigma}{\partial x^k} + \omega_k{}^\sigma{}_\lambda e_p^\lambda$ and $D_k\left(F_A^{ik}e\right) = \frac{\partial\left(F_A^{ik}e\right)}{\partial x^k} - a_k^B\left(F_C^{ik}e\right)C_{BA}^C$ [4].

As a concluding remark, it is worth noticing that all restrictions about the vector fields $\mathscr{J}(X)$ may be removed and that the field equations may be written in the more general form

$$\mathscr{J}\sigma^*(X\lrcorner d\Theta_L) = 0, \qquad \forall X \in D^1(\mathscr{J}(\mathscr{E}\times\mathscr{C}\times J_1Q/G)).$$

SYMMETRIES AND NOETHER THEOREM

The Poincaré–Cartan representation of the field equations is especially useful in the study of symmetries and conserved quantities. To see this point, borrowing from [5], let us introduce the following

Definition 1 *A vector field Z on $\mathscr{J}(\mathscr{E}\times\mathscr{C}\times J_1Q/G)$ is called a* generalized infinitesimal Lagrangian symmetry *if it satisfies the requirement*

$$L_Z(\mathscr{L}\,ds) = d\alpha,$$

for some $(m-1)$-form α on $\mathscr{J}(\mathscr{E}\times\mathscr{C}\times J_1Q/G)$.

Definition 2 *A vector field Z on $\mathscr{J}(\mathscr{E}\times\mathscr{C}\times J_1Q/G)$ is called a* Noether vector field *if it satisfies the condition*

$$L_Z\Theta_L = \omega + d\alpha,$$

where ω is a m-form belonging to the ideal generated by the contact forms and α is any $(m-1)$-form on $\mathscr{J}(\mathscr{E}\times\mathscr{C}\times J_1Q/G)$.

Proposition 5 *If a generalized infinitesimal Lagrangian symmetry Z is a $\mathscr{J}$-prolongation $(Z = \mathscr{J}(X)$ for some $X)$, then it is a Noether vector field.*

73

Proposition 6 *If a Noether vector field Z is a $\mathscr{J}$-prolongation then it is an infinitesimal dynamical symmetry.*

We can associate with any Noether vector field Z a corresponding conserved current, so restating a sort of Noether theorem in the present geometrical setting. In fact, given a Noether vector field Z and a critical section $\sigma : M \to \mathscr{E} \times \mathscr{C} \times J_1 Q/G$, one has

$$d\,\mathscr{J}\,\sigma^* (Z \lrcorner \,\Theta_L - \alpha) = \mathscr{J}\,\sigma^* (\omega - Z \lrcorner \, d\Theta_L) = 0$$

showing that the current $\mathscr{J}\,\sigma^* (Z \lrcorner \,\Theta_L - \alpha)$ is conserved on shell.

In [4] some examples related to the gauge and diffeomorphism invariance of the theory are explicitly dealt with.

REFERENCES

1. Fatibene L and Francaviglia M, *Natural and gauge natural formalism for classical field theories. A geometric perspective including spinors and gauge theories*, Kluwer Academic Publishers, Dordrecht, 2003.
2. R. Cianci, S. Vignolo and D. Bruno, J. Phys. A: Math. Gen. **36**, 8341 (2003).
3. R. Cianci, S. Vignolo and D. Bruno, J. Phys. A: Math. Gen. **37**, 2519 (2004).
4. S. Vignolo and R. Cianci, A new geometrical look at gravity coupled with Yang–Mills fields, *J. Math. Phys.*, 2004, in print.
5. L. Fatibene, M. Ferraris, M. Francaviglia and R. G. McLenaghan, J. Math. Phys. **43**, 3147 (2002).

Interferometric readout for acoustic gravitational wave detectors

L. Conti[1*], M. De Rosa[2†], F. Marin[†], L. Taffarello[**] and M. Cerdonio[*]

*Dip. Fisica, Univ. Padova and INFN sez. Padova, Via Marzolo 8, I-35131 Padova, Italy
†INFN, Sez. Firenze and Dip. Fisica, Univ. Firenze, and LENS, Via Sansone 1, I-50019 Sesto Fiorentino (Firenze), Italy
**INFN sez. Padova, Via Marzolo 8, I-35131 Padova, Italy

Abstract. A review is given of the optical readout for acoustic gravitational wave detectors, ranging from the working principle, to the experimental data and to the future developments. A summary is also given of the scientific results obtained while developing the optical readout for a bar detector, of interest also for the broader interferometer community.

INTRODUCTION

The work described in this paper has been carried out within the AURIGA collaboration. AURIGA (http://www.auriga.lnl.infn.it) is a gravitational wave (gw) bar detector located at the Legnaro National Laboratories of INFN, near Padova (Italy). The detector consists of a massive (2.3 ton), 3 m long cylinder (the bar) made out of Al5056: the detection principle lies in the fact that the gravitational radiation would excite the quadrupole resonant modes of the bar; the bar mode with the largest cross section to the gravitational waves is the first longitudinal one which resonates at about 900Hz for the AURIGA bar. The mechanical signal induced by the passing gw is mechanically amplified and converted into an electromagnetic signal by a capacitive resonant transducer fixed to one of the bar end faces: one of the capacitor plates is a mechanical resonator which itself resonates at the same frequency of the bar first longitudinal mode (this explains why the transducer is called 'resonant'). Thus, since the transducer is much lighter than the bar (by a factor of about 10^3), a mechanical amplification of the signal is also accomplished. Then the signal passes through a resonant low-loss electrical circuit tuned to the mechanical resonances and then it is inductively coupled to a dual stage dc-SQUID amplifier [1].

Fundamental noise sources come from the thermal noise of the oscillators and from the dc-SQUID amplifier: to reduce the thermal noise the detector is cooled to cryogenic temperatures and the material choice and detector assembling is carefully done so as to guarantee low losses. For instance, the quality factor of each of the three modes (the two mechanical resonance and the electric one) is of the order of a million. As

---

[1] email: livia.conti@lnl.infn.it
[2] now at INOA, Complesso Olivetti, Via Campi Flegrei 34, 80078 Pozzuoli (Napoli), Italy

CP751, *General Relativity and Gravitational Physics, 16th SIGRAV Conference*, edited by G. Esposito et al.
© 2005 American Institute of Physics 0-7354-0236-1/05/$22.50

for the amplifier, the AURIGA group has carried out an intense R&D program in the past years that has lead to the achievement of a dual stage dc-SQUID amplifier [2] with noise limited by the thermal one down to 1K also after coupling to the mechanical resonator and with energy resolution a factor of 20-30 better than that ever achieved in a gw detector. In order to reduce the mechanical noise due to floor vibrations the bar is suspended horizontally by a cascade of mechanical filters.

AURIGA is collecting data for its second scientific run since December 2003; as is customary for gw detectors, the sensitivity is quoted as (the square root of) the power spectral density of the total noise expressed in terms of gw amplitude at the detector input, $\sqrt{S_{hh}}$ (units of $Hz^{-1/2}$). The AURIGA sensitivity is shown in fig. 5 together with that of other bar detectors.

THE OPTO-MECHANICAL READOUT

In the past several, different kinds of devices have been proposed to detect the small vibration of a resonant bar determined by the passage of a gravitational wave. One of the most appealing is that proposed by J.P. Richard [3], who suggested to use laser interferometric techniques for this purpose. A more recent version of the Richard idea is the object of an R&D program carried out by the AURIGA collaboration to bring into reality a cryogenic bar detector equipped with optomechanical readout.

In the following we describe the experimental progress toward this end and describe the measurement of the photothermal effect that has been made with the apparatus developed for the optomechanical bar readout. The foreseen performance of a ultracryogenic detector equipped with optomechanical readout is similar to that expected for SQUID based readouts: $\sqrt{S_{hh}}$ at the level of $1 \cdot 10^{-22}/\sqrt{Hz}$ in a bandwidth of some 50-100 Hz around the bar resonance, at a temperature of 0.1 K.

Working principle

The fundamental idea for signal optical transduction is to have a resonant optical cavity of length L formed by a mirror attached to a bar end face and a second fixed mirror. Then a relative motion ΔL of the two mirrors, due to bar vibration induced eventually by a gw, is converted into a change Δv of the optical resonant frequency v according to

$$\frac{\Delta L}{L} = \frac{\Delta v}{v}. \tag{1}$$

The second mirror is attached to a resonant transducer and thus one also achieves a mechanical amplification of the signal.

The idea expressed in eq. (1) has been developed in the scheme shown in fig. 1: this represents the working principle of the room temperature gw bar detector equipped with the optomechanical readout, better explained in the following section. The optical resonant cavity cited above is hereafter named *transducer cavity*. The naive approach of sending a light beam to this cavity and looking at its output as a function of light

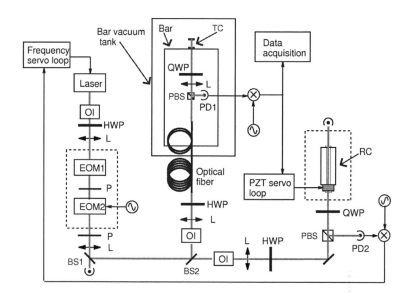

FIGURE 1. Experimental setup for a gw bar detector equipped with opto-mechanical readout. OI: optical isolator; HWP: half-wave plate; L: lens; EOM: electro-optic modulator; P: polarizer; BS: beam-splitter; PD: photodiode; QWP: quarter-wave plate; TC: transducer cavity; RC: reference cavity. From ref. [4].

frequency cannot work, essentially because of light frequency noise. In fact what one really needs are two cavities, one for stabilizing the light frequency and one for signal: for that reason, along with the transducer cavity there is also the *reference cavity*. The basic components of the scheme are therefore a laser source and these two cavities.

Main noise sources for this detection scheme come from the thermal noise of the mechanical resonators (bar and transducer), from the laser frequency noise and from the laser power noise. In the following we report about the strategies against the latter noise sources. As for the thermal noise, the strategy is identical to that used in AURIGA: low loss material (i.e. Al5056), low temperatures and careful design and assembling so as to avoid mechanical losses.

Laser power noise reduction

Laser power noise accounts for the back-acting noise force that affects the detector: power noise results in fluctuation of radiation pressure on transducer-cavity mirrors and this ends up as a noise force driving the two mechanical oscillator systems.

Back-action noise force has two contributions: one is purely classical and depends on the fluctuations in the laser intensity, and one is a quantum effect related to the rate at which photons hit the cavity mirror. This last contribution is fundamental, unavoidable (unless one uses a non-standard technique like squeezing) and sets the lower limit on the

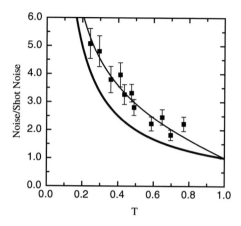

FIGURE 2. Intensity noise spectral density at about 900 Hz, referred to as shot noise, as a function of the transmittivity of the beam-splitter BS1. The experimental data fit with theoretical prediction [5]. The thick curve represents the theoretical limit for infinite gain. From ref. [6]

back-action noise force.

The beam from a 100 mW Nd:YAG cw laser source passes through an optical isolator, a first electro-optic modulator (EOM1), a polarizer and a second electro-optic modulator (EOM2). The two electro-optic modulators and the polarizer are housed into an aluminum box, with actively stabilized temperature. After the box a polarizer is followed by a variable beam-splitter. The beam transmitted by the BS1 is detected by a photodiode (PD1). The signal is then amplified and filtered by the loop electronics and sent to the EOM1 for the laser power noise reduction. The beam reflected by the BS1 is analyzed by a balanced detection scheme for both shot noise calibration and laser power noise measurement.

The calculation of the obtainable noise reduction requires a quantum mechanical approach which takes into account the vacuum fluctuations introduced by the beam-splitter. The experimental results are reported in Fig. 2, together with the theoretical prediction [5]. The thick line gives the theoretical limit with infinite gain. The laser power noise with loop off was measured to be a factor of 200 above the shot noise level for T=0.5.

We have also repeated the measurements after passing the beam reflected by the BS1 through a 2 m long single-mode polarization maintaining fiber. The fiber does introduce an excess of noise but nonetheless the back-action noise force is predicted to be an order of magnitude lower than the transducer thermal noise.

Laser frequency noise reduction

It is clear that the frequency stability of the reference cavity around bar resonance (i.e. at about 1kHz) is of crucial importance for the good performance of the optomechanical

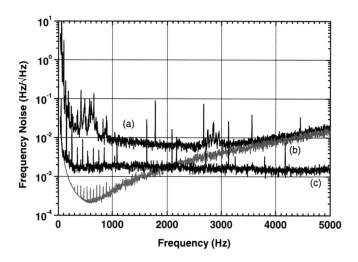

FIGURE 3. (a) Residual frequency noise of the laser when locked to C1, measured with respect to C2. (b) Noise spectrum measured on the C2 detector when the laser is far from resonance. (c) In-loop spectrum. Frequency resolution: 2 Hz. From ref. [7]

readout. The limit is set by the shot noise of the current developed at the photodiodes that detect light from the reference cavity: this sets the limit on the measurement of frequency stability.

We have constructed two identical reference cavities (C1 and C2) [7], in order to estimate an upper limit for the laser frequency noise from their (uncorrelated) relative fluctuations: the cavities are made by a 20 cm long Zerodur spacer with optically contacted mirrors. The finesses are of the order of a few 10^4; the cavities are housed in separate vacuum chambers pumped via ion pumps and are placed on top of a two-stage mechanical suspension. The temperature of the cavities can be varied between room temperature up to about $100°$ C for coarse length tuning via thermal expansion; the temperatures are then actively stabilized at a level of about 10^{-4} K so as to stay within the cavity line-width; moreover, one of the two is equipped with piezoelectric actuators for a fine length tuning by putting pressure on the Zerodur spacer.

We have frequency locked our laser source to one of these cavities (i.e. C1) by means of the FM-sidebands technique [8]: then we have measured the relative fluctuations of the two cavities by bringing the other (i.e. C2) into resonance with the laser beam and by analyzing the error signal produced by the FM-sidebands technique. The experimental results are shown in fig. 3. The achieved noise level, about +9 dB above shot noise at about 1 kHz, should allow efficient performance of the optomechanical readout when coupled to a cryogenic bar.

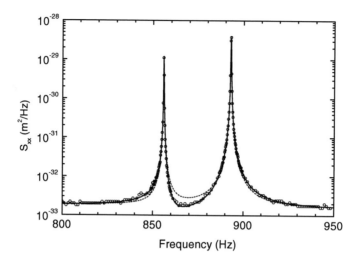

FIGURE 4. Total noise at detector output. Hollow circles: experimental data; solid line: data fit from direct fluctuation-dissipation theorem (also accounting for the noise floor); dashed line: data fitting with normal mode expansion (also accounting for the noise floor). From ref.[4].

THE ROOM TEMPERATURE DETECTOR

In order to test the opto-mechanical readout we have coupled it to a bar identical to that in AURIGA but kept at room temperature [4]. In this experiment we have used a simpler version of reference cavity with Invar spacer: one mirror was fixed to the spacer with a piezoelectric transducer. In this way the frequency tuning is wider and faster, and it is possible to follow the variations of the transducer cavity length caused by room temperature fluctuations. The worse frequency stability is not important here, by virtue of the high thermal noise of mechanical oscillators (bar and transducer). The bar is housed in a vacuum chamber and suspended through a cascade of mechanical filters to reduce vibration noise transmission from the floor. For this test the finesse of the transducer cavity was $2.8 \cdot 10^4$ but we plan to use a 10^5 finesse cavity for the cryogenic test of the full readout; the finesse of the reference cavity was $4.4 \cdot 10^4$.

The experimental setup of the room temperature detector equipped with the opto-mechanical readout is shown in fig. 1.

A zoom at around bar resonance of the total noise at the detector output is shown in fig. 4 The peaks are the thermally-excited modes formed by coupling the resonant transducer to the bar: they emerge from a noise floor resulting from readout. The data are fitted with thermal noise prediction and we find excellent agreement only if we take into account the non-homogeneously distributed losses: the usual normal-mode expansion, which assumes that losses are uniformly distributed, fails to describe our system a bit far from the resonance.

The sensitivity to gravitational waves of the room temperature bar detector equipped with the opto-mechanical readout is shown in fig. 5, together with that of AURIGA and other cryogenic bars. The lower sensitivity of our detector is explained by the room temperature operation. The useful bandwidth of our detector is 10 Hz at +3 dB and 50 Hz at +10 dB from minimum: to our knowledge this is the first wide-band bar detector ever operated.

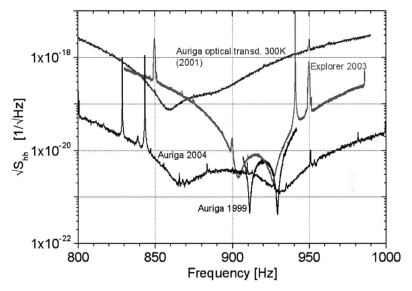

FIGURE 5. Sensitivity of the room temperature bar detector equipped with optomechanical readout compared to that of other cryogenic bars. EXPLORER data from ref. [9].

REFERENCES

1. J.P. Zendri *et al.*, "Status report of the gravitational wave detector AURIGA", in "Gravitational Waves and Experimental Gravity", XXXVIIIth Rencontres de Moriond, J. Dumarchez, J. Trân Thanh Vân Eds., Thê' Gió'i Publishers, Vietnam, 2003.
2. A. Vinante *et al.*, Appl. Phys. Lett. **79**, 2597 (2001).
3. J. P. Richard, J. Appl. Phys. **64**, 2202 (1988).
4. L. Conti *et al.*, J. Appl. Phys. **93**, 3589 (2003).
5. E. Giacobino, F. Marin, A. Bramati and V. Jost, J. Nonlin. Opt. Phys. Mat. **5**, 863 (1996).
6. L. Conti, M. De Rosa and F. Marin, Appl. Opt. **39**, 5732 (2000).
7. L. Conti, M. De Rosa and F. Marin, J. Opt. Soc. Am. **20**, 462 (2003).
8. R. W. P. Drever *et al.*, Appl. Phys. B **31**, 97 (1983).
9. M. Bassan, private communication.

Propagation of Neutrinos and Photons in Gravitational Fields

G. Lambiase, R. Punzi, G. Scarpetta* and G. Papini[†]

*Dipartimento di Fisica E.R. Caianiello, Universitá di Salerno, 84081 Baronissi (SA), Italy.
INFN - Gruppo Collegato di Salerno.
[†]Department of Physics, University of Regina, Regina, Sask, S4S0A2, Canada.

Abstract. We discuss the propagation of bosons and fermions in gravitational fields. The geometrical background is provided by a rotating gravitational source. Use is made of neutrino and photon wave functions that are exact to first order in the metric deviation. These are applied to the lensing of photons and neutrinos and to neutrino helicity transitions and flavor oscillations.

1. INTRODUCTION

The interaction of quantum systems with external gravitational fields is of interest in studies regarding the ultimate structure of spacetime. Obviously, the possibility to carry out experiments at the quantum gravity level is still rather remote. In order to bridge the gap between Planck's length and the typical dimensions at which quantum mechanics normally operates, we are resorting to the study of covariant wave equations. These describe particles, either bosons or fermions, whose behavior can be frequently studied with great precision. For this purpose Schrödinger, Klein–Gordon, Maxwell–Proca and Dirac equations have been discussed in the literature and have led to the observation of the Colella–Overhauser–Werner and Bonse–Wroblewski effects [1]. These equations can be solved exactly to first order in the weak-field approximation, if the solutions of the corresponding free-field equations are known [2]. An interesting property of these solutions is that gravity occurs in the wave functions as a quantum phase. Since phase differences can be measured, it is hoped that interferometers of high sensitivity will be able to observe some of the effects predicted in a not too distant future. In this contribution we apply the covariant Maxwell and Dirac equations [2] to problems like photon and neutrino lensing, neutrino helicity transitions and flavor oscillations. These problems are of interest in astrophysics and cosmology.

2. THE FIELD EQUATIONS FOR FERMIONS AND BOSONS

In this Section we briefly review the procedure followed in [2] to derive solutions of the Dirac and Klein–Gordon equations that are exact to first order in the weak-field approximation. A similar approach is also used in Sect. 2.2 to solve the covariant Maxwell equations.

CP751, *General Relativity and Gravitational Physics, 16th SIGRAV Conference*, edited by G. Esposito et al.
© 2005 American Institute of Physics 0-7354-0236-1/05/$22.50

2.1. Covariant Dirac Equation

The behavior of spin-1/2 particles in the presence of gravitational fields $g_{\mu\nu}$ is determined by the covariant Dirac equation

$$[i\gamma^\mu(x)\mathcal{D}_\mu - m]\Psi(x) = 0, \tag{1}$$

where $\mathcal{D}_\mu = \nabla_\mu + i\Gamma_\mu(x)$, ∇_μ is the covariant derivative, $\Gamma_\mu(x)$ the spin connection and the matrices $\gamma^\mu(x)$ satisfy the relations $\{\gamma^\mu(x), \gamma^\nu(x)\} = 2g^{\mu\nu}$. Both $\Gamma_\mu(x)$ and $\gamma^\mu(x)$ can be obtained from the usual flat spacetime Dirac matrices by using the vierbein fields $e_{\hat\alpha}^\mu$ and the relations

$$\gamma^\mu(x) = e_{\hat\alpha}^\mu(x)\gamma^{\hat\alpha}, \qquad \Gamma_\mu(x) = -\frac{i}{8}[\gamma^{\hat\alpha}, \gamma^{\hat\beta}]e_{\hat\alpha}^\nu e_{\nu\hat\beta;\mu}. \tag{2}$$

A semicolon and a comma are equivalent ways to denote covariant and partial derivatives, respectively. Equation (1) can be solved exactly to first order in the metric deviation $\gamma_{\mu\nu}(x) = g_{\mu\nu} - \eta_{\mu\nu}$, where the Minkowski metric $\eta_{\mu\nu}$ has signature -2. This is achieved by first transforming (1) into the equation

$$[i\tilde\gamma^\nu(x)\nabla_\nu - m]\tilde\Psi(x) = 0, \tag{3}$$

where

$$\tilde\Psi(x) = S^{-1}\Psi(x), \qquad S(x) = e^{-i\Phi_s(x)}, \tag{4}$$

$$\Phi_s(x) = \mathcal{P}\int_P^x dz^\lambda \Gamma_\lambda(z), \qquad \tilde\gamma^\mu(x) = S^{-1}\gamma^\mu(x)S.$$

By multiplying (3) on the left by $(-i\tilde\gamma^\nu(x)\nabla_\nu - m)$, one obtains the equation

$$(g^{\mu\nu}\nabla_\mu\nabla_\nu + m^2)\tilde\Psi(x) = 0, \tag{5}$$

which has the first-order exact solution

$$\tilde\Psi(x) = e^{-i\Phi_G(x)}\Psi_0(x), \tag{6}$$

where $\Phi_G(x)$ is defined as

$$\Phi_G(x) \equiv -\frac{1}{4}\int_P^x dz^\lambda \left[\gamma_{\alpha\lambda,\beta}(z) - \gamma_{\beta\lambda,\alpha}(z)\right]L^{\alpha\beta}(z) + \frac{1}{2}\int_P^x dz^\lambda \gamma_{\alpha\lambda}k^\alpha, \tag{7}$$

$$[L^{\alpha\beta}(z), \Psi_0(x)] = \left((x^\alpha - z^\alpha)k^\beta - (x^\beta - z^\beta)k^\alpha\right)\Psi_0(x), \quad [k^\alpha, \Psi_0(x)] = i\partial^\alpha\Psi_0,$$

and $\Psi_0(x)$ satisfies the usual flat spacetime Dirac equation. $L_{\alpha\beta}$ and k^α are the angular and linear momenta of the free particle. It follows from (6) and (4) that the solution of (1) can be written in the form

$$\Psi(x) = e^{-i\Phi_s}\left(-i\tilde\gamma^\mu(x)\nabla_\mu - m\right)e^{-i\Phi_G}\Psi_0(x), \tag{8}$$

and also as

$$\Psi(x) = -\frac{1}{2m}\left(-i\gamma^\mu(x)\mathscr{D}_\mu - m\right)e^{-i\Phi_T}\Psi_0(x), \tag{9}$$

where $\Phi_T = \Phi_s + \Phi_G$, as well as $\gamma^\mu(x)$, are first-order quantities in $\gamma_{\alpha\beta}(x)$. The factor $-1/2m$ on the r.h.s. of (9) is required by the condition that both sides of the equation agree when the gravitational field vanishes. It is useful to re-derive some known results from the covariant Dirac equation. On multiplying (1) on the left by $(-i\gamma^\nu(x)\mathscr{D}_\nu - m)$ and using the relations

$$\nabla_\mu\Gamma_\nu(x) - \nabla_\nu\Gamma_\mu(x) + i[\Gamma_\mu(x), \Gamma_\nu(x)] = -\frac{1}{4}\sigma^{\alpha\beta}(x)R_{\alpha\beta\mu\nu}, \tag{10}$$

and

$$[\mathscr{D}_\mu, \mathscr{D}_\nu] = -\frac{i}{4}\sigma^{\alpha\beta}(x)R_{\alpha\beta\mu\nu}, \tag{11}$$

one obtains the equation

$$\left(g^{\mu\nu}\mathscr{D}_\mu\mathscr{D}_\nu - \frac{R}{4} + m^2\right)\Psi(x) = 0. \tag{12}$$

In (11) and (12) $R_{\alpha\beta\mu\nu}$ is the Riemann tensor, R the Ricci scalar, and $\sigma^{\alpha\beta}(x) = (i/2)[\gamma^\alpha(x), \gamma^\beta(x)]$. Eq. (12) implies that the gyro-gravitational ratio of a massive Dirac particle is one when $R \neq 0$ [5]. This property can also be extended to the case $R = 0$ [2].

2.2. Covariant Maxwell–de Rahm Equations

Consider now the Maxwell equations with de Rahm term

$$\nabla_\nu\nabla^\nu A_\mu - R_{\mu\sigma}A^\sigma = 0, \tag{13}$$

where the electromagnetic field A_μ satisfies the condition $\nabla_\mu A^\mu = 0$. Equations (13) in the weak-field approximation are

$$\nabla_\nu\nabla^\nu A_\mu \simeq (\eta^{\sigma\alpha} - \gamma^{\sigma\alpha})A_{\mu,\alpha\sigma} + (\gamma_{\sigma\mu,\nu} + \gamma_{\sigma\nu,\mu} - \gamma_{\mu\nu,\sigma})A^{\sigma,\nu} = 0, \tag{14}$$

where the Lanczos–de Donder gauge condition $\gamma_{\alpha\nu,}{}^\nu - \frac{1}{2}\gamma^\sigma{}_{\sigma,\alpha} = 0$ has been used. Equation (14) has the solution [2]

$$A_\mu = e^{-i(\chi + k\cdot x)}a_\mu + \left[\int_P^x dz^\lambda\,\gamma_\mu^\alpha(z)k_\lambda a_\alpha - \frac{1}{2}\int_P^x dz^\lambda(\gamma_{\mu\lambda,}{}^\alpha - \gamma^\alpha{}_{\lambda,\mu})a_\alpha\right]e^{-ik\cdot x}, \tag{15}$$

where $k \cdot x = k_\mu x^\mu \simeq \eta_{\mu\nu}k^\mu x^\nu$,

$$\chi = -\frac{1}{4}\int_P^x dz^\lambda(\gamma_{\alpha\lambda,\beta}(z) - \gamma_{\beta\lambda,\alpha}(z))[(x^\alpha - z^\alpha)k^\beta - (x^\beta - z^\beta)k^\alpha] + \tag{16}$$

$$+\frac{1}{2}\int_P^x dz^\lambda\gamma_{\alpha\lambda}(z)k^\alpha,$$

and k^α is the momentum of the free photon.

3. APPLICATIONS

In what follows the gravitational background is represented by the Lense–Thirring metric which describes the field of a rotating source and has the post-Newtonian form

$$\gamma_{00} = 2\phi, \quad \gamma_{ij} = 2\phi\delta_{ij}, \quad \gamma_{0i} = h_i = \frac{2}{r^3}(\mathbf{J} \wedge \mathbf{r})_i. \tag{17}$$

If M, R and $\omega = (0, 0, \omega)$ are mass, radius and angular velocity of the source, we have $\phi = -\frac{GM}{r}$ and $\mathbf{h} = \frac{4GMR^2\omega}{5r^3}(y, -x, 0)$. The vierbein fields are

$$e_0^0 = 1 - \phi, \quad e_i^0 = 0, \quad e_0^i = h_i, \quad e_k^i = (1 + \phi)\delta_k^i. \tag{18}$$

In the following, we use the convention $e_{\hat{\alpha}}^\mu \simeq \delta_{\hat{\alpha}}^\mu + h_{\hat{\alpha}}^\mu$, where $h_{\hat{\alpha}}^\mu$ obviously contains the gravitational contribution.

3.1. Neutrino Lensing

It is clear from (7) and (8) that, once $\Psi_0(x)$ is chosen to be a plane-wave solution of the flat spacetime Dirac equation, the geometrical phase of a neutrino of momentum k^μ is given by

$$\upsilon(x) = -k_\alpha x^\alpha - \Phi_G(x), \tag{19}$$

where

$$\Phi_G(x) = -\frac{1}{4}\int_P^x dz^\lambda \left[\gamma_{\alpha\lambda,\beta}(z) - \gamma_{\beta\lambda,\alpha}(z)\right]((x^\alpha - z^\alpha)k^\beta - (x^\beta - z^\beta)k^\alpha) + \tag{20}$$

$$\frac{1}{2}\int_P^x dz^\lambda \gamma_{\alpha\lambda}k^\alpha.$$

The components of $\mathbf{p}_\perp$ can be determined from

$$p_i = \frac{\partial\upsilon}{\partial x^i} = -k_i - \Phi_{G,i} = \tag{21}$$

$$= -k_i - \frac{1}{2}\gamma_{\alpha i}(x)k^\alpha + \frac{1}{2}\int_P^x dz^\lambda (\gamma_{i\lambda,\beta}(z) - \gamma_{\beta\lambda,i}(z))k^\beta.$$

In what follows, we study the two relevant cases of propagation along z-axis (parallel to the angular momentum of the source) and x-axis (orthogonal to the angular momentum of the source). In both cases, we consider the ultrarelativistic regime, i.e. $dz^0 \simeq dz(1 + m^2/2E^2)$, $E \simeq k(1 + m^2/2E^2)$.

The deflection angle φ is defined by

$$\tan\varphi \equiv \frac{\sqrt{-g_{ij}p_\perp^i p_\perp^j}}{p_\parallel}, \tag{22}$$

where $\mathbf{p}_\perp$ and $p_\parallel$ are the orthogonal and parallel components of the momentum with respect to the initial direction of propagation. In the weak-field approximation $\tan\varphi \simeq \varphi$ and (22) reduces to

$$\varphi \simeq \frac{|\mathbf{p}_\perp|}{k_\parallel}, \tag{23}$$

where $k_\parallel = p_\parallel$ is the unperturbed momentum and $|\mathbf{p}_\perp| = \sqrt{-\eta_{ij}p_\perp^i p_\perp^j}$, for $p_\perp^i \sim \mathcal{O}(\gamma_{\mu\nu})$.

3.1.1. z-axis propagation

We consider neutrinos starting from $z = -\infty$ with impact parameter b. Without loss of generality, we assume that the neutrinos propagate in the plane $x = b, y = 0$. From (21) and (17), we find

$$p_1 = -2k\left(1 + \frac{m^2}{2E^2}\right)\int_{-\infty}^{z}\phi_{,1}dz, \quad p_2 = 0. \tag{24}$$

One can then evaluate $(p_\perp)^i$, $i = 1, 2$,

$$(p_\perp)^1 = g^{1\mu}p_\mu \simeq -p_1 = -\frac{2GMk}{b}\left(1 + \frac{m^2}{2E^2}\right)\left(1 + \frac{z}{r}\right), \tag{25}$$

$$(p_\perp)^2 = g^{2\mu}p_\mu \simeq h_2E = -\frac{4GMR^2\omega bk}{5r^3}\left(1 + \frac{m^2}{2E^2}\right).$$

From (23), we finally obtain

$$\varphi = \frac{2GM}{b}\left(1 + \frac{m^2}{2E^2}\right)\sqrt{\left(1 + \frac{z}{r}\right)^2 + \left(\frac{2R^2b^2\omega}{r^3}\right)^2}, \tag{26}$$

which is the deflection predicted by general relativity for photons with corrections resulting from the neutrino mass. In the limit $z \to \infty$, Eq. (26) reduces to

$$\varphi = \frac{4GM}{b}\left(1 + \frac{m^2}{2E^2}\right). \tag{27}$$

3.1.2. x-axis propagation

In this case neutrinos start from $x = -\infty$ with impact parameter b. For simplicity, we consider neutrinos propagating in the equatorial plane $z = 0, y = b$.

$$p_2 = -k\left(1 + \frac{m^2}{2E^2}\right)\int_{-\infty}^{x}(2\phi_{,2} + h_{1,2})dx, \quad p_3 = 0. \tag{28}$$

86

One gets

$$(p_\perp)^2 = -\frac{2GMk}{b}\left(1+\frac{m^2}{2E^2}\right)\left(1-\frac{2R^2\omega}{5b}\right)\left(1+\frac{z}{r}\right),\tag{29}$$

$$(p_\perp)^3 = 0.$$

It follows that

$$\varphi = \frac{2GM}{b}\left(1-\frac{2R^2\omega}{5b}\right)\left(1+\frac{m^2}{2E^2}\right)\left(1+\frac{x}{r}\right).\tag{30}$$

In this case, the contribution of the angular momentum of the gravitational source does not vanish in the limit $x \to \infty$. In fact, we find in this limit

$$\varphi = \frac{4GM}{b}\left(1-\frac{2R^2\omega}{5b}\right)\left(1+\frac{m^2}{2E^2}\right),\tag{31}$$

which coincides with the prediction of general relativity.

3.2. Photon Lensing

Let us now assume that the photons propagate in the z-direction, hence $k^\alpha \simeq (k,0,0,k)$, and $ds^2 = 0$ or $dt = dz$, because photons propagate along null geodesics. The photon momentum is given by [4]

$$k_i = \frac{\partial\chi}{\partial x^i},\tag{32}$$

where χ is defined in Eq. (16). It is convenient to write the photon momentum as

$$\mathbf{k} = \mathbf{k}_\perp + k_3\,\mathbf{e}_3,\quad \mathbf{k}_\perp = k_1\,\mathbf{e}_1 + k_1\,\mathbf{e}_2,\tag{33}$$

where $\mathbf{k}_\perp$ is the component of the momentum orthogonal to the propagation direction of photons.

In order to simplify our calculations, we consider the limit $P \to \infty$, i.e. the source of photons is located at very large distances with respect to the dimension of the lens, and the generic point is located in the z direction, i.e. $Q \to z$ and $z \gg x,y$. For non-rotating lenses, i.e. $h_3 = 0$, the components of the photon momentum are [4]

$$k_1 \sim k\frac{2GM}{R^2}x\left(1+\frac{z}{r}\right),\quad k_2 \sim k\frac{2GM}{R^2}y\left(1+\frac{z}{r}\right),\quad k_3 = k(1+\phi),\tag{34}$$

where $R = \sqrt{x^2+y^2}$. The definition of the deflection angle (22), with $p \to k$, gives

$$\tan\varphi \sim \varphi \sim \frac{2GM}{R}\left(1+\frac{z}{r}\right),\tag{35}$$

and, in the limit $z \to \infty$, we obtain the usual result $\varphi_M \sim \frac{4GM}{R}$.

4. NEUTRINO HELICITY TRANSITION

By using the same formalism, one can also calculate the probability that left-handed neutrinos be transformed into right-handed ones by virtue of the gravitational interaction. For this purpose, it is useful to introduce the following conventions and normalization for left/right neutrinos

$$\Psi_0(x) = v_{0L,R} e^{-ik_\alpha x^\alpha} = \sqrt{\frac{E+m}{2E}} \begin{pmatrix} v_{L,R} \\ \frac{\sigma^i k^i}{E+m} v_{L,R} \end{pmatrix} e^{-ik_\alpha x^\alpha}, \tag{36}$$

where $\sigma^i = (\sigma^1, \sigma^2, \sigma^3)$ are the Pauli matrices and $v_{L,R}$ are eigenvectors of $\sigma^i k_i$ corresponding to negative and positive eigenvalues. Note that $\bar{v}_{0L,R}(k) \equiv v_{0L,R}^\dagger(k)\gamma^0$, $v_{0L,R}^\dagger(k)v_{0L,R}(k) = 1$. The transition probability is [3]

$$P_{L \to R} = \left| \int_{\lambda_0}^{\lambda} \langle v_R | \frac{dx^\mu}{d\lambda} \partial_\mu \hat{T} | v_L \rangle d\lambda \right|^2. \tag{37}$$

Here $dx^\mu/d\lambda = k^\mu/m$, and (to lowest order)

$$\partial_\mu \hat{T} = \frac{1}{2m} \left(-i2m\Phi_{G,\mu} - i(\gamma^{\hat{\alpha}} k_\alpha + m)\Phi_{s,\mu} + \gamma^{\hat{\alpha}} (h_{\hat{\alpha},\mu}^\beta k_\beta + \Phi_{G,\alpha\mu}) \right) \tag{38}$$

$$\Phi_{s,\lambda} = \Gamma_\lambda, \quad \Phi_{G,\alpha\mu} = k_\beta \Gamma_{\alpha\mu}^\beta, \quad v_0^\dagger(\gamma^{\hat{\alpha}} k_\alpha + m) = 2E v_0^\dagger \gamma^0.$$

$\Gamma_{\alpha\mu}^\beta$ are the Christoffel symbols. Equations (37) and (38) do not specify the propagation direction of neutrinos. We consider the following two cases:

- For neutrinos propagating along the z-axis, we get

$$P_{L \to R}(-\infty, z) \simeq \left(\frac{m}{E}\right)^2 \left(\frac{GM}{2b}\right)^2 \left[\left(1 + \frac{z}{r}\right)^2 + \left(\frac{2\omega b^2 R^2}{5r^3}\right)^2\right]. \tag{39}$$

The first term results from the mass of the gravitational source, the second from its angular momentum. This last term vanishes for $r \to \infty$ because it is odd, i.e. the contribution from $-\infty$ to 0 exactly cancels out the one from 0 to $+\infty$. In fact, if we consider neutrinos propagating from 0 to $+\infty$, we obtain

$$P_{L \to R}(0, +\infty) \simeq \left(\frac{m}{E}\right)^2 \left(\frac{GM}{2b}\right)^2 \left[1 + \left(\frac{2\omega R^2}{5b}\right)^2\right]. \tag{40}$$

- When propagation takes place along the x-axis, the transition probability is

$$P_{L \to R}(-\infty, x) \simeq \left(\frac{m}{E}\right)^2 \left(\frac{GM}{2b}\right)^2 \left(1 - \frac{2\omega R^2}{5b}\right)^2 \left(1 + \frac{x}{r}\right)^2. \tag{41}$$

Of course, the mass contribution is the same as that for z-axis propagation. On the contrary, the two cases differ substantially in the behavior of the angular momentum term. In this case, in fact, this term is even, hence it does not vanish for $r \to \infty$. If we consider neutrinos generated at $x = 0$ and propagating to infinity, we find

$$P_{L \to R}(0, +\infty) \simeq \left(\frac{m}{E}\right)^2 \left(\frac{GM}{2b}\right)^2 \left(1 - \frac{2\omega R^2}{5b}\right)^2 . \tag{42}$$

5. NEUTRINO OSCILLATIONS

To study the effects of the gravitational field on neutrino oscillations, it is convenient to recast Eq. (9) into a form that contains only first-order contributions. By using (9), we find

$$\Psi(x) = f(x) e^{-i\Phi_T - i k_\alpha x^\alpha} \Psi_0 , \tag{43}$$

where

$$f(x) = \frac{1}{2m} \left[e_{\hat{\alpha}}^\mu \gamma^{\hat{\alpha}} (k_\mu + \Phi_{G,\mu}) + m \right] , \tag{44}$$

and Ψ_0 is given by the phase independent part of (9) in the case of mass eigenstate neutrinos. The relationship between flavor eigenstates (Greek indices) and mass eigenstates (Latin indices) is given by the standard expression

$$|v_\alpha(x)\rangle = \sum_j U_{\alpha j}(\theta) |v_j(x)\rangle , \tag{45}$$

into which (43) must now be substituted. We obtain

$$|v_\alpha(x(\lambda))\rangle = \sum_j U_{\alpha j}(\theta) f_j e^{-i\Phi_T^j(x) - i k_\alpha^j x^\alpha} |v^j\rangle , \tag{46}$$

where U is the mixing matrix

$$U = \begin{pmatrix} \cos\theta & \sin\theta \\ -\sin\theta & \cos\theta \end{pmatrix} . \tag{47}$$

Restricting α to e for simplicity, we define the column matrix

$$\chi = \begin{pmatrix} \langle v_e | v_e(x) \rangle \\ \langle v_\mu | v_e(x) \rangle \end{pmatrix} . \tag{48}$$

To order $\mathcal{O}(\gamma_{\mu\nu} m_i^2 / E^2)$, the evolution equation is [3]

$$i\frac{d\chi}{d\lambda} \simeq \left(\frac{M_f^2}{2} + k^\rho (\Phi_{G,\rho}^{(f)} + \Gamma_\rho^{(f)}) I \right) \chi , \tag{49}$$

where I is the identity matrix,

$$M_f^2 = U^\dagger \begin{pmatrix} m_1^2 & 0 \\ 0 & m_2^2 \end{pmatrix} U , \tag{50}$$

$$\Phi^f_G = U^\dagger \begin{pmatrix} \langle \zeta_1 | k^\rho \Phi_{G,\rho} | \zeta_1 \rangle & 0 \\ 0 & \langle \zeta_2 | k^\rho \Phi_{G,\rho} | \zeta_2 \rangle \end{pmatrix} U, \tag{51}$$

and Γ^f_ρ is related to Γ_ρ in a similar way. As usual, we assume that neutrinos propagate along null geodesics to prevent the mass eigenstates from being measured at different positions or times, thus destroying the interference pattern. The flavor oscillation probabilities are then [3]

$$P_{\nu_e \to \nu_e} = 1 - \sin^2 2\theta \sin^2 \left(\frac{z - z_0}{2L} + \frac{\Delta \Phi_G}{2} \right), \tag{52}$$

$$P_{\nu_\mu \to \nu_e} = \sin^2 2\theta \sin^2 \left(\frac{z - z_0}{2L} + \frac{\Delta \Phi_G}{2} \right),$$

where

$$\Delta \Phi_G = -\frac{\Delta m^2}{4E} [(z - z_0) \gamma_{00}(z_0)]. \tag{53}$$

The phase entering (52) has the form

$$\Omega \equiv \frac{z - z_0}{2L} + \frac{\Delta \Phi_G}{2} \sim \frac{z - z_0}{2L}, \tag{54}$$

because the constant factor $1 - \gamma_{00}(z_0)/2 \simeq (g_{00}(z_0))^{-1/2}$ can be absorbed in a redefinition of the constant $E \to E/\sqrt{g_{00}(z_0)}$, where E is the neutrino energy measured by an inertial observer at rest at infinity. E is, in fact, conserved on the geodesic along which the neutrinos move, because the metric (17) is stationary. Neutrino states are hence assumed to be eigenstates of this quantity. The coordinate difference $z - z_0$ is not the physical distance, the latter being defined as $l = \int_{z_0}^z \sqrt{-g_{33}} \, dz' \sim \int_{z_0}^z (1 - \gamma_{00}/2) dz'$. The phase Ω can be rewritten in terms of the physical quantities by introducing $E_l = E/\sqrt{g_{00}(z)}$, which is the energy measured by a locally inertial observer momentarily at rest in the gravitational field. Since $z - z_0 = \int_{l_0}^l dl/\sqrt{-g_{33}}$, we find

$$\Omega = \frac{\Delta m^2}{4E} (z - z_0) \simeq \frac{\Delta m^2}{4} \int_{l_0}^l \frac{dl}{E_l}, \tag{55}$$

which reflects the fact that the space-time curvature affects the oscillation probability through the gravitational red-shift of the local energy E_l and the proper distance dl.

6. CONCLUSIONS

The formalism presented in this paper can be applied to a variety of problems in astroparticle physics. The lepton asymmetry in the Universe [6] is indeed one such problem. As is well known, lepton asymmetry is generated as a consequence of the active-sterile oscillation of neutrinos, which gives rise to a discrepancy in the neutrino and antineutrino number densities. The lepton number of a neutrino of flavor f is defined by

$$L_f \equiv (n_{\nu_f} - n_{\bar{\nu}_f})/n_\gamma(T),$$

where n_{ν_f} ($n_{\bar{\nu}_f}$) is the number density of neutrinos (antineutrinos), and $n_\gamma(T)$ is the number density of photons at temperature T. As discussed in this paper, the gravitational field may generate transition from left-handed (active) neutrino to right-handed (sterile or anti-) neutrinos. It is then possible, in principle, that this lepton asymmetry be generated by an helicity transition, or that the latter might contribute to it (see Eqs. (40) and (42)). Another issue of interest is related to the variation of the ratio $\nu_e : \nu_\mu : \nu_\tau$ to $\nu_e : \nu_\mu : \nu'_\tau$. In other words, could the τ-neutrino flux measured at Earth be reduced relative to the expected one because of gravity induced helicity and flavor transitions $\nu_\tau \to \nu_X$? Here ν_X is a neutrino with a different flavor or helicity relative to ν_τ. Similar arguments hold, of course, for the other flavors. Work along these directions is currently in progress [3].

REFERENCES

1. R. Colella, A.W. Overhauser, S.A. Werner, *Phys. Rev. Lett.* **34**, 1472 (1975); U. Bonse, T. Wroblewski, *Phys. Rev. Lett.* **51**, 1401 (1983).
2. Y.Q. Cai, G. Papini, *Phys. Rev. Lett.* **66**, 1259 (1991); D. Singh, G. Papini, *N. Cim. B* **115**, 233 (2000); G. Papini, in "Relativity in Rotating Frames", Edited by G. Rizzi and M.L. Ruggiero, Kluwer Academic Publishers, Dordrecht, 2004, Ch. 16, and references therein (see also gr-qc/0304082).
3. G. Lambiase, G. Papini, R. Punzi, G. Scarpetta, *Neutrino Optics and Oscillations in Weak Gravitational Fields*, in preparation.
4. G. Lambiase, G. Papini, R. Punzi, G. Scarpetta, *Optics in Weak Gravitational Fields*, in preparation.
5. C.G. De Oliveira, J. Tiomno, *Nuovo Cimento* **24**, 672 (1962); J. Audretsch, *J. Phys.* **14**, 411 (1981); L. Kannenberg, *Ann. Phys. (N.Y.)* **103**, 64 (1977).
6. A.D. Dolgov, *Phys. Rep.* **370**, 333 (2002).

Virgo and the worldwide search for gravitational waves

F. Acernese*, P. Amico†, S.Aoudia**, N. Arnaud‡, S. Avino*, D.Babusci§, G. Ballardin¶, R. Barillé¶, F. Barone*, L. Barsotti ‖, M. Barsuglia‡, F. Beauville††, M.A. Bizouard‡, C. Boccara‡‡, F. Bondu**, L. Bosi†, S. Braccini‖, C. Bradaschia‖, A. Brillet**, V. Brisson‡, L. Brocco§§, D. Buskulic††, E. Calloni*, E. Campagna¶¶, F. Cavalier‡, R. Cavalieri¶, G. Cella‖, E. Chassande-Mottin**, F. Cleva**, J.-P. Coulon**, E. Cuoco¶, V. Dattilo¶, M. Davier‡, R. De Rosa*, L. Di Fiore*, A. Di Virgilio‖, B. Dujardin**, A. Eleuteri*, D. Enard¶, I. Ferrante‖, F. Fidecaro‖, I. Fiori‖, R. Flaminio¶, J.-D. Fournier**, S. Frasca §§, F. Frasconi¶, A. Freise¶, L. Gammaitoni†, A. Gennai¶, A. Giazotto‖, G.Giordano§, L. Giordano*, R. Gouaty††, D. Grosjean††, G. Guidi¶¶, S.Hebri**, H. Heitmann**, P. Hello‡, P. Heusse‡, L. Holloway¶, S. Kreckelbergh‡, P. La Penna¶, V. Loriette‡‡, G. Losurdo¶¶, M. Loupias¶, J.-M. Mackowski***, E. Majorana§§, C. N. Man**, F. Marchesoni†, F. Marion††, J. Marque¶, F. Martelli¶¶, A. Masserot††, M. Mazzoni¶¶, L. Milano*, C. Moins¶, J. Moreau‡‡, N. Morgado***, B. Mours††, J. Pacheco**, A. Pai §§, C. Palomba §§, F. Paoletti¶, S. Pardi*, A. Pasqualetti¶, R. Passaquieti‖, D. Passuello‖, S.Peirani**, B. Perniola¶¶, F. Piergiovanni¶¶, L. Pinard***, R. Poggiani ‖, M. Punturo†, P. Puppo§§, K. Qipiani*, P. Rapagnani §§, V. Reita‡‡, A.Remillieux***, F. Ricci §§, I. Ricciardi*, P. Ruggi¶, G. Russo*, S. Solimeno*, A. Spallicci**, R. Stanga¶¶, R. Taddei¶, D. Tombolato††, E. Tournefier††, F. Travasso†, D. Verkindt††, F. Vetrano¶¶, A. Viceré¶¶, J.-Y. Vinet**, H. Vocca†, M. Yvert†† and Z. Zhang¶

*INFN - Sezione di Napoli and/or Università di Napoli "Federico II" Complesso Universitario di Monte S. Angelo Via Cintia, I-80126 Napoli, Italia and/or Università di Salerno Via Ponte Don Melillo, I-84084 Fisciano (Salerno), Italia
†INFN Sezione di Perugia and/or Università di Perugia, Via A. Pascoli, I-06123 Perugia - Italia
**Department Artemis - Observatoire de la Côte d'Azur - BP 4229, 06304 Nice Cedex 4 - France
‡Laboratoire de l'Accélérateur Linéaire (LAL),IN2P3/CNRS-Université de Paris-Sud, B.P. 34, 91898 Orsay Cedex - France
§INFN, Laboratori Nazionali di Frascati Via E. Fermi, 40, I-00044 Frascati (Roma) - Italia
¶European Gravitational Observatory (EGO), Via E. Amaldi, I-56021 Cascina (PI) Italia
‖INFN - Sezione di Pisa and/or Università di Pisa, Via Filippo Buonarroti, 2 I-56127 PISA - Italia
††Laboratoire d'Annecy-le-Vieux de physique des particules Chemin de Bellevue - BP 110, 74941 Annecy-le-Vieux Cedex - France
‡‡ESPCI - 10, rue Vauquelin, 75005 Paris - France
§§INFN, Sezione di Roma and/or Università "La Sapienza", P.le A. Moro 2, I-00185, Roma
¶¶INFN - Sezione Firenze/Urbino Via G.Sansone 1, I-50019 Sesto Fiorentino; and/or Università di Firenze, Largo E.Fermi 2, I - 50125 Firenze, and/or Università di Urbino, Via S.Chiara, 27 I-61029 Urbino, Italia

CP751, *General Relativity and Gravitational Physics, 16th SIGRAV Conference*, edited by G. Esposito et al.
© 2005 American Institute of Physics 0-7354-0236-1/05/$22.50

*** SMA - IPNL 22, Boulevard Niels Bohr 69622 - Villeurbanne- Lyon Cedex France

Abstract. Large interferometric detectors of gravitational waves are approaching their design sensitivities and the plans for second generation detectors, expected to start the gravitational wave astronomy, are already under way. The goal of this paper is to give an overview of the status and perspectives of gravitational wave research with ground-based interferometric detectors in the world, with special attention to the French-Italian detector Virgo. The main design features and the status of largest detectors are reviewed, the main upgrades to the first generation detectors foreseen in the next years are outlined.

INTRODUCTION

A community of experimentalists has been working since the sixties to build detectors sensitive enough to detect gravitational waves (GW). They have two main goals: to provide a direct proof that GW exist and compare their properties with the predictions of general relativity and other metric theories of gravity; to open a new window on the universe giving rise to the GW astronomy. Observing the universe in this window may drastically change our view. The gravitational radiation brings information of a relativistic universe unaccessible to the optical telescopes: black holes, cores of supernovae, strongly interacting compact stars could be observed directly. Two main families of GW detectors have been studied and realized: the *resonant detectors* or *bars* [1] and the *interferometric detectors*, subject of this paper.

The idea of using interferometry to detect GW was first formulated by Gertsenshtein and Pustovoit in 1963 [2]. The first complete work on the noise limiting the sensitivity of an interferometric detector is due to Weiss [3], while Forward [4] did the first experimental attempt in 1978.

The theory describes the interaction of GW propagating on an Euclidean space-time with freely falling test masses. The interferometer armlengths are modulated by the passage of GW in a peculiar way, associated to the quadrupolar nature of the waves (for an explanatory description see ref. [5]). The relative displacement of the two end mirrors in a Michelson interferometer induced by GW is

$$\Delta L \sim hL$$

where L is the interferometer armlength and h is the GW strain amplitude. Such an amplitude is very small: two neutron stars coalescing in the Virgo cluster emit GW that reach the Earth with an amplitude $h \sim 10^{-21}$.

The first generation of ground-based interferometric detectors has been developed in the last decade. The construction of a few of them has been completed and they are presently taking data or completing the commissioning:

- **LIGO**: two detectors in the USA at Hanford (WA) and Livingston (LA)4 km long [6]. The Hanford detector hosts two interferometers in the same vacuum system;
- **Virgo**: French-Italian interferometer in Italy (Cascina, PI), 3 km long [7] (fig.1);

---

[1] Presented by G.Losurdo, e-mail: losurdo@fi.infn.it

FIGURE 1. Aerial view of the Virgo detector.

- **GEO600**: British-German project, 600 m long interferometer located in Hannover, Germany [8];
- **TAMA**: 300 m long Japanese interferometer, located near Tokyo [9].

All these detectors are Michelson interferometers, using various optical techniques to enhance their sensitivity. Virgo and LIGO, when running at their design sensitivity, will be the most sensitive ones. In the following we will describe the main design features (common to all detectors) and some peculiar to Virgo. We will analyze the perspectives of GW detection according to the current theoretical estimates on the expected sources. Eventually, we will outline the plans for second-generation interferometers.

DETECTOR DESIGN

The designs of Virgo and LIGO are similar. They are long Michelson interferometers (3 km and 4 km respectively), the mirrors of which are located in an ultra-high vacuum system and suspended in order to be isolated from natural vibrations of the ground. The performance of the vibration isolators determines the low-frequency cutoff of the detector bandwidth. Virgo has developed the most advanced suspension system (see fig.2), able to make the residual seismic vibrations negligible above 4 Hz. For LIGO, the cutoff is about 40 Hz. Above the cutoff the mirrors can be considered as "freely falling". The interferometer response to GW is enhanced by embedding Fabry–Perot cavities in the interferometer arms: the light is stored between two mirrors for a certain time, effectively increasing the optical length and hence the interaction time with the gravitational radiation. Optical cavities work as amplifiers of the interferometer phase sensitivity to GW. The drawback is that they work only at resonance: the interferometer armlength must be maintained constant within the linear range of the cavity (about 10^{-12} m). Therefore, interferometers have to be provided with a sophisticated control system,

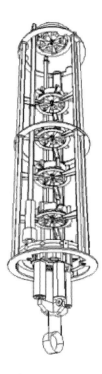

FIGURE 2. The Virgo vibration isolator for suspending the mirrors: the Superattenuator [10]. It is composed by a chain of pendulums, suspended from an inverted pendulum platform actively controlled. The whole suspension is designed in order to have low-frequency normal modes (below 3 Hz), thus achieving an attenuation of the seismic noise of about 10^{-14} at Hz. Control action can be performed at three different stages of the Superattenuator, allowing to control the mirror over a large dynamic range in a hierarchical way.

able to compensate for the large tidal strain or for the seismic excitation of the mirror in the low-frequency region where the vibration isolators do not isolate, and to keep the cavities in resonance without re-injecting control noise in the detection bandwidth. The same control system keeps the interferometer output on the *dark fringe*: in this condition the dependence on the laser amplitude fluctuations is minimized.

Above the low-frequency cutoff, the design sensitivity of an interferometric detector is limited by two fundamental noise sources:

- **shot noise:** the quantum nature of light sets an ultimate limit to the sensing process. The discrete arrival of photons on the photodiode is analogous to stochastic power fluctuations that mimic the effect of GW. The relevance of this noise depends on the input laser power P_0 as $1/\sqrt{P_0}$. Therefore, the interferometer sensitivity can be increased by using more power. The Virgo laser has an input power of about 20 Watts. Such a power would not be enough to reach the sensitivity necessary to detect a coalescence of two neutron stars in the Virgo cluster. To increase the power

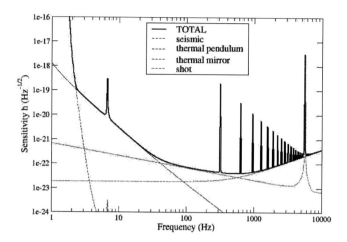

FIGURE 3. Virgo design sensitivity: unlike the *bars*, interferometers are *wideband* detectors. It is possible to search for GW in the frequency range from a few Hz to a few kHz. Therefore, sources of different nature and emitting in a wide range of frequencies can, in principle, be detected.

an optical technique called "power recycling" has been applied [11]: since most of the power would be reflected towards the laser (and thus wasted) one more mirror is used to reflect it back towards the beam splitter. In this way, the power impinging on the beam splitter can be increased by a factor of 50, reducing the shot noise level by a factor of ~ 7. The use of another optical cavity, to be kept in resonance, makes the overall control of the interferometer more difficult.

- **thermal noise:** the Brownian motion of the mirror surface and the suspension wires associated to internal dissipation processes sets a limit to the sensitivity that can be pushed down by cooling the test masses or by using low-dissipation materials [12]. First generation interferometers are run at room temperature. The mirrors are made of high purity fused silica. The mirrors of GEO600 are suspended by means of fused silica fibers to reduce the dissipation in the wires. Virgo and LIGO use suspension wires made of special steel: they are worse with respect to silica fibers for thermal noise issues, but are more robust and reliable.

Figure 3 shows the Virgo design sensitivity [13]: it is limited by thermal noise up to about 500 Hz and by shot noise above. A simplified optical scheme of the Virgo detector is shown in Fig.4.

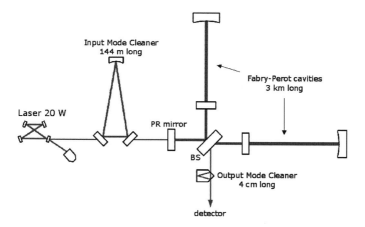

FIGURE 4. Virgo simplified optical scheme: the laser is spatially filtered by a triangular optical cavity (the *mode cleaner*, 144 m long. It is split in the two arms and resonates into the two 3 km long Fabry–Perot cavities. The fraction of the light going to the detector is filtered through a 4 cm long output mode cleaner. The light going towards the laser is reflected back towards the interferometer, with the right phase, by the *power recycling mirror*.

STATUS OF THE DETECTORS

Virgo is in the commissioning phase (the commissioning of the whole detector started in 2003). Fig.5 describes the evolution of the Virgo sensitivity curve over the last months: several orders of magnitude are still missing to reach the design sensitivity, but the progress so far has been quite fast. Virgo is now running in the so-called "recombined configuration": the two 3 km cavities are resonant and the interferometer output is locked on the dark fringe. One more step, the locking of the power recycling cavity, is needed for the full operation. The first Virgo science run is foreseen during 2005.

Among all detectors the two LIGO are the ones in the most advanced state. LIGO commissioning started in 1999. Three science runs have been performed with increasing sensitivity: the first run lasted 17 days and the detector might in principle detect a coalescence of two neutron stars at a distance of 0.2 Mpc. The results of the analysis of the first runs have been published [14, 15, 16, 17] and have established new upper limits on sources of different kinds. The second and third runs lasted 59 days. The detection range for a two neutron stars coalescence event in the third run was 2.2 Mpc (about 1/10 of the range expected with the design sensitivity). Recently, the Hanford interferometer has been operated with a sensitivity very close to the design one: this is very important for the whole GW community, since it demonstrates the feasibility of the project.

WHAT CAN WE SEE WITH FIRST-GENERATION DETECTORS?

It is likely that, during 2005, three independent detectors (Virgo and the two LIGO) will be running with sensitivity enough to detect the coalescence of two neutron stars at a

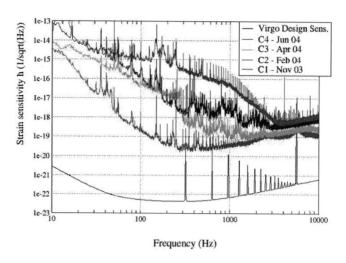

FIGURE 5. The evolution of the Virgo sensitivity since the beginning of the commissioning.

distance of 20 Mpc. The coalescence of two compact stars (neutron stars or black holes) is likely to be the first event to be detected: the emission of GW is very efficient and the waveform is accurately predicted. On the other hand, the rate of such events is uncertain. The best estimates indicate that Virgo/LIGO might expect to detect a few events per year [18, 19].

The probability of detecting a supernova collapse is lower: the efficiency of GW emission in a collapse depends on the asymmetry of the mass distribution and is badly known. The present simulations suggest that Virgo/LIGO might detect a collapse at distances up to 50 kpc: the detection range for first-generation interferometers is thus limited to the Milky Way, where one event every 30/50 years is expected. Moreover, the waveform is not well predicted (though supercomputers are being dedicated to run the simulation codes), preventing the use of optimal filtering techniques for the signal extraction. Supernovae detection will require the analysis of coincidences in the various detectors, as has been done by the *bars* [20].

Periodic GW can be emitted by rotating neutron stars if the mass distribution is non-axisymmetric. The expected signals are rather weak, but the signal-to-noise ratio can be increased by integrating for a long time (weeks or months). Upper limits on the emission expected from some known galactic pulsars are within the reach of the first generation detectors, but the detection of such signals is unlikely.

To sum up, ground-based first-generation interferometers may realize the first detection if nature helps, but they will not see a large number of events. To start a gravitational-wave astronomy more sensitive detectors are needed.

PLANS FOR ADVANCED DETECTORS: TOWARDS A GW ASTRONOMY

Plans to enhance the detector sensitivity and produce a second generation of interferometric detectors are on the way. Several R&D projects have been started and are giving results. To beat the limiting noise of Fig.3 one needs to:

- reduce the thermal noise resulting from the dissipation in the wires suspending the mirrors. This can be achieved by replacing the steel wires with fused silica fibers, as already done in GEO600;
- reduce the thermal noise due to the dissipation in the test mass substrate. Sapphire and silicon test masses are being investigated on purpose and an effort to understand the dissipation mechanisms in the mirror coatings is going on. Moreover, a reduction of mirror thermal noise can be obtained also by using a non-Gaussian laser beam, shaping the wavefront in order to make it less sensitive to the thermally driven test mass surface deformations.
- reduce the shot noise. This can be achieved by increasing the laser power and the storage of power inside the cavities. Tests on lasers ten times more powerful than the Virgo laser are well advanced. Moreover, a special optical technique, called signal recycling, can be used to reduce the shot noise in part of the detector bandwidth. Signal recycling requires one more optical cavity into the interferometer.

On paper, the developments of such R&D programs promise to enhance the sensitivity by ~ 10 times on the whole detector bandwidth. Second-generation detectors might observe ten times farther and, thus, access a volume of universe 10^3 times larger (with a detection rate increased by the same factor for sources at cosmological distances).

At the moment, a complete (and approved) proposal for *Advanced LIGO* exists [21]. Besides the upgrades described above, Advanced LIGO will replace the vibration isolators with completely new ones, able to widen the lower detector bandwidth cutoff towards 10 Hz. The upgrade of LIGO to Advanced LIGO will start in 2007.

The Virgo collaboration is starting a similar effort to define a detector upgrade. Virgo is already provided with vibration isolators suitable for advanced detectors. The required upgrades are presently being discussed.

CONCLUSIONS

The worldwide search for GW has been boosted by the construction of the wideband interferometric detectors. In 2005 three kilometric detectors, the two LIGO and Virgo, might be running with a sensitivity close to the design one. They might do the first detection, but to start a GW astronomy one has to wait for the second generation of interferometers. Those upgraded detectors will be realized at the end of the decade. Together with LISA, the 5 million km space interferometer, they might provide a new picture of the universe.

REFERENCES

1. Visco, M. (2004), these proceedings.
2. Gersenshtein, M., and Pustovoit, V., *Sov. Phys. - JETP*, **16**, 433 (1963).
3. Weiss, R., *Quart. Prog. Rep. Res. Lab. Electron. MIT*, **105**, 54 (1972).
4. Forward, R., *Phys. Rev. D*, **17**, 379 (1978).
5. Saulson, P., *Fundamentals of Interferometric Gravitational Wave Detectors*, World Scientific, 1994.
6. Abramovici, A. *et al.*, *Science*, **256**, 325 (1992), URL http://ligo.caltech.edu.
7. Bradaschia, C. *et al.*, *Nucl. Instrum. Meth. Phys. Res. A*, **289**, 518 (1992), URL http://virgo.infn.it.
8. Danzmann, K. *et al.*, GEO600: Proposal for a 600 m Laser Interferometric Gravitational Wave Antenna, Tech. Rep. 190, Max Planck Institut für Quantenoptik, Garching, Germany (1994), URL http://www.geo600.uni-hannover.de.
9. Tsubono, K., "300 m Laser Interferometer Gravitational Wave Detector (TAMA300) in Japan," in *Gravitational Wave Experiments*, edited by E. Coccia, G. Pizzella, and F. Ronga, World Scientific, Singapore, 1995, pp. 112–114, URL http://tamago.mtk.nao.ac.jp.
10. Ballardin, G. *et al..*, *Rev. Sci. Instrum.*, **72**, 3643 (2001).
11. Meers, B., *Phys. Rev. D*, **38**, 2317 (1988).
12. Saulson, P., *Phys. Rev. D*, **42**, 2437 (1990).
13. Punturo, M., The Virgo Sensitivity Curve, Tech. Rep. VIR-NOT-PER-1390-51, INFN Perugia (2003), URL http://www.virgo.infn.it/senscurve.
14. Abbot, B. *et al..*, *Phys. Rev. D*, **69**, 082004 (2004).
15. Abbot, B. *et al..*, *Phys. Rev. D*, **69**, 102001 (2004).
16. Abbot, B. *et al..*, *Phys. Rev. D*, **69**, 122001 (2004).
17. Abbot, B. *et al..*, *Phys. Rev. D*, **69**, 122004 (2004).
18. Grishchuk, L. *et al..*, *Physics-Uspekhi*, **44**, 1 (2001), arXive: astro-ph/0008481.
19. Burgay, M. *et al..*, *Nature*, **426**, 531 (2003).
20. Astone, P. *et al..*, *Phys. Rev. D*, **68**, 022001 (2003).
21. Advanced ligo proposal (2003), URL http://ligo.caltech.edu/advLIGO.

Analytical modelling of gravitational collapse

Giulio Magli

Dipartimento di Matematica, Politecnico di Milano, Piazza Leonardo da Vinci, 32, 20133 Milano, Italy, e-mail magli@mate.polimi.it

Abstract. Analytical modelling of gravitational collapse is an open problem in General Relativity. No satisfactory non-spherical model is known, and even in spherical symmetry our knowledge is still fragmentary. In the present paper status and perspectives in this field of research are briefly reviewed, putting special emphasis on the companion problem of obtaining exact solutions of the Einstein field equations in matter.

1. INTRODUCTION

Studying gravitational collapse in the framework of General Relativity means trying to gain insight into problems like the dynamics of black hole formation, the emission of gravitational waves in violent astrophysical events, and the *Cosmic Censorship Conjecture*.This conjecture, first formulated by Roger Penrose [1], asserts (roughly speaking) that complete gravitational collapse should always lead to singularities "covered" by an event horizon in physically realistic situations.

In recent years relevant advances have been made in the field of numerical modelling of gravitational collapse. However, numerical relativity alone cannot, of course, be of much help in addressing topics like the Cosmic Censorship. A famous and important example is Choptuik's numerical discovery of critical behavior in the gravitational collapse of scalar fields, a discovery which still awaits for a complete theoretical explanation. In the present contribution I will briefly review the status of analytical modelling of gravitational collapse and its relationship with the Cosmic Censorship.

What is known *analytically* in gravitational collapse is essentially restricted to spherical symmetry, one exception being given by the Szekeres "quasi-spherical" spacetimes [2], which however have a causal structure similar to their spherical counterparts. The situation with respect to the true astrophysical problem of collapse, in which departures from sphericity and, consequently, emission of gravitational waves have to be expected, is therefore very bad. Standing this situation, the recent research in this field developed in the direction of obtaining at least an acceptable understanding of spherical symmetry in order to proceed via perturbation theory in the approach to non-sphericity.

The analytical framework for spherically symmetric fields can be summarized as follows. The general non-static line element in comoving coordinates t, r, θ, φ can be written in terms of three functions ν, η, R of r and t:

$$ds^2 = -e^{2\nu}dt^2 + \eta^{-1}dr^2 + R^2(d\theta^2 + \sin^2\theta d\varphi^2) \,. \tag{1}$$

The collapsing body is characterized via its energy-momentum tensor which has the

CP751, *General Relativity and Gravitational Physics, 16ᵗʰ SIGRAV Conference*, edited by G. Esposito et al.

form $T_\nu^\mu = \text{diag}(-\varepsilon, P_r, P_t, P_t)$. Here the radial stress P_r and the tangential stress P_t have to be understood as expressed, whenever possible, in terms of the stress-strain relations for the material. In the case of perfect fluids (isotropic pressure $P = P_r = P_t$) these relations reduce to the equation of state having the form $P = P(\varepsilon)$.

Once an equation of state has been chosen, the Einstein field equations become a closed system; in spherical symmetry there are three independent equations for the three variables ν, η and R. It has proven, however, to be very useful to write the field equations as a system of four compatible equations. This is done by introducing the *mass function* [3] defined as

$$m(r,t) = \frac{R}{2}\left(1 - R'^2\eta + \dot{R}^2 e^{-2\nu}\right) , \tag{2}$$

where a dash and a dot denote derivatives with respect to r and t, respectively. The mass function makes it possible to write the field equations as

$$m' = 4\pi\varepsilon R^2 R' , \tag{3}$$

$$\dot{m} = -4\pi P_r R^2 \dot{R} , \tag{4}$$

$$R'\dot{\eta} = -2\eta(\dot{R}' - \dot{R}\nu') , \tag{5}$$

$$P_r' = -(\varepsilon + P_r)\nu' - 2(P_r - P_t)(R'/R) . \tag{6}$$

A solution of the above equations can be considered as physically valid only if it satisfies the weak energy condition $T_{\mu\nu}u^\mu u^\nu \geq 0$ for any non-spacelike u^μ. In addition, a solution describes the collapse of an initially regular distribution of matter only if the spacetime admits a spacelike hypersurface ($t = 0$ say) which carries regular initial data for the fields (this means that the metric, its inverse, and the second fundamental form all have to be continuous at $t = 0$). Furthermore, in what follows, we consider only solutions which can be interpreted as models of collapsing stars, i.e. isolated objects rather than "universes". This implies that the metric can be matched smoothly to the Schwarzschild vacuum solution, a non-trivial requirement which turns out to be very restrictive especially in the case of perfect fluid solutions.

2. THE KNOWN EXACT SOLUTIONS AND THEIR PHYSICAL CONTENT

In the book by Kramer, Stephani, Herlt and Mac Callum [4] one can learn that most of the known spherically symmetric solutions have vanishing shear. As we shall see later, the physical content of shearfree solutions is poor (that is to say, vanishing of the shear is a quite unphysical condition) and therefore the situation with perfect fluids is very unhappy. However, the authors of the book cited above do not consider anisotropic solutions, while there are many anisotropic shearing solutions that have given a quite good understanding of gravitational collapse in recent years (as a matter of fact, anisotropy of stresses is what one should aspect in physically realistic situations). I will start the review from the simplest case, which is also - trivially - a perfect fluid solution: dust.

2.1. Dust solutions

The general exact solution of the Einstein field equations for zero stresses (dust) is very well known [5, 6, 7]. From equation (6), one gets $v = 0$ (more precisely, e^v is an arbitrary function of t only which can be rescaled to unity without loss of generality). Then one gets $\eta^{-1} = R'^2/(1+f)$ where $f(r)$ is an arbitrary function. Thus, the general dust metric is

$$ds^2 = -dt^2 + \frac{R'^2}{1+f}dr^2 + R^2(d\theta^2 + \sin^2\theta d\varphi^2) . \tag{7}$$

The evolution of R can be obtained by pointing out that the mass is constant in time ($m = F(r)$) by virtue of equation (4). Thus, eq. (2) can be written as a Kepler-like equation ($\dot{R}^2 = f + 2F/R$), which is integrable in parametric form for $f \neq 0$ and in closed form for $f = 0$ (this case is called marginally bound). Finally the density can be read off from (3) as $\varepsilon = F'/(4\pi R^2 R')$.

Although the dust solutions are known since the work by Lemaitre, the study of the nature of their singularities started only in 1984 [8]. Since then, a great effort has been paid to understand the nature of the central singularity in this solution and today the complete spectrum of possible endstates of the dust evolution in dependence of the initial data is known. I recall here what happens in the case of marginally bound solutions ($f = 0$) [9]: the solutions can be uniquely characterized by the expansion of the function $F(r)$ at $r = 0$ or, and that is the same, by the expansion of $\varepsilon_0 = F'/4\pi r^2$. If the initial density profile is linear or parabolic, the singularity is naked; if it is of fourth order or greater, the singularity is covered (a black hole) while if it is cubic there is a transition between naked singularities and blackholes occurring at a sort of "magic number" $\xi = F_3/F_0^{5/2} = -(26 + 15\sqrt{3})/2$.

The above results can be extended to the general case of collapsing dust clouds, so that the final fate of the dust solutions is completely known [9]. The final fate depends on a parameter which is a combination of coefficients of the expansions of F and f near $r = 0$, and a structure similar to that of marginally bound collapse arises.

2.2. Solutions with vanishing radial stresses

The general solution for dust is relatively simple because of mass conservation, which allows the reduction to quadratures. However, the mass is constant in time whenever the radial stress vanishes, and, as a consequence, the complete solution with vanishing radial stresses can be found [10, 11]. The solution is at best displayed using a system of coordinates first introduced by Ori [12]. One of the new coordinates is the mass m, the other coordinate is the "area radius" R. In such coordinates the metric reads as

$$ds^2 = -\Gamma^2\left(1 - \frac{2m}{R}\right)dm^2 + 2\sqrt{1+f}\frac{\Gamma}{hu}dRdm - \frac{1}{u^2}dR^2 + R^2(d\theta^2 + \sin^2\theta d\varphi^2) , \tag{8}$$

where

$$u = -\sqrt{-1 + \frac{2m}{R} + \frac{1+f}{h^2}} , \quad \Gamma = g(m) + \int \frac{h}{u^2\sqrt{1+f}}\frac{\partial u}{\partial m}dR . \tag{9}$$

Here the function $g(m)$ is arbitrary and the function $h(m,R)$ describes the stress-strain relation of the material. When h depends only on m, the solution reduces to the dust one.

In recent years, the problem of understanding the nature of the singularities for these solutions has been solved. The first class which has been investigated is the so-called *Einstein cluster* [13, 14, 15]. This is a spherically symmetric cluster of rotating particles. The motion of the particles is sustained by the angular momentum whose average effect is to introduce a non vanishing tangential stress in the energy-momentum tensor. The final state "at fixed dust background" (i.e. at fixed choices of initial density and velocity) depends on the expansion of the angular momentum M near $m = 0$. It can be shown that regularity of initial data implies $M \approx \beta m^{y/3}$ where $y \geq 4$ (the "strange" exponent divided by 3 is due to the choice of coordinates, and reduces to an integer if the standard r coordinate is restored) [16, 17]. On considering, for simplicity, the marginally bound case, one finds that for $y = 4$ the singularity does not form (the system bounces back), while for $y = 5, 6$ a naked singularity is always formed. At $y = 7$ a transition takes place and the evolution of the model is such that only the critical branch is changed, un-covering a part of the blackhole region in the corresponding dust spacetime; the non-critical branch is the same as in dust spacetimes. Finally for $y > 7$ the evolution always leads to the same end state of the dust solution which has the same initial distribution of density and velocity.

At least in one special case, the singularity forming at the center of the Einstein cluster has been shown to be timelike [16]. This is a quite strong "counterexample" to the Cosmic Censorship conjecture, since the singularity is visible "forever". By contrast,the naked singularities arising in inhomogeneous dust collapse are always null-like.

From the physical point of view, the Einstein cluster is interesting since it mimics the effects of rotation without introducing deviations from spherical symmetry. However, although such a system can be formally obtained by the choice of a particular function of state within the general exact solution, this state function is essentially a "centrifugal potential" and therefore does not fulfill the physical requirements which are typical of the equations of state of matter continua, like those to be expected in strongly collapsed matter states (e.g. in neutron stars). As a consequence, the results obtained for the Einstein cluster cannot be straightforwardly extended to the general case of tangential stresses, which is of course interesting in its own right [18].

Recently, a new approach to the problem of characterizing the final fate of collapse based on non-linear o.d.e. techniques has been developed [19]. Using this approach, the complete spectrum of singularities in gravitational collapse with tangential stresses has been found [20] as well as the spectrum of a more general class of solutions with both radial and tangential stresses [21]. Once again the final state depends on the value of a parameter which in turn depends on the first non-vanishing terms of the Taylor expansion of the data near the center, as in dust models. I will skip here all technical details of such calculations, but what I want to stress is that these results show that the existence of naked singularities in gravitational collapse is not a phenomenon resulting from the unphysical assumption of vanishing stresses. Actually *naked singularities exist - and can even be timelike - in the presence of general tangential stresses*.

Of course, one could claim that proving existence does not prove genericity (ad therefore physical validity) of a phenomenon. It is very difficult to assess the genericity of the data leading to naked singularities. A non-censorist like who writes would tend to say as follows: "pick up a function, any function, and check what is the probability of finding equal to zero the first three derivatives at zero..." (this is what a censorist needs to cancel out naked singularities from dust and tangential stresses solutions). Of course, however, I understand that other people need not share this view.

The other objection to physical validity, i.e. that regarding the stability of naked singularities, requires the non-spherical framework and, as I have said, belongs to that field of study which is recently - and slowly - coming up [22]. A third objection is, of course, the fact that the most widely used model in astrophysical modelling, that of fluids, has not been addressed. I am therefore going to review what is actually known on perfect fluid collapse.

2.3. Shearfree Perfect fluid solutions

A spherically symmetric solution can be invariantly characterized by the canonical decomposition of the velocity field. Standing the complexity of the field equations for perfect fluids, one can try to superimpose simplifying assumptions on the fluid flow lines, and, in particular, to impose the vanishing of shear. This implies (after fixing an arbitrary function of r)

$$\eta = \frac{r^2}{R^2} \, .$$

Note that, since this is a *superimposed* condition on the Einstein field equations, there is no *a priori* guarantee that the fluid satisfies an equation of state in closed form.

The shearfree condition is mathematically very convenient since it allows the introduction of a system of coordinates which is simultaneously comoving and isotropic. In addition, equation (5) can be reduced to an equation relating v and R which is immediately integrated to

$$e^{2v} = \frac{\dot{R}^2}{R^2 H^2(t)} \, ,$$

where $H(t)$ is (one third) the expansion of the fluid flow lines. The metric for shearfree perfect fluids thus reads as

$$ds^2 = -\frac{\dot{R}^2}{R^2 H^2(t)} dt^2 + \frac{1}{\eta}(dr^2 + r^2(d\theta^2 + \sin^2\theta d\varphi^2)) \, , \tag{10}$$

where only η has to be determined. On using $L := \sqrt{\eta}$ nd $x := r^2$ as independent variable, it can be shown that L must be a solution of

$$L_{xx} = l(x)L^2 \, ,$$

where $l(x)$ is arbitrary and a subscript denotes derivative. The remaining field equations make it possible to express the energy density and the pressure of the fluid in terms of L and its derivatives as follows:

$$8\pi\varepsilon = 3H^2 + 12L_x(L - xL_x) + 8xlL^3 \, ,$$

$$8\pi P = -3H^2 + \frac{L}{L}\left[2H\dot{H} + 4\dot{L}_x(L - 2xL_x)\right] + 4L_x(3xL_x - 2L).$$

The fact that the equation for L does not contain time derivatives has the consequence that its solutions (and therefore the metric) depends on two arbitrary functions of time. However, these functions turn out to be severely constrained by the requirements of reality of the solution and physical reasonableness. In fact, as mentioned before, we are interested here only in solutions describing an isolated object which collapses to a singular state starting from an initially regular data set.

Although the problem of shearfree dynamics is mathematically reduced to the integration of a single equation, this equation is non-linear and only a few solutions are known. Most papers on shearfree spacetimes have focused on the mathematical difficulties of the integration problem so that little attention has been paid to the physical issues and, especially, to the nature of the singularities (notable exceptions are [23, 24]). Recently, the nature of singularities has been clarified at least in the simplest case $l = const.$[25]. It turns out that all solutions satisfy strong cosmic censorship if the weak energy condition is assumed. What happens is that either the singularity is *synchronous* (all "points" r become singular simultaneously) or the boundary becomes singular first. In both cases, there is only one comoving time at which the singularity can emit light rays, and it can be shown that it does not. There are very strong hints that *all* physically valid shearfree models behave in this way, i.e., that non-vanishing shear is a necessary condition for nakedness. However, I want to stress again that vanishing of shear is a very strong superimposed condition on the dynamics as well.

2.4. Shearing perfect fluid solutions

From the previous discussion it is clear that if we want to understand perfect fluid collapse we cannot avoid the general case, when, however,very few exact solutions are available. This led many researchers to consider another form of simplifying assumption: self-similarity. With this assumption the dimensionless variables v, η and R/r depend only on the "similarity variable" $z := r/t$, and the Einstein field equations become *ordinary* differential equations (see [27] and references therein).

Being governed by ordinary differential equations, self-similar spherical collapse can be analyzed with the techniques of dynamical systems theory [26]. The analysis of the singularities forming in self-similar perfect fluids spacetimes[28, 29] once again leads to a spectrum of endstates in which both naked singularities and blackholes can form, depending on the values of the parameters characterizing the solution.

The presence of naked singularities in perfect fluids is therefore, as expected, certain, and it has been recently proved also with the use of explicit examples [30]. In the last two years, a completely new approach to perfect fluids collapse has also been proposed [32, 31]. In these papers it has been shown that the Einstein field equations for barotropic fluids with linear equation of state can be *always* reduced to a single, although of course non-linear, second-order partial differential equation for the mass. On using this result, it is possible to show that analytic mass functions always lead to naked singularities. I stress however that the genericity of this last result is still unclear, since the mathematical notion of analyticity depends, of course, on the choice of frame.

3. DISCUSSION

Our understanding of gravitational collapse in spherical symmetry is still incomplete but - at least in my opinion - it is reasonable to think that a sector of data leading to naked singularities typically exists for general equations of state if standard physical requirements are satisfied. Therefore, the Cosmic Censorship conjecture in spherical symmetry looks simply to be false if matter models based on the theory of continua are considered. It remains the possibility to reject such models by default adopting the point of view that only matter models which do not exhibit singularities at the non relativistic level have to be considered [33], such as scalar fields [35] or kinetic models [34]. However one could note, as H. Seifert already pointed out many years ago, that our understanding of cosmology is essentially based on a model, the Friedmann-Robertson-Walker model, which of course exhibits the big-bang singularity at the classical (non-relativistic) level.

REFERENCES

1. R. Penrose, Nuovo Cimento **1**, 252 (1969).
2. P. Szekeres, Phys. Rev. D **12**, 2941 (1975).
3. C. W. Misner and D. H. Sharp, Phys. Rev. B **136**, 571 (1964).
4. D. Kramer, H. Stephani, E. Herlt, and M. MacCallum, *Exact solutions of the Einstein's field equations*, (Cambridge Univ. Press, 1980).
5. G. Lemaitre, Ann. Soc. Sci. Bruxelles A **53**, 51 (1933); Gen. Rel. Grav. **29**, 641 (1997).
6. R. C. Tolman, Proc. Nat. Acad. Sci. **20**, 169 (1934).
7. H. Bondi, Mon. Not. Roy. Astr. Soc. **142**, 333 (1969).
8. D. Christodoulou, Comm. Math. Phys. **93**, 171 (1984).
9. T. P. Singh and P. S. Joshi, Class. Quantum Grav. **13**, 559 (1996).
10. G. Magli, Class. Quantum Grav. **14**, 1937 (1997).
11. G. Magli, Class. Quantum Grav. **15**, 3215 (1998).
12. A. Ori, Class. Quantum Grav. **7**, 985 (1990).
13. A. Einstein, Ann. Math. **40**, 922 (1939).
14. B. K. Datta, Gen. Rel. Grav. **1**, 19 (1970).
15. H. Bondi, Gen. Rel. Grav. **2**, 321 (1971).
16. T. Harada, H. Iguchi and K. Nakao, Phys. Rev. D **58**, R041502 (1998).
17. S. Jhingan and G. Magli, Phys. Rev. D **58**, R041502 (1998).
18. S. Goncalves, S. Jhingan and G. Magli, Phys. Rev. D **65**, 064011 (2002).
19. R. Giambò, F. Giannoni and G. Magli, Class. Quantum Grav. **19**, L15 (2002).
20. R. Giambò, F. Giannoni, G. Magli and P. Piccione, Class. Quantum Grav. **20**, L75 (2003).
21. R. Giambò, F. Giannoni, G. Magli and P. Piccione, Commun. Math. Phys. **235**, 545 (2003).
22. H. Iguchi, K. Nakao and T. Harada, Phys. Rev. D **57**, 7262 (1998).
23. E.N. Glass, J. Math. Phys. **20**, 1508 (1979).
24. R.A. Sussman, J. Math. Phys. **28**, 1118 (1987).
25. E. Brinis, S. Jhingan and G. Magli, Class. Quantum Grav. **17**, 4481 (2000).
26. O. I. Bogoyavlenski, *Methods in the Qualitative Theory of Dynamical Systems in Astrophysics and Gas Dynamics*, (Springer–Verlag, 1985).
27. B. J. Carr, in *Proceedings of the 7th Canadian conference on General Relativity and Relativistic Astrophysics*, (ed.) D. Hobill.
28. A. Ori and T. Piran, Phys. Rev. Lett. **59**, 2137 (1987).
29. P. S. Joshi and I. H. Dwivedi, Commun. Math. Phys. **146**, 333 (1992).
30. R. Goswami and P. S. Joshi, Class. Quantum Grav. **21**, 3645 (2004).
31. R. Giambò, F. Giannoni, G. Magli and P. Piccione, Class. Quantum Grav. **20**, 4943 (2003).
32. R. Giambò, F. Giannoni, G. Magli and P. Piccione, Gen. Rel. Grav. **36**, 1279 (2004).
33. R. M. Wald, gr-qc/9710068
34. G. Rein, A. D. Rendall and J. Schaeffer, Phys. Rev. D **58**, 044007 (1998).
35. D. Christodoulou, Ann. Math. **149**, 183 (1999).

The Chaplygin gas, a model for dark energy in cosmology

V. Gorini, U. Moschella[*], A. Kamenshchik[†] and V. Pasquier[**]

[*]*Dipartimento di Scienze Matematiche, Fisiche e Chimiche, Università dell'Insubria, Via Valleggio 11, 22100 Como, Italy and INFN sez. di Milano, Italy.*
[†]*Dipartimento di Scienze Matematiche, Fisiche e Chimiche, Università dell'Insubria, Via Valleggio 11, 22100 Como, Italy*
L.D. Landau Institute for Theoretical Physics of Russian Academy of Sciences, Kosygin str. 2, 119334 Moscow, Russia.
[**]*Service de Physique Theórique, C.E. Saclay, 91191 Gif-sur-Yvette, France*

Abstract. We review the essential features of the Chaplygin gas cosmological models and provide some examples of appearance of the Chaplygin gas equation of state in modern physics. A possible theoretical basis for the Chaplygin gas in cosmology is discussed. The relation with scalar field and tachyon cosmological models is also considered.

INTRODUCTION

Recent years observations of the luminosity of type Ia distant supernovae [1, 2, 3] point towards an accelerated expansion of the universe, which implies that the pressure p and the energy density ρ of the universe should violate the strong energy condition, i.e. $\rho + 3p < 0$.

The matter responsible for this condition to be satisfied at some stage of the cosmological evolution is referred to as "dark energy" (for a review see [4, 5, 6]). There are different candidates for the role of dark energy.

The most traditional candidate is a nonvanishing cosmological constant, which can also be thought of as a perfect fluid satisfying the equation of state $p = -\rho$. However, it remains to be understood why the observed value of the cosmological constant is so small in comparison with the Planck mass scale.

Moreover, there also arises in this connection the so called "cosmic coincidence conundrum". It amounts to the following question: why are the energy densities of dark energy and of dust-like matter at the present epoch of the same order of magnitude? This seems to be a problem, because it would imply that at the time of recombination these two densities were different by many orders of magnitude (see for instance [7]).

A less featureless candidate to provide dark energy is represented by the so called quintessence scalar field [8]. Scalar fields are traditionally used in inflationary models to describe the transition from the quasi-exponential expansion of the early universe to a power law expansion. It has been a natural choice to try to understand the present acceleration of the universe by also using scalar fields [9, 10]. However, we now deal with the opposite task, i.e. we would like to describe the transition from a universe filled

CP751, *General Relativity and Gravitational Physics, 16th SIGRAV Conference*, edited by G. Esposito et al.

with dust-like matter to an exponentially expanding one.

Scalar fields are not the only possibility but there are (of course) alternatives. Among these one can point out the so called k-essence models, where one deals again with a scalar field, but with a non-standard kinetic term [11]. The tachyonic models of dark energy have a similar structure, where the kinetic term of the tachyon field has a form suggested by string theory (see the review [12] and references therein). One can also mention models where the role of dark energy is played by quantum corrections to the effective action of a scalar field [13].

Here we consider a recently proposed class of simple cosmological models based on the use of peculiar perfect fluids [14]. In the simplest case, we study the model of a universe filled with the so called Chaplygin gas, which is a perfect fluid characterized by the following equation of state:

$$p = -\frac{A}{\rho}, \tag{1}$$

where A is a positive constant.

Chaplygin introduced this equation of state [15] as a suitable mathematical approximation for calculating the lifting force on a wing of an airplane in aerodynamics. The same model was rediscovered later in the same context [16, 17].

The convenience of the Chaplygin gas is connected with the fact that the corresponding Euler equations have a very large group of symmetry, which implies their integrability. The relevant symmetry group has been recently described in modern terms [18].

The negative pressure following from the Chaplygin equation of state could also be used for the description of certain effects in deformable solids [19], of stripe states in the context of the quantum Hall effect and of other phenomena.

It is worth mentioning a remarkable feature of the Chaplygin gas, i.e. that it has positive and bounded squared sound velocity

$$v_s^2 = \frac{\partial p}{\partial \rho} = \frac{A}{\rho^2},$$

which is a non-trivial fact for fluids with negative pressure (this follows from $\rho^2 \geq A$, see formula (22) below).

Beyond cosmology, the Chaplygin gas equation of state has recently raised a growing attention [20] because it displays some interesting and, in some sense, intriguingly unique features.

Indeed, Eq. (1) has a nice connection with string theory and it can be obtained from the Nambu-Goto action for d-branes moving in a $(d+2)$-dimensional spacetime in the light-cone parametrization [21]. Also, the Chaplygin gas is the only fluid which, so far, admits a supersymmetric generalization [22]. We ourselves came across this fluid [23] when studying the stabilization of branes [24] in black hole bulks [25]. An "anti-Chaplygin" state equation, i.e. Eq. (1) with negative constant A, arises in the description of wiggly strings [26, 27].

Inspired by the fact that the Chaplygin gas possesses a negative pressure we have undertaken the simple task of studying a FRW cosmology of a universe filled with this type of fluid [14]. Further theoretical developments of the model were given in

[28, 29, 30, 31]. One of its most remarkable properties is that it describes a transition from a decelerated cosmological expansion to a stage of cosmic acceleration. The inhomogeneous Chaplygin gas can do more: it is able to combine the roles of dark energy and dark matter [29].

Another model that has been discussed in some detail [30] is the generalized Chaplygin gas that has two free parameters:

$$p = -\frac{A}{\rho^\alpha}, \quad 0 < \alpha \leq 1 \tag{2}$$

A further possibility is to use [31] a more realistic two-fluid cosmological model including both the Chaplygin gas and the usual dust-like matter; this was also studied by using the statefinder parameters [32, 33]. While the model looks less economical than pure Chaplygin, it is more flexible from the point of view of the comparison with observational data. Moreover, the two-fluid model can suggest a solution of the cosmic coincidence conundrum [31].

The cosmological models of the Chaplygin class have at least three significant features: they describe a smooth transition from a decelerated expansion of the universe to the present epoch of cosmic acceleration; they attempt to give a unified macroscopic phenomenological description of dark energy and dark matter; and, finally, they represent, perhaps, the simplest deformation of traditional ΛCDM models.

Taking into account these attractive features, it is important to try to explain what could be the microscopic origin of the presence of the Chaplygin gas in our universe. An interesting attempt [34] makes use of a field theory approach to the description of a $(3+1)$-dimensional brane immersed in a $(4+1)$-dimensional bulk [35]. Phenomenologically, the Chaplygin gas manifests itself as the effect of the immersion of our four-dimensional world into some multidimensional bulk. The appearance of the Chaplygin gas in such a context does not depend on the details of the theory.

Another interesting feature of the Chaplygin gas is that it can be considered as the simplest model within the family of tachyon cosmological models. In other words, the Chaplygin gas cosmological model can be identified with a tachyon field theoretical model with a constant value of the field potential [36]. Using this fact as a starting point, we have developed a technique of construction of tachyon models which are in general correspondence with the present observational data, describing the contemporary epoch of cosmic acceleration [37]. On the other hand, these models have a rather rich dynamics, opening various scenarios for the future of the universe. In particular, it is possible that the present cosmological acceleration be followed by a catastrophically decelerated expansion culminating in a *Big Brake* cosmological singularity. This type of cosmological singularity is characterised by an infinite value of cosmic deceleration which is achieved in a finite span of time.

A significant amount of work has been devoted to the comparison of the Chaplygin cosmological predictions with observational data[38-64]. In this context, one can mention the following directions of investigation:
1. Supernova of type Ia observations.
2. Cosmic microwave background radiation.
3. Growth of inhomogeneities and the large scale structure of the universe.
4. Statistics of gravitational lensing.

5. X-ray luminosity of galaxy clusters.

6. Age restrictions from high-redshift objects.

7. New methods of diagnostics of the dark energy equation of state for future supernovae observations.

The restrictions coming from the studies of the large-scale structure of the universe are crucial for the viability of the Chaplygin gas models [40, 46, 47, 50, 52, 53, 57]. One can safely say that a careful investigation of the non-linear regime of the growth of inhomogeneities is necessary for the purpose of coming to definite conclusions concerning the compatibility of the Chaplygin cosmologies with the observable large-scale structure of the universe.

Finally, we remark that until recently the standard ΛCDM model was considered as the most natural candidate for the role of dark energy from the observational point of view. Thus, the usual discussions of such theoretical problems as the relation between the cosmological constant and the Planck mass scale and the cosmic coincidence conundrum coexisted with a tacit agreement that there is nothing that can fit the data better than this simplest model. One can now detect first signs of a possible change of the situation. An example is paper [65], where some essential arguments in favour of a non-constant ratio between the pressure and the energy density of dark energy were put forward. Moreover, the study of conditions under which future SNAP (Supernova / Acceleration Probe) and cosmic microwave background radiation observations would be able to rule out the simplest ΛCDM model is also attracting attention [66]. In this context, further study of the Chaplygin gas cosmological model and its relatives looks promising.

The structure of this paper is the following: in sections and we present some theoretical foundations of the Chaplygin gas in modern physics; section contains the discussion of the Chaplygin cosmological model; in section we consider a two-fluid cosmological model in terms of the statefinder parameters; section presents the relationships between the Chaplygin cosmological model and some corresponding scalar field and tachyon field models.

BRANES AND THE CHAPLYGIN'S EQUATION OF STATE

The rather special status of the Chaplygin fluid may be appreciated by reviewing how its equation of state reappears in a modern theoretical physics context.

To this end one is led to consider the Nambu-Goto action for a d-brane moving in a $(d+2)$-dimensional spacetime in the light-cone parametrization [21]. However, to keep the discussion elementary, we restrict our attention to the 3-dimensional case [23] i.e. the string case. When written in the light-cone gauge, the Hamiltonian for such a string has the following structure:

$$H = \frac{1}{2} \int [\Pi^2 + (\partial_\sigma x)^2] d\sigma, \tag{3}$$

where σ is a spatial world-sheet coordinate, x is a transversal spatial coordinate and Π is its conjugate momentum. The Hamilton equations following from (3) are very simple:

$$\partial_\tau x = \Pi, \tag{4}$$

$$\partial_{\tau\tau}^2 x - \partial_{\sigma\sigma}^2 x = 0. \qquad (5)$$

At this point one wants to interpret the functions

$$\rho(x) = (\partial_\sigma x)^{-1}, \qquad (6)$$
$$v = \Pi, \qquad (7)$$

as the density and the velocity fields of a certain fluid associated with the string.

This is substantiated by a special instance of the hodograph transformation [21], that makes one move from the independent variables τ and σ to the variables $t = \tau$ and x. There hold the following relations:

$$\frac{\partial}{\partial \tau} = \frac{\partial}{\partial t} + \Pi \frac{\partial}{\partial x} = \frac{\partial}{\partial t} + v \frac{\partial}{\partial x}, \qquad (8)$$

$$\frac{\partial}{\partial \sigma} = (\partial_\sigma x) \frac{\partial}{\partial x} = \frac{1}{\rho} \frac{\partial}{\partial x}. \qquad (9)$$

At this point one can easily see that the density and velocity that we have defined always satisfy the continuity equation

$$\frac{\partial \rho}{\partial t} + \frac{\partial(\rho v)}{\partial x} = 0. \qquad (10)$$

Furthermore Eqs. (8) and (9) can be used to show that Eq. (5) for the string is equivalent to the Euler equation

$$\rho \left(\frac{\partial}{\partial t} + v \frac{\partial}{\partial x} \right) v + \frac{\partial p}{\partial x} = 0 \qquad (11)$$

for the fluid, provided that the pressure field satisfies the Chaplygin equation of state (1) with $A = 1$.

The Chaplygin equation of state also arises in connection with the Randall-Sundrum model [24]. In this model one thinks of our four dimensional Minkowski spacetime to be a brane in a higher dimensional manifold. The difference of the Randall-Sundrum model w.r.t. the standard Kaluza-Klein models is that the higher dimensional manifold has now a (non-factorizable) warped structure as given by its metric

$$ds^2 = e^{-2|y|/l}(dt^2 - dx_1^2 - dx_2^2 - dx_3^2) - dy^2, \qquad (12)$$

where y is an additional fifth coordinate and $e^{-2|y|/l}$ is the warping factor. Eq. (12) is the metric of a portion of a five-dimensional anti-de Sitter spacetime of radius l.

At $y = 0$ one has the so called orbifold boundary conditions. Here the Christoffel symbols have finite jumps while the components of the curvature tensor contain δ-like terms. To compensate them, one should introduce a brane located at $y = 0$ whose tension is $\lambda = 6/l$. This tension can be interpreted as a nonvanishing cosmological constant on the 4-dimensional brane spacetime, or equivalently, as a fluid living on the brane whose state equation is $p = -\rho$. The 5-dimensional anti-de Sitter curvature makes the graviton essentially trapped on the 4-brane.

It is possible to consider other geometries for the brane, and this in general requires other kinds of matter on it for its stabilization. We have considered [23] a foliation of the $(n+2)$-dimensional anti-de Sitter spacetime by static universes with topology $R \times S^n$, always imposing orbifold boundary conditions. In this case [23] the matter on the brane is a fluid satisfying the following state equation:

$$p = -\frac{(n-1)\rho}{n} - \frac{4n}{\rho l^2}. \tag{13}$$

In the three-dimensional case $(n = 1)$ this reduces again to the Chaplygin gas state equation.

A POSSIBLE THEORETICAL BASIS FOR THE CHAPLYGIN GAS IN COSMOLOGY

As we shall see in the next section, the Chaplygin gas cosmological model has interesting features. Before discussing the latter in detail, we would like to address the question whether there is any fundamental mechanism to produce a Chaplygin gas source term at the RHS of the Einstein equations.

An interesting attempt in this direction [34] makes use of a $(3+1)$-brane immersed in a $(4+1)$-bulk, following a recent stream of ideas [35]. Consider the embedding of a $(3+1)$-dimensional brane in a $(4+1)$-dimensional bulk described by coordinates $x^M = (x^\mu, x^4)$, where the index μ runs over $0,1,2,3$. Denote the bulk metric by g_{MN}. Then, the induced metric on the brane is given by

$$\tilde{g}_{\mu\nu} = g_{\mu\nu} - \theta_{,\mu}\theta_{,\nu}, \tag{14}$$

where $\theta(x^\mu)$ is a scalar field describing the embedding of the brane into the bulk.

The action on the brane has the following structure:

$$
\begin{aligned}
S_{brane} &= \int d^4x \sqrt{-\tilde{g}}(-f + \cdots) \\
&= \int d^4x \sqrt{-g}\sqrt{1 - g^{\mu\nu}\theta_{,\mu}\theta_{,\nu}}(-f + \cdots),
\end{aligned} \tag{15}
$$

where the constant f gives a brane tension and the dots $\cdots$ stay for other possible contributions. Equation (15) follows from the identity

$$\det(a_{ij} - b_i b_j) = \det(a_{ij})\left(1 - b_m(a^{-1})_{mn}b_n\right),$$

whose proof is straightforward.

The energy-momentum tensor following from the tension-like action (15) is then

$$T_{\mu\nu} = f\left(\frac{\theta_{,\mu}\theta_{,\nu}}{\sqrt{1 - g^{\mu\nu}\theta_{,\mu}\theta_{,\nu}}} + g_{\mu\nu}\sqrt{1 - g^{\mu\nu}\theta_{,\mu}\theta_{,\nu}}\right). \tag{16}$$

113

This expression corresponds to a perfect fluid energy-momentum tensor

$$T_{\mu\nu} = (\rho + p)u_\mu u_\nu - p g_{\mu\nu},$$

provided one makes the following identifications: the four-velocity

$$u_\mu = \frac{\theta_{,\mu}}{\sqrt{g^{\mu\nu}\theta_{,\mu}\theta_{,\nu}}};$$

the pressure and the energy density:

$$p = -f\sqrt{1 - g^{\mu\nu}\theta_{,\mu}\theta_{,\nu}} \tag{17}$$

$$\rho = f\frac{1}{\sqrt{1 - g^{\mu\nu}\theta_{,\mu}\theta_{,\nu}}}. \tag{18}$$

It follows that the pressure and energy density exactly satisfy the Chaplygin's equation of state (1) with $A = f^2$.

FRW COSMOLOGY WITH THE CHAPLYGIN GAS

We consider a homogeneous and isotropic universe with the metric

$$ds^2 = dt^2 - a^2(t)dl^2, \tag{19}$$

where dl^2 is the metric of a 3-manifold of constant curvature ($K = 0, \pm 1$), and the expansion factor $a(t)$ evolves according to the Friedmann equation

$$\frac{\dot{a}^2}{a^2} = \rho - \frac{K}{a^2}. \tag{20}$$

Energy conservation

$$d(\rho a^3) = -p\, d(a^3) \tag{21}$$

together with the equation of state (1) give the following relation:

$$\rho = \sqrt{A + \frac{B}{a^6}}, \tag{22}$$

where B is an integration constant. By choosing a positive value for B we see that for small a (i.e. $a^6 \ll B/A$) the expression (22) is approximated by

$$\rho \sim \frac{\sqrt{B}}{a^3} \tag{23}$$

which corresponds to a universe dominated by dust-like matter. For large values of the cosmological radius a it follows that

$$\rho \sim \sqrt{A}, \quad p \sim -\sqrt{A}, \tag{24}$$

which, in turn, corresponds to an empty universe with a cosmological constant $\sqrt{A}$ (i.e a de Sitter universe). In the flat case it is also possible to find the exact solution as follows:

$$t = \frac{1}{6\sqrt[4]{A}} \left(\ln \frac{\sqrt[4]{A + \frac{B}{a^6}} + \sqrt[4]{A}}{\sqrt[4]{A + \frac{B}{a^6}} - \sqrt[4]{A}} - 2\arctan\sqrt[4]{1 + \frac{B}{Aa^6}} + \pi \right). \tag{25}$$

Note that $\rho = \sqrt{A}$ solves the equation

$$\rho + p = \rho - \frac{A}{\rho} = 0. \tag{26}$$

The circumstance that this equation has a nonzero solution lies at the heart of the possibility of interpreting the model as a "quintessential" one. Let us estimate the constant A by comparing our expressions for pressure and energy with observational data. An indirect and naive way to do it is to consider the nowadays accepted values for the contributions of matter and cosmological constant to the energy density of the universe. To use these data we decompose pressure and energy density as follows:

$$p = p_\Lambda + p_M = -\Lambda, \tag{27}$$
$$\rho = \rho_\Lambda + \rho_M = \Lambda + \rho_M. \tag{28}$$

An application of Eq. (1) gives

$$A = \Lambda(\Lambda + \rho_M). \tag{29}$$

If the cosmological constant contributes seventy percent to the total energy we get $\sqrt{A} \approx 1.2\,\Lambda$. We now observe that, in the context of a Chaplygin cosmology, once an expanding universe starts accelerating it cannot decelerate any more. Indeed Eqs. (20) and (21) imply that

$$\frac{\ddot{a}}{a} = -\frac{1}{2}(\rho + 3p). \tag{30}$$

Condition $\ddot{a} > 0$ is equivalent to

$$a^6 > \frac{B}{2A}, \tag{31}$$

which is obviously preserved by time evolution in an expanding universe. It thus follows that the observed value of the (effective) cosmological constant will increase up to 1.2Λ.

There is a relation of our Chaplygin cosmology with cosmologies based on fluids admitting a bulk viscosity proportional to a power of the density [67, 68, 69, 70]. In the flat $K = 0$ case the FRW equations for Chaplygin fit in this scheme as a special case [68] and indeed a transition from power law to exponential expansion was already noticed [67, 71]. However, since the state equation for the corresponding fluid is different from our Eq. (1) this coincidence of the solutions is destroyed by any small perturbation, for instance by a small spatial curvature or by adding another matter source.

Considering now the subleading terms in Eq. (22) at large values of a (i.e. $a^6 \gg B/A$), one obtains the following expressions for the energy and pressure:

$$\rho \approx \sqrt{A} + \sqrt{\frac{B}{4A}}\, a^{-6}, \tag{32}$$

$$p \approx -\sqrt{A} + \sqrt{\frac{B}{4A}}\, a^{-6}. \tag{33}$$

Eqs. (32) and (33) describe the mixture of a cosmological constant $\sqrt{A}$ with a type of matter known as "stiff" matter, described by the following equation of state:

$$p = \rho. \tag{34}$$

Note that a massless scalar field is a particular instance of stiff matter. Therefore, in a generic situation, a Chaplygin cosmology can be looked upon as interpolating between different phases of the universe: from a dust dominated universe to a de Sitter one passing through an intermediate phase which is the mixture just mentioned above. The interesting point, however, is that such an evolution is accounted for by using one fluid only.

For open or flat Chaplygin cosmologies ($K = -1, 0$), the universe always evolves from a decelerating to an accelerating epoch. For the closed Chaplygin cosmological models ($K = 1$), the Friedmann equations (20) and (30) tell us that it is possible to have a static Einstein universe solution $a_0 = (3A)^{-\frac{1}{4}}$ provided the following condition holds:

$$B = \frac{2}{3\sqrt{3A}}. \tag{35}$$

When $B > \frac{2}{3\sqrt{3A}}$ the cosmological radius $a(t)$ can take any value, while if

$$B < \frac{2}{3\sqrt{3A}}$$

there are two possibilities: either

$$a < a_1 = \frac{1}{\sqrt{3A}}\left(\sqrt{3}\sin\frac{\varphi}{3} - \cos\frac{\varphi}{3}\right) \tag{36}$$

or

$$a > a_2 = \frac{2}{\sqrt{3A}}\cos\frac{\varphi}{3}, \tag{37}$$

where $\varphi = \pi - \arccos 3\sqrt{3A}B/2$. The region $a_1 < a < a_2$ is not accessible. Further information on the dynamics of the Chaplygin gas cosmological model can be found in the literature [72, 73].

For the generalized Chaplygin gas [14, 30] the dependence of the energy density on the cosmological radius is

$$\rho = \left(A + \frac{B}{a^{3(1+\alpha)}}\right)^{\frac{1}{1+\alpha}}. \tag{38}$$

This type of matter at the beginning of the cosmological evolution behaves like dust and at the end of the evolution like a cosmological constant, while during the intermediate stage it could be treated as a mixture of two-fluids: the cosmological constant and a perfect fluid with equation of state $p = \alpha\rho$. The generalized Chaplygin gas cosmological models have an additional free parameter α to play with and are convenient for the comparison with observational data. However, the prospects of the construction of a physical theory explaining the origin of these models seem even less evident than those for the true Chaplygin gas with $\alpha = 1$.

THE STATEFINDER PARAMETERS AND A TWO-FLUID COSMOLOGICAL MODEL

Since models trying to provide a description (if not an explanation) of the cosmic acceleration are proliferating, there exists the problem of discriminating between the various contenders. To this aim a new proposal introduced in [32] may turn out useful, which exploits a pair of parameters $\{r, s\}$, called "statefinder". The relevant definition is as follows:

$$r \equiv \frac{\dddot{a}}{aH^3}, \quad s \equiv \frac{r-1}{3(q-1/2)}, \tag{39}$$

where $H \equiv \frac{\dot{a}}{a}$ is the Hubble constant and $q \equiv -\frac{\ddot{a}}{aH^2}$ is the deceleration parameter. The new feature of the statefinder is that it involves the third derivative of the cosmological radius.

Trajectories in the $\{s, r\}$-plane corresponding to different cosmological models exhibit qualitatively different behaviours. ΛCDM model diagrams correspond to the fixed point $s = 0$, $r = 1$. The so-called "quiessence" models [32] are described by vertical segments with r decreasing from $r = 1$ down to some definite value. Tracker models [74, 75] have typical trajectories similar to arcs of parabola lying in the positive quadrant with positive second derivative.

The current location of the parameters s and r in these diagrams can be calculated in models (given the deceleration parameter); it may also be extracted from data coming from future SNAP (SuperNovae Acceleration Probe)-type experiments [32]. Therefore, the statefinder diagnostic combined with forthcoming SNAP observations may possibly be used to discriminate among different dark energy models.

Here, we consider the one-fluid pure Chaplygin gas model and a two-fluid model where dust is also present [31]. We show that these models are different from those considered in [32].

To begin with, let us rewrite the formulae for the statefinder parameters in a form convenient for our purposes. Since

$$\dot{p} = \frac{\partial p}{\partial \rho}\dot{\rho} = -3\sqrt{\rho}\,(\rho + p)\frac{\partial p}{\partial \rho}, \tag{40}$$

we easily get:

$$r = 1 + \frac{9}{2}\left(1 + \frac{p}{\rho}\right)\frac{\partial p}{\partial \rho}, \quad s = \left(1 + \frac{\rho}{p}\right)\frac{\partial p}{\partial \rho}. \tag{41}$$

For the Chaplygin gas one has simply that

$$v_s^2 = \frac{\partial p}{\partial \rho} = \frac{A}{\rho^2} = -\frac{p}{\rho} = 1 + s \tag{42}$$

and therefore

$$r = 1 - \frac{9}{2}s(1+s). \tag{43}$$

Thus, the curve $r(s)$ is an arc of parabola. It is easy to see that

$$v_s^2 = \frac{A}{A + \frac{B}{a^6}}. \tag{44}$$

When the cosmological scale factor a varies from 0 to ∞ the velocity of sound varies from 0 to 1 and s varies from -1 to 0. Thus in our model the statefinder s takes negative values; this feature is not shared by quiessence and tracker models [32].

As s varies in the interval $[-1,0]$, r first increases from $r = 1$ to its maximum value and then decreases to the ΛCDM fixed point $s = 0, r = 1$ (see Fig. 1).

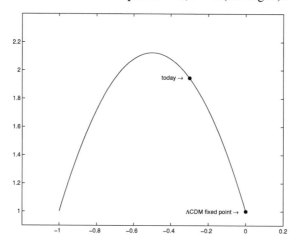

FIGURE 1.   s-r evolution diagram for the pure Chaplygin gas

If $q \approx -0.5$ the current values of the statefinder (within our model) are $s \approx -0.3$, $r \approx$ 1.9. In [32] an interesting numerical experiment based on 1000 realizations of a SNAP-type experiment, probing a fiducial ΛCDM model is reported. Our values of the statefinder lie outside the three-sigma confidence region displayed in [32]. Based on this fact it can be expected that future SNAP experiments should be able to discriminate between the pure Chaplygin gas model and the standard ΛCDM model.

Now consider a more "realistic" cosmological model which, besides a Chaplygin's component, contains also a dust component. For a two-component fluid Eqs. (41) take the following form:

$$r = 1 + \frac{9}{2(\rho + \rho_1)}\left[\frac{\partial p}{\partial \rho}(\rho + p) + \frac{\partial p_1}{\partial \rho_1}(\rho_1 + p_1)\right], \tag{45}$$

$$s = \frac{1}{p + p_1} \left[\frac{\partial p}{\partial \rho}(\rho + p) + \frac{\partial p_1}{\partial \rho_1}(\rho_1 + p_1) \right]. \tag{46}$$

If one of the fluids is dust, i.e. $p_1 = p_d = 0$, the above formulae become

$$r = 1 + \frac{9(\rho + p)}{2(\rho + \rho_d)} \frac{\partial p}{\partial \rho}, \quad s = \frac{\rho + p}{p} \frac{\partial p}{\partial \rho}. \tag{47}$$

If the second fluid is the Chaplygin gas, proceeding exactly as before we obtain the following relation:

$$r = 1 - \frac{9}{2} \frac{s(s+1)}{1 + \frac{\rho_d}{\rho}}. \tag{48}$$

To find the term ρ_d/ρ we write down the dependence of the dust density on the cosmological scale factor:

$$\rho_d = \frac{C}{a^3}, \tag{49}$$

where C is a positive constant. Eq. (44) gives $Aa^6 + B = -\frac{B}{s}$ and therefore

$$\frac{\rho_d}{\rho} = \frac{C}{\sqrt{Aa^6 + B}} = \kappa\sqrt{-s}, \tag{50}$$

where the constant $\kappa = C/\sqrt{B}$ is the ratio between the energy densities of dust and of the Chaplygin gas at the beginning of the cosmological evolution. Thus

$$r = 1 - \frac{9}{2} \frac{s(s+1)}{1 + \kappa\sqrt{-s}}. \tag{51}$$

Graphs of the function (51) for different choices of κ are plotted in Fig. 2.

In this case there are choices of the parameters so that the current values of the statefinder are close to the ΛCDM fixed point. For $\kappa = 1$ we have $s = -0.09$ and $r = 1.2835$; by increasing κ we get closer and closer to the point $(0, 1)$. Already for $\kappa = 2$ we get $s = 0.035$, $r = 1.11$ while for $\kappa \gtrsim 5$ the statefinder essentially coincides with the ΛCDM fixed point (see Fig. 2).

Thus our two-fluid cosmological models (with κ say bigger than 5) cannot be discriminated from the ΛCDM model on the basis of the statefinder analysis.

However, even if the Chaplygin component closely mimics today the cosmological constant, this neither spoils the interest of the two-fluid model nor makes it equivalent to ΛCDM; for instance, one advantage of the model is that it may suggest a solution to the cosmic coincidence conundrum: here the initial values of the energies of dust and of the Chaplygin gas can be of the same order of magnitude. In particular the value $\kappa = 1$ is not excluded by current observations. This may be seen by using the results of [40] and taking into account the relation

$$\kappa = \frac{\Omega_m}{(1 - \Omega_m)\sqrt{1 - v_s^2}}, \tag{52}$$

where $\Omega_m = \frac{\rho_d}{\rho_d + \rho}$ and where ρ, ρ_d and v_s are evaluated at the present epoch.

Further details concerning application of the statefinder diagnostic to the study of Chaplygin gas models can be found in [31, 33].

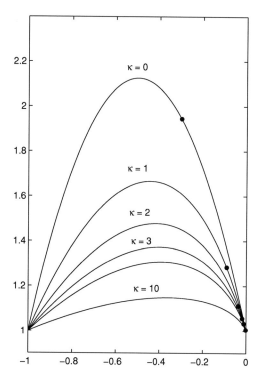

FIGURE 2. s-r evolution diagram for the Chaplygin gas mixed with dust. Dots locate the current value of the statefinder

THE CHAPLYGIN GAS, SCALAR FIELDS, TACHYONS AND THE FUTURE OF THE UNIVERSE

It is well-known that for isotropic cosmological models, given the dependence of the cosmological radius on time it is always possible to construct a potential for a minimally coupled scalar field model, which would reproduce this cosmological evolution (see e.g. [9]), provided rather general reasonable conditions are satisfied. Sometimes, it is possible to construct an explicit scalar field potential, which, provided some special initial conditions are chosen, can reproduce the evolution arising in some perfect fluid cosmological model [69, 37].

Consider the Lagrangian

$$L(\phi) = \frac{1}{2}\dot{\phi}^2 - V(\phi) \tag{53}$$

and set the energy density of the field equal to that of the Chaplygin gas:

$$\rho_\phi = \frac{1}{2}\dot{\phi}^2 + V(\phi) = \sqrt{A + \frac{B}{a^6}}. \tag{54}$$

120

The corresponding "pressure" coincides with the Lagrangian density:

$$p_\phi = \frac{1}{2}\dot{\phi}^2 - V(\phi) = -\frac{A}{\sqrt{A + \frac{B}{a^6}}}. \tag{55}$$

It immediately follows that

$$\dot{\phi}^2 = \frac{B}{a^6\sqrt{A + \frac{B}{a^6}}} \tag{56}$$

and

$$V(\phi) = \frac{2a^6\left(A + \frac{B}{a^6}\right) - B}{2a^6\sqrt{A + \frac{B}{a^6}}}. \tag{57}$$

We restrict ourselves to the flat case $K = 0$. Then Eq. (56) also implies that

$$\phi' = \frac{\sqrt{B}}{a(Aa^6 + B)^{1/2}}, \tag{58}$$

where prime means differentiation w.r.t. a. This equation can be integrated and it follows that

$$a^6 = \frac{4B\exp(6\phi)}{A(1 - \exp(6\phi))^2}. \tag{59}$$

Finally, by substituting the latter expression for the cosmological radius in Eq. (57) one obtains the following potential, which has a surprisingly simple form:

$$V(\phi) = \frac{1}{2}\sqrt{A}\left(\cosh 3\phi + \frac{1}{\cosh 3\phi}\right). \tag{60}$$

Note that the potential does not depend on the integration constant B and therefore it reflects only the state equation (1) as it should.

The cosmological evolution of the model with a scalar field with potential (60) coincides with that of the Chaplygin gas model provided the initial values $\phi(t_0)$ and $\dot{\phi}(t_0)$ satisfy the relation $\dot{\phi}^4(t_0) = 4(V^2(\phi(t_0)) - A)$.

We have mentioned in the Introduction, that one of the most popular candidates for the role of dark energy is a tachyon field described by the effective action [76, 77]

$$S = -\int d^4x \sqrt{-g} V(T) \sqrt{1 - g^{\mu\nu} T_{,\mu} T_{,\nu}}, \tag{61}$$

where $V(T)$ is a tachyon potential. For a spatially homogeneous model

$$L = -V(T)\sqrt{1 - \dot{T}^2}.$$

The energy density of the tachyon field is

$$\rho = \frac{V(T)}{\sqrt{1 - \dot{T}^2}} \tag{62}$$

while the pressure is

$$p = -V(T)\sqrt{1 - \dot{T}^2}. \tag{63}$$

It is easy to see [36] that for the constant tachyon potential $V(T) = V_0$ the pressure (63) and the energy density (62) are connected by the Chaplygin state equation (1) with $A = V_0^2$.

Thus, we see that the Chaplygin gas model coincides with the simplest tachyon cosmological model. It is interesting to construct also other tachyon potentials reproducing the dynamics of some perfect fluid models. For example [78, 79] a tachyon model with a potential

$$V(T) = \frac{4\sqrt{-k}}{9(1+k)T^2}, \quad -1 < k < 0 \tag{64}$$

has a solution

$$T = \sqrt{1 + kt}, \tag{65}$$

corresponding to the cosmological evolution

$$a = a_0 t^{\frac{2}{3(1+k)}}, \tag{66}$$

which, in turn, could be obtained in the model with a perfect fluid obeying the state equation

$$p = k\rho. \tag{67}$$

It is not difficult to show [37], that for $k > 0$ it is impossible to reproduce the power-law cosmological evolution (66) using a tachyon action. Instead, one can use a "pseudo" - tachyon action with the Lagrangian

$$L = V(T)\sqrt{\dot{T}^2 - 1} \tag{68}$$

with the potential

$$V(T) = \frac{4\sqrt{k}}{9(1+k)T^2}.$$

In this case the corresponding solution (65) preserves its form.

We have considered [37] a more complicated toy tachyon model. Studying a two-fluid cosmological model, where one of the fluids is the cosmological constant and the other fluid obeys the state equation $p = k\rho, -1 < k < 1$, one gets the following expression for the cosmological evolution

$$a(t) = a_0 \left(\sinh \frac{3\sqrt{\Lambda}(1+k)t}{2} \right)^{\frac{2}{3(1+k)}}. \tag{69}$$

The same evolution can be reproduced in the tachyon model with a potential

$$V(T) = \frac{\Lambda}{\sin^2 \left(\frac{3\sqrt{\Lambda(1+k)}T}{2} \right)} \sqrt{1 - (1+k)\cos^2 \left(\frac{3\sqrt{\Lambda(1+k)}T}{2} \right)}. \tag{70}$$

The solution of the tachyon equation of motion corresponding to the evolution (69) has the form

$$T(t) = \frac{2}{3\sqrt{\lambda(1+k)}} \arctan \sinh \frac{3\sqrt{\Lambda}(1+k)t}{2} \tag{71}$$

and could be obtained provided some special initial conditions are chosen.

Considering all possible initial conditions we get a rich family of cosmological evolutions, which are rather different from (69), representing a simple two-fluid model [37]. Here we encounter two "surprises". First, when $k > 0$, for the description of the dynamics of the model it is necessary to consider regions of the phase plane $(T, \dot{T})$, where the tachyon action (61) with the potential (70) is not well-defined and should be substituted by a pseudo-tachyon action. Second, (again for the case $k > 0$) there are two types of trajectories :
a) infinitely expanding universes;
b) universes, hitting a cosmological singularity of a special type which we call *Big Brake*, and which is characterised by the following behaviour of the cosmological radius

$$\ddot{a}(t_B) = -\infty, \; \dot{a}(t_B) = 0, \; 0 < a(t_B) < \infty. \tag{72}$$

Here t_B means the final moment of time when the Big Brake is achieved.

In the current literature devoted to the future of the universe the following scenarios are usually considered:
a) an infinite asymptotically de Sitter expansion,
b) present accelerated expansion followed by contraction and the achieving of a Big Crunch cosmological singularity (see, e.g. [80, 81])
c) an accelerated expansion culminating in a Big Rip cosmological singularity arising in the phantom dark energy cosmological models (see, e.g. [82, 83, 84]). At the Big Rip singularity the cosmological radius and the Hubble parameter achieve an infinite value in a finite interval of time.

On the basis of the above considerations, it seems reasonable to envisage an additional scenario, in which the present cosmic acceleration is followed by a decelerated expansion culminating in the hitting of a Big Brake cosmological singularity. The latter, as we have seen, can be described in terms of a rather simple tachyon model [37].

ACKNOWLEDGMENTS

A.K. is grateful to CARIPLO Science Foundation and to University of Insubria for financial support. His work was also partially supported by the Russian Foundation for Basic Research under the grant No 02-02-16817 and by the scientific school grant No. 2338.2003.2 of the Russian Ministry of Science and Technology.

REFERENCES

1. A. Riess et al., *Astron. J.* **116**, 1009 (1998).

2. S.J. Perlmutter et al., *Astroph. J.* **517**, 565 (1999).
3. N.A. Bachall, J.P. Ostriker, S. Perlmutter and P.J. Steinhardt, *Science* **284**, 1481 (1999).
4. V. Sahni and A.A. Starobinsky, *Int. J. Mod. Phys.* **A9**, 373 (2000).
5. P.J.E. Peebles and B. Ratra, *Rev. Mod. Phys.* **75**, 559 (2003).
6. T. Padmanabhan, *Phys.Rept.* **380**,235 (2003).
7. V. Sahni, *Class. Quantum Grav.* **19**, 3435 (2002).
8. R.R. Caldwell, R. Dave and P.J. Steinhardt, *Phys. Rev. Lett.* **80**, 1582 (1998).
9. A.A. Starobinsky, *JETP Lett.* **68**, 757 (1998).
10. T.D. Saini, S. Raychaudhury, V. Sahni, A.A. Starobinsky, *Phys. Rev. Lett.* **85**, 1162 (2000).
11. C. Armendariz-Picon, V. Mukhanov and P.J. Steinhardt, *Phys. Rev. Lett.* **85**, 4438 (2000).
12. A. Sen, Remarks on tachyon driven cosmology, hep-th/0312153.
13. L. Parker and A. Raval, *Phys. Rev.* **D60**, 063512 (1999).
14. A.Yu. Kamenshchik, U. Moschella and V. Pasquier, *Phys. Lett.* **B511**, 265 (2001).
15. S. Chaplygin, *Sci. Mem. Moscow Univ. Math. Phys.* **21**, 1 (1904).
16. H.-S- Tsien, *J. Aeron. Sci.* **6**, 399 (1939).
17. T. von Karman, *J. Aeron. Sci.* **8**, 337 (1941).
18. D. Bazeia and R. Jackiw, *Ann. Phys. (N Y)* **270**, 246 (1998).
19. K. Stanyukovich, Unsteady Motion of Continuos Media (Pergamon, Oxford, UK 1960).
20. R. Jackiw, A Particle Field Theorist's Lectures on Supersymmetric, Non-Abelian Fluid Mechanics and d-Branes, physics/0010042.
21. M. Bordemann, J. Hoppe, *Phys. Lett.* **B317**, 315 (1993).
22. R. Jackiw and A. P. Polychronakos, *Phys. Rev.* **D62**, 085019 (2000).
23. A. Kamenshchik, U. Moschella and V. Pasquier, *Phys. Lett.* **B487**, 7 (2000).
24. L. Randall, R. Sundrum, *Phys. Rev. Lett.* **83**, 4690 (1999).
25. M. Banados, M. Henneaux, C. Teitelboim, J. Zanelli, *Phys. Rev.* **D48** 1506 (1993).
26. B. Carter, *Phys. Lett.* **B224**, 61 (1989).
27. A. Vilenkin, *Phys. Rev.* **D41**, 3038 (1991).
28. J.C. Fabris, S.V.B. Gonsalves and P.E. de Souza, *Gen. Rel. Grav.* **34**, 53 (2002).
29. N. Bilic, G.B. Tupper and R. Violler, *Phys. Lett.* **B535** (2002).
30. M.C. Bento, O. Bertolami and A.A. Sen, *Phys. Rev.* **D66**, 043507 (2002).
31. V. Gorini, A.Yu. Kamenshchik and U. Moschella, *Phys. Rev.* **D67**, 063509 (2003).
32. V. Sahni, T.D. Saini, A.A. Starobinsky and U. Alam, *JETP Lett.* **77**, 201 (2003).
33. U. Alam, V. Sahni, T.D. Saini and A.A. Starobinsky, *Mon. Not. Roy. Astron. Soc.* **344**, 1057 (2003)
34. N.Bilic, G.B. Tupper and R. Viollier, Dark matter, dark energy and the Chaplygin gas, astro-ph/0207423.
35. R. Sundrum, *Phys. Rev.* **D59**, 085009 (1999).
36. A. Frolov, L. Kofman and A. Starobinsky, *Phys. Lett.* **B545**, 8 (2002).
37. V. Gorini, A. Kamenshchik, U. Moschella and V. Pasquier, *Phys. Rev.***D69**, 123512 (2004).
38. J.C. Fabris, S.V.B. Goncalves and P.E. De Souza, *Gen. Rel. Grav.* **34**, 2111 (2002).
39. J.C. Fabris, S.V.B. Goncalves and P.E. De Souza, Fitting the supernova type Ia data with the Chaplygin gas, astro-ph/0207430.
40. P.P. Avelino, L.M.G. Beca, J.P.M. de Carvalho, C.J.A.P. Martins and P. Pinto, *Phys. Rev.* **D67**, 023511 (2003).
41. A. Dev, J.S. Alcaniz and D. Jain, *Phys. Rev.* **D67**, 023515 (2003).
42. M. Makler, S. Quinet de Oliveira and I. Waga, *Phys. Lett.* **B555**, 1 (2003).
43. J.S. Alcaniz, D. Jain and A. Dev, *Phys. Rev.* **D67**, 043514 (2003).
44. M.C. Bento, O. Bertolami and A.A. Sen, *Phys. Rev.* **D67**, 063003 (2003).
45. D. Cartugan and F. Finelli, *Phys. Rev.* **D68**, 103501 (2003).
46. H. Sandvik, M. Tegmark, M. Zaldariaga and I. Waga, The end of unified dark matter?, astro-ph/0212114.
47. R. Bean and O. Dore, *Phys. Rev.* **D68**, 023515 (2003).
48. R. Colistete, Jr., J.C. Fabris, S.V.B. Goncalves and P.E. de Souza, Bayesian analysis of the Chaplygin gas and cosmological consstant models using the SNE Ia data, astro-ph/0303338.
49. M.C. Bento, O. Bertolami and A.A. Sen, *Phys. Lett.* **B575**, 172 (2003).
50. L.M.G. Beca, P.P. Avelino, J.P.M. de Carvalho and C.J.A.P. Martins, *Phys. Rev.* **D67**, 101301 (2003).
51. L. Amendola, F. Finelli, C. Buriana and D. Carturan, *JCAP* **0307**, 005 (2003).

52. R.R.R. Reis, I. Waga, M.O. Calvao and S.E.Joras, *Phys. Rev.* **D68**, 061302 (2003).
53. P.P. Avelino, L.M.G. Beca, J.P.M. Carvalho, C.J.A.P. Martins and E.J. Copeland, **Phys. Rev. D69**, 041301 (2004).
54. M. Makler, S. Quinet de Oliveira and I. Waga, *Phys. Rev.* **D68**, 123521 (2003).
55. P.P. Avelino, L.M.G. Beca, J. P. M. de Carvalho and C.J.A.P. Martins, *JCAP* **0309**, 002 (2003).
56. N. Bilic, R.J. Lindebaum, G.B. Tupper and R.D. Viollier, The Chaplygin gas and the evolution of dark matter and dark energy in the universe, astro-ph/0307214;
57. T. Multamaki, M. Manera and E. Caztanaga, *Phys. Rev.* **D69**, 023004 (2004).
58. J.S. Alcaniz and J.A.S. Lima, Measuring the Chaplygin gas equation of state from angular and luminosity distances, astro-ph/0308465.
59. N. Bilic, R. J. Lindenbaum, G.B. Tupper and R.D. Viollier, Inhomogeneous Chaplygin gas cosmology, astro-ph/0310181.
60. A. Dev, D. Jain and J.S. Alcaniz, Constraints on Chaplygin quartessence from the CLASS gravitational lens statistics and supernova data, astro-ph/0311056.
61. Y. Gong and C.-K. Duan, Constraints on alternative models to dark energy, gr-qc/0311060.
62. Y. Gong and C.-K. Duan, Supernova constraints on alternative models to dark energy, astro-ph/0401530.
63. S. Nesseris and L. Perivolaropoulos, A comparison of cosmological models using recent supernova data, astro-ph/0401556.
64. O. Bertolami, A.A. Sen, S. Sen and P.T. Silva, Latest supernova data in the framework of generalized Chaplygin gas model, astro-ph/0402387.
65. U. Alam, V. Sahni, T.D. Saini and A.A. Starobinsky, Is there supernova evidence for dark energy metamorphosis ?, astro-ph/0311364.
66. J. Kratochvil, A. Linde, E.V. Linder and M. Shmakova, Testing the cosmological constant as a candidate for dark energy, astro-ph/0312183.
67. J.D. Barrow, *Phys. Lett.* **B180**, 335 (1986).
68. J.D. Barrow, *Nucl. Phys.* **B310**, 743 (1988).
69. J.D. Barrow, *Phys. Lett.* **B235**, 40 (1990).
70. J.D. Barrow and P. Saich, *Phys. Lett.* **B249**, 406 (1990).
71. R. Treciokas and G.F.R. Ellis, *Commun. Math. Phys.* **23**, 1 (1971).
72. I.M. Khalatnikov, *Phys. Lett.* **B563**, 123 (2003).
73. M. Szydlowski and W. Czaja, *Phys. Rev.* **D69**, 023506 (2004).
74. B. Ratra and P.J.E. Peebles, *Phys. Rev.* **D37**, 3406 (1988).
75. P. Brax and J. Martin, *Phys. Rev.* **D61**, 103502 (2000).
76. A.Sen, Rolling Tachyon, *JHEP* **0204**, 048 (2002) 048.
77. M.R. Garousi, *Nucl. Phys.* **B584**, 284 (2000).
78. A. Feinstein, *Phys. Rev.* **D66**, 063511 (2002).
79. T. Padmanabhan, *Phys. Rev.* **D66**, 021301 (2002).
80. R. Kallosh and A. Linde, *JCAP* **02**, 002 (2003).
81. U. Alam, V. Sahni and A.A. Starobinsky, *JCAP* **04**, 002 (2003).
82. R.R. Caldwell, M. Kamionkowski and N.N. Weinberg, *Phys. Rev. Lett.* **91**, 071301 (2003).
83. M.P. Dabrowski, T. Stachowiak and M. Szydlowski, Phantom cosmologies, hep-th/0307128.
84. M. Sami and A. Toporensky, Phantom field and the fate of the universe, gr-qc/0312009.

The Double Pulsar binary J0737-3039: a two-clocks relativistic system

Andrea Possenti*, Marta Burgay*, Nichi D'Amico*†, Andrew Lyne**,
Michael Kramer**, Maura McLaughlin**, Duncan Lorimer**, Dick
Manchester‡, Fernando Camilo§, John Sarkissian¶, Paulo Freire‖ and Bhal
Chandra Joshi††

*INAF - Oss. Astronomico di Cagliari, Loc. Poggio dei Pini, Strada 54, 09012 Capoterra, Italy
†Università degli Studi di Cagliari, Dipartimento di Fisica, SP Monserrato-Sestu km 0.7, 09042
Monserrato, Italy
**University of Manchester, Jodrell Bank Observatory, Macclesfield, Cheshire, SK11 9DL, UK
‡Australia Telescope National Facility, CSIRO, P.O. Box 76, Epping, New South Wales 2121,
Australia
§Columbia Astrophysics Laboratory, Columbia University, 550 West 120 th Street, New York
10027, USA
¶Australia Telescope National Facility, CSIRO, Parkes Observatory, P.O. Box 276, Parkes, New
South Wales 2870, Australia
‖NAIC, Arecibo Observatory, Puerto Rico, USA
††National Center for Radio Astrophysics, P.O. Bag 3, Ganeshkhind, Pune 411007, India

Abstract. The double-pulsar system J0737−3039 is the most fascinating pulsar discovery of last decade. Its orbital parameters and close distance allow to perform unprecedented tests of theories of Gravity and of the physics of highly condensed matter. The discovery of this system enhances of about an order of magnitude the estimate of the merger rate of double neutron-star systems, opening new possibilities for the current generation of gravitational-wave detectors. The high orbital inclination offers the opportunity of using the radio beams from one of the two sources as a probe for studying the magnetosphere of the other. We report the status of the observations and discuss short- and long-term perspectives.

THE DISCOVERY

PSR J0737−3039A, a millisecond pulsar with a spin period of 22.7 ms, was first detected in April 2003 [1] in a 4-minute pointing of the Parkes High-Latitude Pulsar Survey (Burgay et al. in preparation). Few days of follow-up observations were enough for determining both the Keplerian parameters (binary period P_b ~2.4 hr, eccentricity $e \sim 0.09$ and projected semi-major axis $a \sin i \sim 1.42$ lt-s) and a remarkably high value of the advance of the periastron, $\dot{\omega}$ ~17°/yr (the previously highest observed value being 5.33°/yr for PSR J1141-6545, [2]). That puts PSR J0737−3039A (hereafter A) on top of the list of the most relativistic binary pulsars.

If interpreted within the framework of general relativity, the value of $\dot{\omega}$ implied a total mass for the binary of about 2.58 $M_\odot$ giving a maximum mass for the pulsar of ~ 1.34 $M_\odot$ and a minimum mass for the companion of 1.24 $M_\odot$. The maximum mass for the pulsar perfectly agrees with the other measurements of neutron star masses [3], while

CP751, General Relativity and Gravitational Physics, 16th SIGRAV Conference, edited by G. Esposito et al.

the mass of the companion is only slightly lower than average. This piece of evidence strongly suggested that the discovered binary was a new double neutron-star system.

The ultimate confirmation of the above picture came few months later when, analyzing the follow-up observations of PSR J0737−3039A, a strong signal with a repetition period of about 2.8 seconds occasionally appeared [4]. The newly discovered pulsar, henceforth called PSR J0737−3039B (or simply B), had the same dispersion measure as PSR J0737−3039A (or A), and showed orbital Doppler variations that identified it, without any doubt, as the companion to the millisecond pulsar. After 30 years of worldwide searches and when the total pulsar counting had reached more then 1600 known sources, the first Double Pulsar system had been eventually discovered.

In the following some of the implications of this discovery are discussed as well as short- and long-term perspectives.

DOUBLE NEUTRON-STAR-SYSTEM FORMATION

A scenario (see [5] for a review) for the formation of double neutron-star systems was proposed since the time of the discovery of the first binary pulsar [6]. It starts with two main-sequence stars. The initially more massive first evolves off the main-sequence and eventually explodes in a supernova to form a neutron star. If the binary survives the explosion, the new born compact object spins down for the next few millions years, possibly shining as a normal radiopulsar. Under favorable values of the orbital parameters of the binary, when the secondary star leaves the main sequence a phase of mass transfer from the companion onto the neutron star may occur. During this stage the system is visible as an X-ray binary and the neutron star is spun up to short rotational periods while its surface magnetic field gets reduced by some orders of magnitude [7].

Close double neutron star binaries (like J0737−3039) can be explained only if a large amount of matter from the secondary star is ejected from the system [8], instead of being accreted onto the first born neutron star (for instance in a common envelope phase). This allows for the binary separation to decrease dramatically, leaving a very compact binary including a rapidly rotating neutron star and a helium star. If the latter is massive enough, nuclear evolution leads to a second supernova explosion in the system. If the system does not ionize, the final result is a double neutron star binary whose components follow eccentric orbits and should have significantly different values of both the rotational rate and the surface magnetic field (the second born neutron star not having being recycled).

These predictions could not be tested in the five previously known binary pulsars whose companion is supposed to be a neutron star [9], because of lack of a detectable radio signal from the recycled pulsar's companion. The J0737−3039 system opens the possibility of performing this check: in fact the observed rotational rates (P_A=22.7 ms, P_B=2.7 s) and surface magnetic fields ($B_A = 6.3 \times 10^9$ Gauss and $B_B = 1.2 \times 10^{12}$ Gauss[1]) of the two pulsars are nicely consistent with the above general model.

Regular timing observations, coupled with an interferometric determination of the

---

[1] Due to the interaction of A's wind which penetrates deeply into B's magnetosphere, some care should be taken when interpreting B's magnetic field strength.

distance of the system and with scintillation and polarimetric measurements may soon allow to infer the full geometry of the binary, together with an estimate of the transversal velocity vector of its center of mass. This will permit a determination of the dynamics of J0737−3039 with unprecedented accuracy, putting constraints on the origin and on the progenitor masses and evolution, besides assessing the magnitude of the kick associated with the second supernova explosion [10].

TEST OF GENERAL RELATIVITY

When a new pulsar is discovered, the parameters immediately available are its approximate period, dispersion measure and position within about half a beam-width of the radio-telescope. Precise parameters of the source can be obtained with regular monitoring of the new pulsar according to a procedure known as *timing*. For most binary systems including a radiopulsar, the determination of the position in the sky, of the rotational parameters and of the five Keplerian parameters allows to satisfactorily reproduce the orbital motion and to predict the times of arrival of the pulse from the given source within the measurement errors.

That is not expected to be the case for close double neutron star binaries. In fact, by virtue of their strong gravitational fields and rapid motions, these binaries exhibit large relativistic effects [11]. When they are large enough, the system can be used for testing the predictions of General relativity and of other theories of gravity. In this perspective, PSR J0737−3039A/B promises to be the best source ever. The tests can be performed when a number of relativistic corrections to the Keplerian description of the orbit (the so-called post-Keplerian, hereafter PK, parameters) can be measured. In this formalism, for point masses with negligible spin contributions, the PK parameters in each theory should only be functions of the a priori unknown neutron star masses and the classical Keplerian parameters. With the two masses as the only free parameters, the measurement of three or more PK parameters over-constrains the system, and thereby provides a test-ground for theories of gravity [12]. In a theory that describes the binary system correctly, the PK parameters define lines in a mass-mass diagram that all intersect at a single point. Such tests have been possible to date in only two double neutron star systems, PSR B1913+16 [13] and PSR B1534+12 [14]. For PSR B1913+16, the relativistic periastron advance, $\dot{\omega}$, the orbital decay due to gravitational wave damping, $\dot{P}_b$, and the gravitational redshift/time dilation parameter, γ, have been measured, providing a total of three PK parameters. For PSR B1534+12, Shapiro delay, caused by passage of pulses through the gravitational potential of the companion, is also visible, since the orbit is seen nearly edge-on. This results in two further PK parameters, r (range) and s (shape) of the Shapiro delay. However, the observed value of $\dot{P}_b$ requires correction for kinematic effects, so that PSR B1534+12 provides four PK parameters usable for precise tests [14].

With an intense campaign of regular timing observations started immediately after the discovery, we measured in only 6 months A's $\dot{\omega}$ and γ and have also detected the Shapiro delay in the pulse arrival times of A due to the gravitational field of B [15]. After adding further 6 months of observation, we have now got also the first meaningful determination of the orbital period decay $\dot{P}_b = -1.2 \pm 0.3 \times 10^{-12}$ s/s. This provides five

128

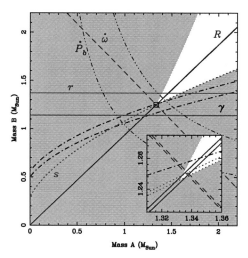

FIGURE 1. The observational constraints upon the masses M_A and M_B. The colored regions are those which are ruled out by the Keplerian mass functions of the two pulsars. Further constraints are shown as pairs of lines enclosing permitted regions as predicted by general relativity: (a) the measurement of the advance of periastron $\dot{\omega}$, giving the total mass $M_A+M_B=2.588\pm0.003$ M$_\odot$ (dashed lines); (b) the measurement of R= $M_A/M_B = x_B/x_A = 1.069 \pm 0.006$ (solid lines); (c) the measurement of the gravitational red-shift/time dilation parameter γ (dot-dash lines); (d) the measurement of Shapiro parameter r (solid horizontal lines) and Shapiro parameter s (dotted lines); (e) the measurement of the orbital decay (dot-dot-dot-dash lines). Inset is an enlarged view of the small square which encompasses the intersection of the three tightest constraints, with the scales increased by a factor of 16. The permitted regions are those between the pairs of parallel lines and we see that an area exists which is compatible with all constraints.

measured PK parameters[2], resulting in a M_A-M_B plot (Figure 1) through which we can test the predictions of general relativity. We can see that a region satisfying all constraints exists. In particular the current data, spanning only 12 months of observations, indicate an agreement of the observed with the expected Shapiro parameter s of $s_{obs}/s_{exp}=$ 1.00007$\pm$0.00220, where the uncertainties are likely to decrease quickly. The position of the allowed region in Fig. 1 also determines the inclination of the orbit to the line-of-sight. It turns out that the system is observed nearly edge-on with an inclination angle of about 88°.

However, the detection of B as a pulsar opens up opportunities to go beyond what is possible with previously known double neutron star binary systems. First, we can rule out all regions in the M_A-M_B plot plane that are forbidden by the individual mass functions of A and B due to the requirement $\sin i \leq 1$ (where i is the inclination of the orbital plane). Second, with a measurement of the projected semi-major axes of the orbits of A and B, we obtain a precise measurement of the mass ratio, $R(M_A, M_B) = M_A/M_B = x_B/x_A$, providing a further constraint in the M_A-M_B plot (Fig. 1). This relation

---

[2] Assuming that correction to $\dot{P}_b$ due to the motion of the binary system in the galactic gravitational potential are negligible or can be isolated by proper motion and distance measurements.

is valid for any theory of gravity; most importantly, the R-line is independent of strong-field (self-field) effects, providing a new constraint for tests of gravitational theories [11]. With five PK parameters already available for tests, this additional constraint makes this system the most over-determined double neutron star binary.

By virtue of space-time curvature near massive objects, the spin axes of both pulsars will precess about the total angular momentum vector, changing the orientation of the pulsars as seen from Earth [16]. Within the framework of general relativity, the periods of such a geodesic precession should be only 75 yr for A and 71 yr for B. This would lead to measurable changes in the profiles of A and B over a short time span. Deviations from the value predicted for $\dot{\omega}$ by general relativity may be caused by contributions from spin-orbit coupling [17], which is about an order of magnitude larger than for PSR B1913+16. This potentially will allow us to simultaneously measure the spin rate, the mass and the moment of inertia of a neutron star. This would lead to constrain the equation of state for the nuclear matter.

A REVISED DOUBLE NEUTRON STAR COALESCENCE RATE

The merging of a double-neutron-star system should produce a burst of emission of gravitational waves. By virtue of the energy budget and of the expected typical frequency of these events, they are among the primary targets for the current generation of ground-based gravitational wave detectors, which should be able to detect them up to a distance of about 20 Mpc. Hence, a key question is the occurrence rate of these double-neutron-star coalescences in a volume of universe of that radius. This rate can in turn be estimated on the basis of the rate of events in the Galaxy.

Among the double-neutron-star systems previously known, only three had tight enough orbits so that the two neutron stars will merge within a Hubble time. Two of them (PSR B1913+16 and PSR B1534+12) are located in the Galactic field, while the third (PSR B2127+11C) is found on the outskirts of a globular cluster. The contribution of globular cluster systems to the Galactic merger rate is estimated to be negligible [18]. Furthermore, recent studies [19] have demonstrated that the current estimate of the Galactic merger rate $\mathscr{R}$ relies mostly on PSR B1913+16. Thus, one can start by comparing the observed properties of PSR B1913+16 and PSR J0737−3039. The latter will merge because of the emission of gravitational waves in ~ 85 Myr, a time-scale that is a factor 3.5 shorter than that for PSR B1913+16. In addition, the estimated distance of PSR J0737−3039A/B (500 - 600 pc, based on the observed dispersion measure and a model for the distribution of ionized gas in the interstellar medium [20]) is an order of magnitude less than that of PSR B1913+16. These properties have a substantial effect on the prediction of rate of merging events in the Galaxy.

For a given class k of binary pulsars in the Galaxy, apart from a beaming correction factor, the merger rate $\mathscr{R}_k$ is calculated as $\mathscr{R}_k = N_k/\tau_k$ [19]. Here τ_k is the binary pulsar lifetime and N_k is the scaling factor defined as the number of binaries in the Galaxy belonging to the given class. The shorter lifetime of J0737−3039 system [$\tau_{1913}/\tau_{0737} = $ (365 Myr)/(185 Myr)] implies a doubling of the ratio $\mathscr{R}_{0737}/\mathscr{R}_{1913}$. A much more substantial increase results from the computation of the ratio of scaling factors N_{0737}/N_{1913}.

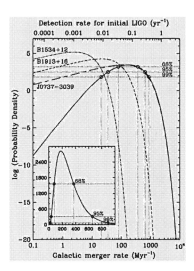

FIGURE 2. Probability density function representing the expectation that the actual double neutron star merger rate in the Galaxy (*bottom axis*) and the predicted initial LIGO merger rate (*top axis*) takes on particular values, given the observations. The solid line shows the total probability density along with those obtained for each of the three known coalescing binary systems in the galactic field, (*dashed lines*). *Inset:* Total probability density, and corresponding 68%, 95%, and 99% confidence limits, shown in a linear scale. From [21].

The luminosity $L_{400}=30$ mJy kpc^2 of PSR J0737−3039A is much lower than that of PSR B1913+16. For a planar homogeneous distribution of pulsars in the Galaxy, the ratio $N_{0737}/N_{1913} \sim L_{1913}/L_{0737} \sim 6$. Hence we obtain $\mathscr{R}_{0737}/\mathscr{R}_{1913} \sim 12$. Including the moderate contribution of the longer-lived PSR B1534+12 system to the total rate, we obtain an increase factor for the total merger rate of about an order of magnitude.

Extensive simulations [21] give results consistent with this simple estimate and show that the peak of the merger rate increase factor resulting from the discovery of J0737−3039 system lies in the range 5-7 and is largely independent of the adopted pulsar population model. For the most favorable distribution model available (model nr. 15 of [19]), the updated cosmic detection rate for first generation gravitational-wave detectors such as VIRGO, LIGO and GEO can be as high as 1 every 3-5 years at 95% confidence level (Figure 2). Hence, with the discovery of PSR J0737−3039A/B the double-neutron-star coalescence rate estimates enter an astrophysical interesting regime. Within a few years of gravitational-wave detectors operations, it should be possible to directly test these predictions and, in turn, place better constraints on the cosmic population of double-neutron-star binaries.

PROBING PULSAR MAGNETO-SPHERE

It is now clear that PSR J0737−3039B was missed in the original observation of A because pulses from the B pulsar are only very bright for two short intervals along an orbit. The discovery observation happened to be at a time when B was not pulsing. The first burst of strong emission is centered near orbital longitude[3] 210°, covering about 40° of longitude ($\sim$ 13 min), whereas the second, centered near longitude 280°, covers about 25° of longitude ($\sim$ 8 min). Even more surprising, not only does the pulse intensity change with orbital phase, but the pulse shape changes as well! Furthermore, an eclipse of the A pulsar at superior conjunction (90° orbital longitude) occurs. The eclipse is very short with a duration of less than 30 s centered slightly after conjunction.

The short eclipse of A[4] and the unprecedented peculiarities of the radio emission from B immediately called for some kind of interactions occurring between the energetic flux released from the recycled pulsar A and the magneto-sphere of the canonical (and much less energetic) pulsar B. In fact the energy density at the B pulsar of the wind resulting from spin-down of the A pulsar is about 2.1 erg cm^{-3}. This is about two orders of magnitude greater than the energy density (or equivalently, pressure) of the B pulsar wind at its light cylinder. Hence, the A wind should penetrate deep into the B-pulsar magneto-sphere. The penetration depth is determined by pressure balance between the A wind and the B dipolar magnetic field giving an effective Alfvén radius of about 0.45 R_{LC}, where R_{LC} is the light cylinder radius of B, if it were undisturbed. This implies that 90% (in volume) of the B-pulsar magneto-sphere may be blown away by the A wind, making it unlikely that the radio emission region of B is located near the light cylinder.

The heuristic energetic considerations presented above clearly support the hypothesis that the radio emission peculiarities of the B pulsar can be ascribed to the powerful impact of the A-wind upon the B-magneto-sphere. Clearly this unique MHD-wind vs pulsar magneto-sphere "collider" may give us the possibility of unveiling so far unaccessible features of the physics of a pulsar magneto-sphere and of the pulsar emission mechanism itself, such as the degree of magnetization and the isotropy of the wind, the energetic distribution (among accelerated particles, γ-rays and electro-magnetic radiation at the pulsar spin frequency) of the pulsar flux, the plasma density in the closed magneto-sphere and the height (over the neutron star surface) of the coherent radio emission region. Two recent observations [22], [23] nicely demonstrate the potentiality of the J0737–3039 system. We present them in the following.

Radioemission from B modulated by radiation from A

The ratio P_B/P_A of the intrinsic barycentric rotation period of B to that of A is 122.182. Since the periods are incommensurate, any influence of A's electro-magnetic radiation on B's emission will manifest itself as a drifting behavior caused by the beating of the two periodicities. On using the intrinsic periods of the two pulsars and their orbital

---

[3] The angle of longitude is here always expressed with respect to the position of the ascending node.

[4] Given that the relative orbital velocity of the two stars is about 620 km s^{-1}, a 30-s eclipse corresponds to a width of the eclipsing region at the B pulsar of about 18,000 km or about 0.14 R_{LC}. The best estimate of the orbital inclination (87.7 degrees) corresponds to an impact parameter of about 35,400 km or 0.26 R_{LC} at the B pulsar, much greater than the width of the eclipsing region. This suggests that the eclipse results from a grazing impact on an eclipsing region of radius about 0.27 R_{LC}.

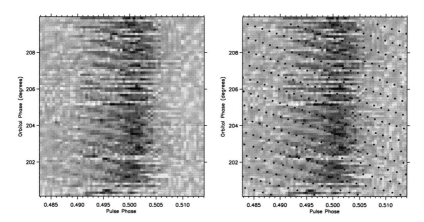

FIGURE 3. *Left:* Single pulses of B at 820 MHz for orbital phases 200° – 210°. as seen at the Green Bank 100-m dish on MJD 52997. Only 3% of the pulse period of B is shown. Drifting features are obvious as darker streams in the grey-scale plot. *Right:* Dots denote the arrival at the center of B of emission from an arbitrary rotational phase of A, retarded by the propagation time across the orbit. From [22].

elements, we can accurately predict the frequency of A's pulsed radiation as seen by B. At orbital phases 195° – 210° the beat frequency between A and B varies from 0.200 to 0.196 cycles/period.

Indeed, in Fig. 3 we can see single pulses from pulsar B for the aforementioned range of orbital phases with the predicted phases of arrival at the center of mass of B of a signal with A's periodicity. A drifting pattern in B emission is obvious [22] and it can be measured that its periodicity is just $0.196^{+0.005}_{-0.002}$ cycles/period (i.e. a modulation period $P_3 = 5.10 P_B$). This, and the fact that the separation of successive features within a given pulse is equal to P_A, shows that the drifting results from the direct influence of a signal with A's periodicity on the B pulse emission mechanism. This signal must be electro-magnetic, as any particle beam would spread in phase as it propagates across the 3 lt-s distance from A to B. The fact that the observed modulation is at 44 Hz with only a single pulse in each 23-ms period suggests that it is not the beamed radiation from A, which has two pulses per period, that excites the B emission. To sum up, there is evidence that the observed modulation results from the influence of the 44-Hz magnetic dipole radiation on the magneto-sphere of B and that the modulation is caused by the electro-magnetic field itself, rather than its intensity or pressure, which have an 88-Hz periodicity. This result provides observational support to the idea that, close to the pulsar, most of the spin-down energy is carried by the Poynting flux of the magnetic-dipole radiation rather than by energetic particles, e.g. [24].

Radioemission from A modulated by B at eclipse

McLaughlin et al. [23] divided each 2.8-s window of B's rotational phase into four equal regions, and calculated an average light curve for the eclipse of A in each region,

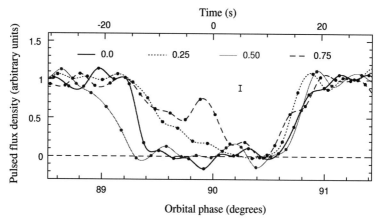

FIGURE 4. Averaged light curves for four regions of B pulse phase, with a smooth curve calculated with a spline fitting procedure drawn through the individual flux density measurements. The four regions are centered on pulse phases 0.0, 0.25, 0.50 and 0.75 of pulsar B, with, for example, the curve for 0.0 covering the 0.25 of pulse phase centered on B's radio pulse. The vertical bar indicates the typical one-sigma measurement error in each flux density. From [23].

as shown in Figure 4. All the four light curves vary smoothly.

As can be inferred from the upper axis of Figure 4, the eclipse durations range from 20 s to 34 s for the four B phase windows, calculated, as in [25], to be the full width at half maximum of the eclipse. Ingress occurs first at B phase 0.5 and then at 0.0, when, assuming B is an orthogonal rotator, the magnetic axis of B is aligned with the line of sight to A. It occurs later and more gradually for the other two phase windows centered on phases 0.25 and 0.75, when the magnetic axis of B is at right angles to the line of sight. The orbital phases of egress are similar for all B pulse phases. Only at around orbital phase 90.5° is flux density of A at all phases of B consistent with being zero.

All these observations demonstrate that the eclipse modulation and duration vary considerably with B phase and, hence, that the properties of the occulting medium are in fact very different in different regions of B's magneto-sphere. The observed phenomenology might be understood in the context of a model invoking synchrotron absorption of the radio emission from A in a magneto-sheath surrounding B's rotating magneto-sphere [26],[15]. If this is the case, both Arons et al. [26] and Lyutikov [15] reach the very interesting conclusion that the density of A's wind must be at least four orders of magnitude greater than is expected given currently accepted models of pair creation.

ACKNOWLEDGMENTS

We gratefully acknowledge the technical assistance with hardware and software provided by Osservatorio Astronomico di Cagliari, Jodrell Bank Observatory, CSIRO ATNF and Swinburne Center for Astrophysics and Super-computing and National Radio

Astronomy Observatory. The Parkes radio telescope is part of the Australia Telescope which is funded by the Commonwealth of Australia for operation as a National Facility managed by CSIRO. The National Radio Astronomy Observatory is facility of the National Science Foundation operated under cooperative agreement by Associated Universities, Inc. NDA, AP and MB received support from the Italian Ministry of University and Research (MIUR) under the national program *Cofin 2003*. IHS holds an NSERC UFA and is supported by a Discovery Grant. DRL is a University Research Fellow funded by the Royal Society. FC acknowledges support from NSF grant AST-02-05853 and a NRAO travel grant.

REFERENCES

1. Burgay, M., et al., *Nature*, **426**, 531 (2003).
2. Kaspi, V. M., et al., *ApJ*, **543**, 321 (2000).
3. Thorsett, S. E., and Chakrabarty, D., *ApJ*, **512**, 288 (1999).
4. Lyne, A. G., et al., *Science*, **303**, 1089 (2004).
5. Bhattacharya, D. and van den Heuvel, E. P. J., *Phys. Rep.*, **203**, 1 (1991).
6. Hulse, R. A., and Taylor, J. H., *ApJ*, **195**, L51 (1975).
7. Bisnovatyi-Kogan, G. S., and Komberg, B. V., *Sov. Astron.*, **18**, 217 (1974).
8. van den Heuvel, E. P. J., and de Loore, C., *A&A*, **25**, 387 (1973).
9. Taylor, J. H., *Phil. Trans. R. Soc. London Ser. A*, **341**, 117 (1992).
10. Willems, B., Kalogera, V., and Henninger, M., *ApJ*, in press, astro-ph/0404423, (2004).
11. Damour, T., and Deruelle, N., *Ann.Inst.H. Poincaré*, **44**, 263 (1986).
12. Damour, T., and Taylor, J. H., *Phys. Rev. D*, **45**, 1840 (1992).
13. Taylor, J. H., and Weisberg, J. M., *ApJ*, **345**, 434 (1989).
14. Stairs, I. H., Thorsett, S. E., Taylor, J. H., and Wolszczan, A., *ApJ*, **581**, 501 (2002).
15. Lyutikov, M., submitted, astro-ph/0403076, (2004).
16. Damour, T., and Ruffini, R., *Acedemie des Sciences Paris Comptes Rendus Ser. Scie. Math.*, **279**, (1974).
17. Barker, B. M., and O'Connell, R. F., *Phys. Rev. D*, **12**, 329 (1975).
18. Phinney, E.S., *ApJ*, **380**, L17 (1991).
19. Kim, C., Kalogera, V., and Lorimer, D. R., *ApJ*, **584**, 98 (2003).
20. Taylor, J. H., and Cordes, J. M., *ApJ*, **411**, 674 (1993).
21. Kalogera, V., et al., *ApJ*, **601**, L179 (2004).
22. McLaughlin, M., et al., *ApJ*, **613**, L57 (2004a).
23. McLaughlin, M., et al., *ApJ*, in press, astro-ph/0408297, (2004b).
24. Michel, F. C., *Rev. Mod. Phys.*, **54**, 1 (1982).
25. Kaspi, V. M., et al., *ApJ*, **613**, L137 (2004).
26. Arons, J., Backer, D. C., Spitkovvsky, A., and Kaspi, V. M., in *Binary Radio Pulsar*, ASP Conf. Ser., in press, astro-ph/0404159, (2004).

Rotation Effects and The Gravito-Magnetic Approach

Angelo Tartaglia

Dip. Fisica del Politecnico and INFN, Turin
angelo.tartaglia@polito.it

Abstract. The paper summarizes a review of the current state of theoretical and experimental research on gravito-magnetic theory and effects. Three slightly different approaches to the problem are described. The observation of the orbital precession both of satellites around the Earth and of pulsars in binary systems is mentioned. On the experimental side GPB is considered. Last, a number of other possibilities are listed.

INTRODUCTION

The idea that gravity and electromagnetism could have much in common is as old as modern physics, and inspired many scientists, including Faraday and of course Maxwell, during the whole XIX century [1]. Immediately after the publication of the General Relativity theory evidence was presented for a strong correspondence between Maxwell and Einstein equations at least in weak-field conditions. From that time date the origins of gravito-electromagnetism, which is the subject of this presentation.

DEFINING GRAVITOMAGNETISM

I shall present in the following three slightly different ways to introduce gravito-magnetism. As we shall see the different approaches can be led to converge, though their application fields may not be completely coincident.

The usual approach to gravito-electromagnetism starts by perturbing a flat space-time metric, then linearizing Einstein's equations [1] [2]. The metric tensor is assumed to be

$$g_{\mu\nu} = \eta_{\mu\beta} \left(\delta_\nu^\beta + h_\nu^\beta \right), \tag{1}$$

where $\eta_{\mu\nu}$ is the Minkowski metric tensor and $h_\nu^\mu \ll 1$.

It is usually simpler to work with

$$\bar{h}_{\mu\nu} = h_{\mu\nu} - \frac{1}{2}\eta_{\mu\nu}h \tag{2}$$

where $h = \eta^{\mu\alpha}h_{\alpha\mu}$. The gauge freedom is commonly exploited by imposing the so-called de Donder, or Lorenz-type gauge condition, i.e. $\bar{h}_{;\alpha}^{\mu\alpha} = 0$.

CP751, *General Relativity and Gravitational Physics, 16th SIGRAV Conference*, edited by G. Esposito et al.

Under these assumptions and conventions it is possible, as any textbook on General Relativity shows, to write the Einstein equations in linearized form, i.e. retaining only terms of first order in the h. The final equations are

$$\eta^{\alpha\beta}\bar{h}_{\mu\nu,\alpha\beta} = -16\pi\frac{G}{c^2}T_{\mu\nu}. \tag{3}$$

The similarity between equations (3) and the equations for the electromagnetic four-potential in flat space-time

$$\eta^{\alpha\beta}A^{\mu}{}_{,\alpha\beta} = \mu_0 j^{\mu} \tag{4}$$

is evident.

If the source of gravity is assumed to be a fluid, one can establish a further correspondence

$$T^{0\nu} = \rho U^0 U^{\nu} \Leftrightarrow j^{\nu} = \left(\sigma c, j^1, j^2, j^3\right), \tag{5}$$

where ρ is the local matter-energy density; U's are the four velocities in the source; the electromagnetic corresponding object is the current density four-vector, with the local charge density σ.

When the velocities within the source are "small", i.e. $v^i \ll c$, one has

$$\begin{aligned}
U^i &\ll 1, & U^0 &\cong 1, \\
T^{0i} &< T^{00}, & T^{ij} &\ll T^{00}, \\
\bar{h}_{0i} &< \bar{h}_{00}, & \bar{h}_{ij} &\ll \bar{h}_{00}.
\end{aligned} \tag{6}$$

Under these conditions, equations (3) are approximately equal to

$$\eta^{\alpha\beta}\bar{h}^{0\mu}{}_{,\alpha\beta} = -16\pi\frac{G}{c^2}T^{0\mu}, \quad T^{0\mu} = \rho U^0 U^{\mu} \cong \rho U^{\mu} = j_g^{\mu}, \tag{7}$$

where j_g^{μ} is, of course, the matter density four-vector in the source.

Looking at eq. (7) we may identify $\bar{h}^{0\mu}$ with the components of a four vector which plays the same role as the vector potential of electromagnetism:

$$\bar{h}^{0\mu} = -\mathscr{A}^{\mu}. \tag{8}$$

This new gravito-electromagnetic four potential may then be used to define a gravito-electric and a gravito-magnetic field formally satisfying the Maxwell equations:

$$E_{(i)\atop g} = -\frac{\partial A^0}{\partial x^i} - \frac{1}{c}\frac{\partial A_i}{\partial t},, \quad B_{(k)\atop g} = \varepsilon_k^{ij}\frac{\partial}{\partial x^i}A_j. \tag{9}$$

The explicit form for the $\mathscr{A}$ can be taken from the analogous solutions of the Maxwell equations in terms of retarded potentials:

$$\begin{aligned}
\bar{h}^{00} &= -\kappa\frac{\int \left(\rho U^{02}\right)\big|_{ret}d^3x'}{|\vec{r}-\vec{r}'|} \cong -\kappa\frac{\int \rho\big|_{ret}d^3x'}{|\vec{r}-\vec{r}'|}, \\
\bar{h}^{0i} &= -\kappa\frac{\int \left(\rho U^0 U^i\right)\big|_{ret}d^3x'}{|\vec{r}-\vec{r}'|} \cong -\kappa\frac{\int \left(\rho U^i\right)\big|_{ret}d^3x'}{|\vec{r}-\vec{r}'|}.
\end{aligned} \tag{10}$$

Finally the free motion of a massive particle moving at a four-velocity u^μ can be described in terms of a Lorentz force whose general expression would be

$$F^i = mc^2 \left[\frac{\partial \bar{h}^{0i}}{c \partial t} + \frac{\partial \bar{h}^{00}}{\partial x^i} + 2\eta^{ij} \left(\frac{\partial \bar{h}^{0k}}{\partial x^j} - \frac{\partial \bar{h}^{0j}}{\partial x^k} \right) u^k \right]. \tag{11}$$

It should be stressed that the u need not be small for eq. (11) to hold.

A slightly different approach to the problem of gravito-electromagnetism is as follows. As is well known, the free fall of a test particle in a gravitational field is described by the geodesic equation

$$\frac{d^2 x^i}{ds^2} = -\Gamma^i_{\mu\nu} u^\mu u^\nu. \tag{12}$$

When $u^i << u^0$ one has $u^0 \cong 1$ and eq. (12) becomes approximately

$$\frac{d^2 x^i}{dt^2} = -c^2 \Gamma^i_{00} - 2c\Gamma^i_{0j} v^j. \tag{13}$$

Now we notice that $\Gamma^\mu_{0\nu}$'s do indeed transform as a rank-2 tensor in three-space: keeping the time coordinate unaltered (apart from scale changes), we are allowed to treat them as tensors in the whole space-time. Explicitly one has

$$\mathscr{C}_{\mu\nu} = -\Gamma_{\mu 0 \nu} = -\frac{1}{2} \left(\frac{\partial g_{\mu 0}}{\partial x^\nu} + \frac{\partial g_{\mu\nu}}{\partial x^0} - \frac{\partial g_{0\nu}}{\partial x^\mu} \right). \tag{14}$$

This tensor is in general neither symmetric nor antisymmetric. We may of course pick out a symmetric and an antisymmetric part:

$$\begin{aligned} \mathscr{D}_{\mu\nu} &= \tfrac{1}{2} \left(\mathscr{C}_{\mu\nu} + \mathscr{C}_{\nu\mu} \right) = -\frac{\partial g_{\mu\nu}}{\partial x^0}, \\ \mathscr{F}_{\mu\nu} &= \tfrac{1}{2} \left(\mathscr{C}_{\mu\nu} - \mathscr{C}_{\nu\mu} \right) = \frac{\partial g_{0\nu}}{\partial x^\mu} - \frac{\partial g_{0\mu}}{\partial x^\nu}. \end{aligned} \tag{15}$$

If we now identify the whole $g_{0\mu}$ with a four-vector potential, $\mathscr{F}_{\mu\nu}$ looks like the electromagnetic tensor. The equation of motion (13) can now be recast in the form

$$\frac{d^2 x^i}{ds^2} = \mathscr{F}^i_0 + \mathscr{D}^i_0 + 2\mathscr{F}^i_j u^j + 2\mathscr{D}^i_j u^j. \tag{16}$$

Reading off terms in eq. (16) we may identify a gravito-electric field

$$E_i = \mathscr{F}_{i0} + \mathscr{D}_{i0} = \frac{\partial g_{00}}{\partial x^i} - 2\frac{\partial g_{0i}}{\partial x^0}, \tag{17}$$

a gravito-magnetic field

$$\mathscr{F}_{ij} = \frac{\partial g_{0j}}{\partial x^i} - \frac{\partial g_{0i}}{\partial x^j}, \tag{18}$$

and finally a term that produces the equivalent of a viscous force; let us then call $\mathscr{D}_{\mu\nu}$ "viscosity" tensor.

The situation is similar to the one illustrated in the previous section, but for the fact that now no assumption about the field being weak has been made and the whole elements of the metric tensor are used. The linearization is in the force expression and requires the test particles to move slowly. Furthermore, there is a restriction in the admitted coordinate transformations which are not spoiling the tensorial behavior of the gravito-electromagnetic $\mathscr{F}$ and the "viscous" $\mathscr{D}$. When the metric tensor is stationary, $\mathscr{D}$ vanishes and we are in a pure gravito-electromagnetic situation.

The force formula, in the stationary case, may be written by introducing a mass matrix

$$\mathfrak{m} = m \begin{pmatrix} 1 & 0 & 0 & 0 \\ 0 & 2 & 0 & 0 \\ 0 & 0 & 2 & 0 \\ 0 & 0 & 0 & 2 \end{pmatrix}, \tag{19}$$

and hence

$$\begin{aligned} F &= \mathfrak{m} \cdot \mathscr{F} \cdot u, \\ F^\mu &= \mathfrak{m}^\mu_\alpha \mathscr{F}^\alpha_\beta u^\beta. \end{aligned} \tag{20}$$

We must in any case recall that the electromagnetic analogy is imperfect both because, once we have chosen our time coordinate, we are not allowed to change it, if we do not wish to recast everything from the beginning; and because, when trying to write down the field equations, on identifying $g_{0\mu}$'s with a vector potential we obtain identities:

$$\mathscr{F}^{\mu\nu}_{;\nu} \equiv 0. \tag{21}$$

We are able to describe only a sourceless field because covariant derivatives of the metric tensor are identically zero. This inconvenient is solved when it is possible to substitute ordinary partial derivatives for covariant ones. In practice this is equivalent to treating the gravitational interaction as described by a field sitting in a Minkowski background, which is approximately possible when the field is weak, so we come back to the linear version of the Einstein equations.

Another feature of the gravito-electromagnetic field is that, when the coupling constant with the gravito-electric part is set to 1, the gravito-magnetic coupling constant is 2, as implied from eqs. (11) and (13).

The third approach I would like to present may be called a "curvature" approach. Consider a Fermi–Walker observer, for example an observer in a freely falling non-rotating (with respect to fixed stars) space-craft. Suppose this observer is performing local experiments in a small four-volume: for example the interior of his/her spacecraft's capsule, for short enough times. The objects on which the experiments are performed are slowly moving (with respect to c). Under these conditions the geodesic deviation equation is written as

$$\frac{d\xi^\alpha}{ds^2} = R^\alpha_{00i}\xi^i + 2R^\alpha_{i0j}u^i\xi^j, \tag{22}$$

where the ξ are distances from the origin (observer) and the R are components of the Riemann tensor. Everything is linearized in distances and velocities; no limits are set on the strength of the field.

Equation (22) is indeed an equation of motion for a test particle and again we may read off a gravito-electric and a gravito-magnetic part (accelerations):

$$E^i_g = c^2 R^i_{00j}\xi^j,$$
$$\eta^i_{j0k}B^k_g = cR^i_{j0l}\xi^l. \tag{23}$$

Here $\eta_{\alpha\beta\gamma\lambda}$ is the completely antisymmetric tensor density corresponding to the Levi–Civita tensor of flat space-time.

ARE GRAVITOMAGNETIC EFFECTS OBSERVABLE?

Of course the question is posed concerning practical, not foundational problems. Let us start estimating the respective size of gravito-electric and gravito-magnetic fields in a stationary background. The comparison is between $\mathscr{F}_{0i}$ and $\mathscr{F}_{ij}$, i.e. between $\left|\frac{\partial g_{00}}{\partial x^i}\right|$ and $\left|\frac{\partial g_{0j}}{\partial x^i} - \frac{\partial g_{0i}}{\partial x^j}\right|$. When calculating the ratio between gravito-magnetic and gravito-electric contributions to the force formula for a typical axially symmetric stationary metric, one obtains

$$\alpha = \frac{F_B}{F_E} \simeq \frac{a}{r}, \tag{24}$$

where $a = \frac{J}{Mc}$ is the specific angular momentum of the source of the gravitational field (length-dimensioned ratio of the angular momentum J to the mass M).

Within the Solar system the highest α ratio is obtained at the surface of the Sun; the order of magnitude is

$$\alpha \approx 10^{-5}. \tag{25}$$

The situation would be more favorable at the surface of some object with approximately the same a as the sun ($\sim 10^3$ m) and a much smaller radius. Collapsed objects can have these requisites with an α as high as 10^{-1} or more.

The Most Renowned Effect: Lense Thirring

Lense and Thirring worked out what is otherwise known as "frame dragging" by a spinning massive source, in 1918 [3]. The Lense Thirring effect is due to the gravito-magnetic field of the source, which in the case of the Earth (or the Sun or other celestial bodies) has a beautiful dipole form:

$$B_r = \frac{G}{c^3}\frac{J\cos\theta}{r^3}, \quad B_\theta = \frac{G}{c^3}\frac{J\sin\theta}{r^3}. \tag{26}$$

A spinning gyroscope is expected to interact with the gravito-magnetic field just as a magnetic dipole interacts with a magnetic field. Actually a complete standard treatment of the motion of a point-like gyroscope in a gravitational field evidences a precession rate corresponding to [4]

$$\vec{\Omega} = -\frac{1}{2}\vec{v} \wedge \vec{a} + \frac{3}{2}\vec{v} \wedge \vec{E}_g - \vec{B}_g. \tag{27}$$

The first term on the right is purely kinematical: it is the Thomas precession and depends on the three-velocity $\vec{v}$ and acceleration $\vec{a}$ of the gyroscope. When the latter is in free fall the Thomas precession contribution to the total precession rate is zero. The second term in eq. (27) represents the geodetic precession due to the interaction with the local gravito-electric field. Finally, the third term in eq. (27) is the proper Lense Thirring precession rate.

An orbiting body is indeed a gyroscope. When the central mass produces a gravito-magnetic field and the orbit is other than equatorial, the frame dragging will produce a precession of the whole plane of the orbit and this effect can be observed.

Ciufolini and coworkers [5] [6] have studied the behavior of the orbits of LAGEOS I and LAGEOS II terrestrial satellites both in the precession of the line of nodes and in the precession of the perigee. The expected precession rates are respectively

$$\vec{\Omega} = 2\frac{G}{c^2}\frac{\vec{J}}{a^3\left(1-e^2\right)^{3/2}} \text{ (line of nodes)},$$

$$\vec{\omega} = 2\frac{G}{c^2}J\frac{\hat{u}_j - 3\cos\alpha\,\hat{n}}{a^3\left(1-e^2\right)^{3/2}} \text{ (perigee)}$$

a is the semi-major axis of the orbit, e is the eccentricity, α the colatitude of the orbital plane; $\hat{u}_j$ a unit vector along $\vec{J}$; $\hat{n}$ a unit vector perpendicular to the plane of the orbit.

Numerical estimates for LAGEOS I and LAGEOS II are shown in the table below.

Table1. Angular precession rates

Satellite	Precession of the line of nodes	Precession of the perigee
LAGEOS I	31×10^{-3} "/year	32×10^{-3} "/year
LAGEOS II	31.5×10^{-3} "/year	-57×10^{-3} "/year

The results in the table are presently verified within 18%.

Work to improve these performances is under way, and many details are discussed in another presentation of this conference. A proposed new mission (LARES [7]) could bring the uncertainty below 1%.

Another observational possibility is with pulsars. The first discovered binary pulsar, PSR 1913+16 (Hulse and Taylor 1974 [8]), is a good laboratory for relativistic effects. The estimated relativistic precession rate of the orbit (both geodesic and gravito-magnetic) is of order $1°$ /year, and the periastron advance rate is $\sim 4.2°$/year. These effects have indeed been observed and the study of the pulse profile makes it possible to pick out the gravito-magnetic contribution.

The recently discovered double pulsar system J0737-3039 is extremely interesting and promises even better results.

Direct Measurement: GPB

Gravity Probe B (GPB) is the name of the experiment built by NASA and the Stanford University in order to measure both the geodetic and gravito-magnetic precession of an orbiting gyroscope. The gyroscopes are actually four, borne on a satellite in a polar orbit around the Earth. The calculated precession rates are respectively 6.6144 "/year for the geodetic term, and 0.0409 "/year for the frame dragging effect. The precessions are in mutually orthogonal planes.

The GPB mission flew on April 20th 2004. At the moment of closing the present paper (October 2004) the actual scientific data taking has begun. Approximately one year will be necessary before reliable conclusions can be drawn concerning the sought for relativistic effects. A precision of better than 1% is expected. The updated state of the mission, whose principal investigator is Francis Everitt, may be seen on the web site http://einstein.stanford.edu.

OTHER POSSIBILITIES AND PROPOSALS

The Clock Effect

The proper time differential for an observer in an axially symmetric stationary system is

$$d\tau = dt\sqrt{g_{00} + g_{rr}\dot{r}^2 + g_{\theta\theta}\dot{\theta}^2 + g_{\phi\phi}\dot{\phi}^2 + 2g_{0\phi}\dot{\phi}}, \qquad (28)$$

where dots stand for coordinate time derivatives. The last term under the square root changes its sign according to the rotation sense of the observer with respect to the one of the source. This fact introduces asymmetries in the way clocks tick and is the base of the gravitomagnetic clock effect. Considering two clocks on identical circular orbits, but moving in opposite directions, their respective coordinate revolution times will differ by [9] [10]

$$\delta T = 4\pi\frac{a}{c}. \qquad (29)$$

Now a is again the specific angular momentum of the central body. The result in eq. (29) is exact; when reading proper times on the clocks the difference after one complete revolution would only approximately be the one in (29). This gravitomagnetic clock effect is indeed big also in the Solar system. In fact it is

$$\delta T_{Earth} = 10^{-7}s, \quad \delta T_{Sun} = 10^{-5}s . \qquad (30)$$

Unfortunately, so far nobody has put forth a credible observational or experimental configuration or set up able to measure the effect.

Anisotropies in Light Propagation

From eq. (28) an anisotropic behavior in the times of flight of light passing by a rotating mass can be derived. On considering the equatorial plane of a rotating mass and using the same approach as for the Shapiro delay, the total coordinate time of flight of a light signal from the source to the observer can be expressed as the sum of three contributions: $t_f = t_0 + t_M + t_J$. The three terms are (b is the impact parameter of the light ray)

$$t_0 = \frac{y_2 - y_1}{c}, \text{Flat geometric term}$$

$$t_M = 2G\frac{M}{c^3} \ln \frac{y_2 + \sqrt{b^2 + y_2^2}}{y_1 + \sqrt{b^2 + y_1^2}}, \text{Mass or curvature effect (Shapiro)} \tag{31}$$

$$t_J = \mp 2\frac{GJ}{c^4 b} \left(\frac{y_2}{\sqrt{b^2 + y_2^2}} - \frac{y_1}{\sqrt{b^2 + y_1^2}} \right). \text{Angular momentum effect}$$

The y axis is along the line of sight, with origin on the spinning mass; index 1 is for the source, 2 for the observer; rays are assumed to be straight. The double sign in the last term shows the anisotropy between right and left with respect to the rotation axis of the central body.

On considering times of flight from the very central mass ($y_1 = 0$) and assuming $y_2 \gg b$, the t.o.f. anisotropy is

$$\delta t = 4\frac{GJ}{c^4 b}. \tag{32}$$

In the case of the Sun the time of flight difference would be $\delta t \sim 10^{-10}$ s. For a collapsed object the difference could be as high as $\sim 10^{-8}$ s. In this respect, the double pulsar system J0737-3039 could be the right place to look for the effect.

The time of flight anisotropy would indeed appear as an anisotropic frequency shift when the signal source passes behind the massive spinning body [11]. When it is possible to consider the motion of the signal source as being along a straight line orthogonal to the line of sight, the relative frequency shift would indeed be

$$\frac{\delta \nu}{\nu} = 4\frac{GM}{c^3}\frac{\nu_0}{b} - \frac{GM}{c^3}\frac{b\nu_0}{r^2} \mp 4\frac{GJ}{c^4}\frac{\nu_0}{b^2}. \tag{33}$$

Now ν_0 is the proper signal frequency at the source, and r is the distance from the source to the observer.

On treating the time delay above, a straight geometric trajectory has been assumed. This assumption may appear somewhat arbitrary since we know that gravitational fields actually bend light rays. The bending is indeed there; furthermore, the influence of the angular momentum of the central body will produce an asymmetric bending. Studying null geodesics in an axially symmetric stationary space-time one finds, at the lowest order in the angular momentum contribution, that the bending angle α is

$$\alpha = 4\frac{G M}{c^2 b} \left(1 - \frac{\vec{J} \cdot \hat{n}}{cMb} \right). \tag{34}$$

The gravito-magnetic term contains the unit vector $\hat{n}$, which is orthogonal to the plane containing the light ray and the center of the spinning body; the plane is oriented according to the position and sense of the ray, so that the gravito-magnetic correction has opposite signs on the right and on the left. The geometric asymmetry in the bending angles is

$$|\delta \alpha| = 8\frac{GJ}{c^3 b^2} \cos \theta. \tag{35}$$

Now θ is the co-latitude of the plane of the rays. In the case of the Sun it is $|\delta\alpha| \sim 10^{-14}$, exceedingly small to be of importance within the solar system; this is, a posteriori, the reason why the geometric contribution has been neglected when calculating the Shapiro delay. Once more, the situation could be more interesting in the neighborhood of a neutron star, where it could be $|\delta\alpha| \sim 10^{-6} \sim 1$".

Other proposals

Many different proposals have been put forward in a decades long pursuit of an experimental verification of gravito-magnetic effects. Here I would like to mention only a limited number.

A pendulum at the South pole. Braginsky, Polnarev and Thorne [12] many years ago proposed an experiment of the same class as GPB, to be performed on the Earth. The frame drag on a pendulum located at the pole should be measured. The expected precession rate would be ~ 0.281"/year, more than 6 times bigger than in the GPB case. The system would however be much noisier than on a freely falling satellite.

Anisotropic weight of spinning objects.A gyroscope in the inhomogeneous gravito-magnetic field of the Earth, just as a magnetic dipole in a non-uniform magnetic field, can be subject to a repulsive or attractive force in addition to the gravito-electric one. On introducing an interaction energy $W = -\vec{s} \cdot \vec{B}_g$ the force would be $\vec{F} = -\vec{\nabla}W$; $\vec{s}$ is the angular momentum of the gyroscope. The force may also be written in terms of an interaction of the angular momentum tensor of the gyroscope with the curvature tensor. With the Papapetrou notation it reads as $F^i = -\frac{1}{2}R^i_{0jk}S^{jk}$. On Earth at the poles it is given by (vertical component)

$$\delta F_z = \mp \frac{G}{c^2}\frac{sJ}{R^4} \sim 10^{-17}\text{s}. \tag{36}$$

Gravitational Faraday law. The gravito-electromagnetic analogy brings about, when it is viable, an induction law similar to the Faraday law for electromagnetism: $\vec{\nabla} \wedge \vec{E}_g = -\frac{\partial \vec{B}_g}{\partial t}$. Imagine an annular tube filled in with a superfluid, then set into rotation about a diameter in the gravito-magnetic field of the Earth [13]. The interaction, just as in an alternator, would induce alternating accelerations whose amplitude would be in the order of 10^{-11}m/s^2. An acceleration like this could indeed produce detectable effects in the superfluid, provided one could build in a reliable manner the equivalent of a Josephson junction for superfluids.

CONCLUSIONS

For length reasons it has not been possible to account for the great number of existing new and old ideas and proposals to detect gravito-magnetic effects, hence I have not mentioned, for instance, lunar laser ranging, anisotropic Doppler effect, differential magnetization of rotating cylinders, effects on polarized light, coupling of the particles' spin with curvature and the gravito-magnetic field, proposed experiments with rotating superconductors. As we have seen, only a limited number of observations on satellites' orbits and pulsars in binary systems and the GPB experiment are presently under way. However, the continuous progress in technologies and in particular in atomic clocks and time measurement techniques are bringing many of the ideas, so far considered as pure intellectual exercises, within the range of feasibility.

REFERENCES

1. Ruggiero, ML., and Tartaglia, A., *Nuovo Cimento B* **117**, 743-768 (2002).
2. Misner, C.W., Thorne, K. S., and Wheeler, J. A., *Gravitation*, San Francisco: Freeman Ed., 1973.
3. Lense, J., and Thirring, H, *Phys. Z.* **19**, 156 (1918). English translation in Mashhoon, B., Hehl, F. W., and Theiss, D. S., *Gen. Rel. Grav.* **16**, 711 (1984).
4. Straumann, N., *General Relativity and Relativistic Astrophysics,* Berlin-Heidelberg-New York: Springer Verlag, 1991, sec 5.5
5. Ciufolini, I., Pavlis, E., Chieppa, F., Lucchesi, D., and Vespe, F., *Class. Quantum Grav.* **14**, 2701 (1997).
6. Ciufolini, I., Pavlis, E., Chieppa, F., Fernandes Vieira, E., and Perez Mercader, J., *Science* **279,** 2100 (1998)
7. Ciufolini, I., *Class. Quantum Grav.* **17,** 2369 (2000).
8. Hulse, R. A., and Taylor, J. H., *Astrophys. J.* **195,** L51 (1975).
9. Cohen, J. M., and Mashhoon, B., *Phys. Lett. A* **181,** 353 (1993).
10. Tartaglia, A., *Class. Quantum Grav.* **17,** 2381 (2000).
11. Tartaglia, A., and Ruggiero, M. L., *Gen. Rel. Grav.* **36,** 293 (2004).
12. Braginsky, V. B., Polnarev, A. G., and Thorne, K. S., *Phys. Rev. Lett.* **53,** 863 (1984).
13. Tartaglia, A., and Ruggiero, M. L., *Eur. J. Phys.* **25,** 203-210 (2004).

The EXPLORER and NAUTILUS Gravitational Wave Detectors and Beyond

Massimo Visco

CNR, Istituto di Fisica dello Spazio Interplanetario, Via del Fosso del Cavaliere 100, Roma, Italy
Istituto Nazionale di Fisica Nucleare, Sez. di Roma 2, Viale della Ricerca Scientifica 1, Roma, Italy

Abstract. Explorer and Nautilus are the two detectors for gravitational waves that during the last few years have gathered data with the highest duty cycle. In this paper the present status, performances and recent results, obtained with these experiments, will be presented. Furthermore, one perspective of resonant antennas development, when the large-scale interferometric detectors are starting to operate, will be discussed.

INTRODUCTION

A good gravitational wave (GW) detector must have high sensitivity and such a stability to allow steady performance over long periods. The resonant bars have shown the possibility to be on the air for several years with continuous performance [1]. We will discuss in particular the performances of the EXPLORER and NAUTILUS detectors, designed and operated by the ROG collaboration. EXPLORER is installed at the CERN Laboratories in Geneva, NAUTILUS at the INFN Laboratories of Frascati. EXPLORER was the first detector to start long term operation at the beginning of the '90s and NAUTILUS was the first antenna of the ultra-cryogenic generation (with a thermodynamic temperature of about 100 mK) and began to operate in 1995. Since the beginning the two apparata have obtained long periods of continuous measurements, many in coincidence between themselves and with other detectors operating worldwide. The total observation time is much longer than that of any other detector (see Fig. 1).

The aim to have reliable and sensitive instruments was reached also developing innovative technologies. All the acquired experiences help to design new generations of resonant detectors like the spherical ones.

SENSITIVITY OF RESONANT DETECTORS

A resonant detector can be schematically described as made of four main parts:

- A large mass, whose amplitude of oscillation is changed by the impinging GW. Most of the present detectors use an aluminum cylinder of about 2000 kg.
- A transducer that transforms the mechanical signal in an electrical one. It is usually based on a capacitance or an inductance modulated by the antenna oscillation.

CP751, *General Relativity and Gravitational Physics, 16th SIGRAV Conference*, edited by G. Esposito et al.
© 2005 American Institute of Physics 0-7354-0236-1/05/$22.50

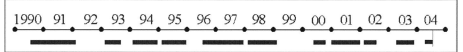

EXPLORER

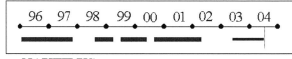

NAUTILUS

FIGURE 1. Explorer and Nautilus working periods over the last 14 years.

- A low noise amplifier to increase the amplitude of the electrical signal. Most of the present detectors use dcSQUID amplifiers.
- A data handling system.

Three main sources of noise, acting on the detector, limit the performances of a gravitational resonant detector:

- the environmental seismic noise,
- the thermal noise of the bar and the transducer,
- the electronic noise of the first stage of amplification.

A detector reaches satisfactory performances if all these sources of noise are under control.

Good mechanical filters are necessary to reduce the environmental seismic noise. The detectors are cooled in order to reduce the contribution of the thermal noise. The readout is periodically upgraded by following the technological progress.

The sensitivity of a resonant bar can be expressed by means of the input spectral strain sensitivity $S_h(\nu)$ (see for example Fig.2). $S_h(\nu)$ reaches its smallest value S_h^{min}, i.e. the detector has its best sensitivity, near the frequencies of the two resonant modes. S_h^{min} does not depend, to first approximation, on the electronic noise, but only on some physical parameters of the system: the mass M of the detector; the Brownian noise, which is proportional to the ratio between the thermodynamical temperature T and mechanical quality factor Q, i.e.

$$S_h^{min} \propto \left(\frac{T}{Q}\right)\frac{1}{M}. \tag{1}$$

On other hand, the signal bandwidth $\Delta\nu$ around the two resonant frequencies depends on the ratio between the electronic noise, expressed by the noise temperature T_n of the amplifier, and the contribution of Brownian noise given once again by the ratio between T and Q. A coupling factor β taking into account the percentage of mechanical energy transferred in the electrical circuit must be introduced in the expression:

$$\Delta v \propto \sqrt{\frac{\beta T_n}{T/Q}}. \tag{2}$$

S_h^{min} gives the maximum sensitivity for monochromatic or stochastic sources. If broad band pulses of GW are considered, the sensitivity can be calculated by integrating $S_h(v)$ over the bandwidth of the signal. An increase of the sensitivity for burst signals can therefore be obtained both decreasing the peak noise S_h^{min} and enlarging the bandwidth Δv. To decrease S_h^{min} it is necessary to have a detector with a larger mass or cooled at a lower temperature, which is equivalent to build a new experiment. A larger bandwidth can be obtained developing new amplifiers characterized by lower noise and new transducers that assure a better energy transfer from the antenna vibrations to the electronic system.

The bandwidth plays a crucial role for the search of different kinds of GW signals. In a search for periodic signals, a larger bandwidth increases the possibility of finding an observable source. The minimum detectable signal in a measurement of relic background radiation, that is performed by cross correlating the output of two detectors, scales with the square root of the overlapping bandwidths. In the searches for short bursts, a larger bandwidth gives a sharper time resolution. In a long term perspective, a larger bandwidth will provide more information on the incoming waveform, allowing a partial reconstruction of the signal and the possibility to introduce search for short time signals based on wave forms different from the delta shape.

The strategy of most groups working on resonant detectors, in order to obtain better measurements, consists, for the immediate future, of trying to enlarge the signal bandwidth by modifying the readout. On the other hand, on a long time scale, we shall design new apparata with larger mass like the large spherical detectors that, as we will show later, have also other interesting proprieties.

PRESENT EXPERIMENTAL CONFIGURATION AND PERFORMANCE

During the operating life of our experiments, phases of data taking have been alternated to periods devoted to upgrading the experiments. The last upgrades on EXPLORER and NAUTILUS took place respectively in 1999 and 2002. The two experiments now have a similar readout based on a small gap capacitive transducer [2] and an advanced commercial dc-SQUID amplifier. These two devices are connected through a superconducting transformer that provides the electrical matching.

The present science run with the two experiments in coincidence has started on June, 2003. The most evident improvement, obtained during this run, is represented by the useful signal bandwidth of 10 Hz, that means about 10 times larger than that we had in the past. The performances of the two detectors are stable and the sensitivity to short bursts of GW is around $2 \cdot 10^{-19}$ for NAUTILUS and $4 \cdot 10^{-19}$ for EXPLORER. In Fig. 3 is reported, for the two experiments, the hourly average of the sensitivity for short

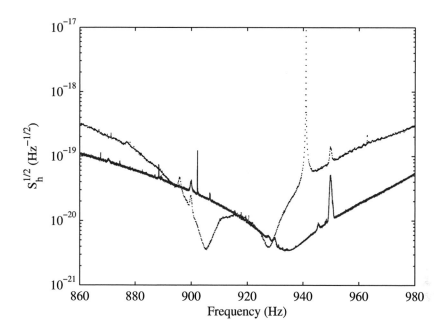

FIGURE 2. Explorer (dotted line) and Nautilus (continuous line) strain sensitivity on June 13, 2003.

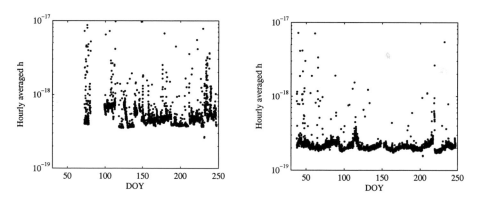

FIGURE 3. The hourly averaged value of the sensitivity to short burst for EXPLORER (left) and NAUTILUS (right), no veto is applied.

GW conventional bursts during 2003. Fig. 4 shows that the data taken follow a Gaussian distribution. No signal due to no modeled noise is present in NAUTILUS distribution, very few extra events are present in the case of EXPLORER.

Both experiments are now equipped with cosmic ray telescopes, that can detect

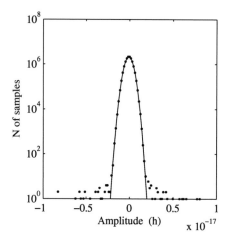
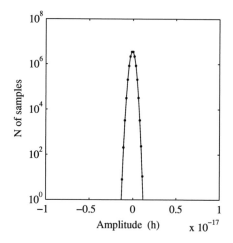

FIGURE 4. Distribution of the filtered output of the two antennas relative to 12 hours of the day September 4th, 2004. EXPLORER data are on the left.

charged particles crossing the antenna. Coincidences analysis between the signal at the output of the antenna and those resulting from extended air showers in the cosmic ray detector were carried out by showing expected and unexpected effects [3, 4].

MAIN RESULTS OBTAINED

Several searches, for different possible sources of signals, were carried out by using the data gathered by EXPLORER and NAUTILUS.

Monochromatic signals

The possible sources of monochromatic GW are generally weak, but they are quite interesting as the presence of the signal for long periods gives two advantages, first of all it is possible to increase the signal-to-noise ratio with the integration time, second, in case of detection, further observations can be done looking for a confirmation.

The spinning down of the source, that entails a change in the frequency of the emission, together with the frequency modulation obtained from the motion of the source relative to the observer, implies that, without any previous hypothesis, all sky search is a complicate procedure requiring huge computational time.

On using 95.7 days of EXPLORER data, an upper limit of $h_c = 3 \cdot 10^{-24}$ in the frequency range $(921.32 \div 921.38)$ Hz for signals coming from the Galaxy center was given [5]. An all-sky search, using 2 days of EXPLORER data has been done, giving an upper limit of $h_c = 2 \cdot 10^{-23}$ in the frequency range $(921.00 \div 921.76)$ Hz [6]. A further phase of all-sky search is in progress.

Stochastic background

During 1997 the two detectors were tuned, for a certain period, at a frequency of $v_c = 907.2\,Hz$. Cross-correlating the data during 10 hours, in a band of 0.1Hz, an upper limit on the quantity that measures the energy density of the Universe $\Omega_{GW}(v_c) < 60$ was given [7]. The present configuration of EXPLORER and NAUTILUS produces a superposition of the two bands (see Fig.2) of a few Hertz, much larger than the past. The measure of the stochastic background depends on the square root of this band, hence with the present configuration it would be possible to improve the previous upper limit by reaching $\Omega < 1$.

Burst signals

This kind of signal is the main target for resonant gravitational antennas. The data gathered with EXPLORER and NAUTILUS were widely used to search for coincidences between candidate events selected at the output of each antenna.

We recall the observations made within the IGEC (International Gravitational Event Collaboration): this collaboration among all five operating resonant detectors has developed a protocol in order to perform coincidence analysis using lists of event candidates for burst signals generated by each antenna [1]. On using the data of the detectors from 1997 to 2000, a new conservative upper limit $h > 2 \cdot 10^{-18}$, with 95% of probability, has been set on the rate of conventional short bursts of GW.

During 2001, for several reasons (maintenance, upgrade etc), NAUTILUS and EXPLORER were the only two resonant detectors in operation and they operated at their best sensitivity ever. The analysis on the data showed an excess of coincidence when a sidereal time analysis was applied. The coincidences found during the sidereal hours 3-5, in addition to be in number larger than the background, also appear to be strongly correlated in energy (Fig.5). The amplitude of these events is around $h \cong 10^{-18}$, therefore the observed excess is compatible with every previous upper limit [8, 9].

An analysis on the data collected during the present run of EXPLORER and NAUTILUS is underway with the aim of dismissing or confirming this effect.

RESONANT DETECTORS IN THE AGE OF INTERFEROMETERS

The large interferometer LIGO has produced the first scientific data and VIRGO is about to start operation. The interferometric detectors are characterized by a bandwidth larger than that of the resonant ones and their target sensitivity is better than that of the present resonant bars. Therefore, one could wonder if both kinds of apparatus should be maintained in operation.

A first general consideration that suggests a positive answer is that, at present, resonant detectors are the only ones that assure long time operation, allowing a continuous monitoring of the sky. A further point is that the first detection of a GW may not be too far and the opportunity to make an observation by two totally different instruments could

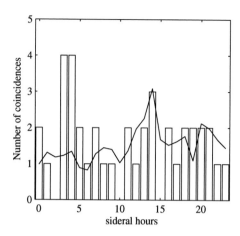

FIGURE 5. The histogram reports the numbers of observed coincidences and the background vs sidereal time.

help to better understand the observed phenomena. And finally there is still the possibility to both substantially increase the sensitivity of present resonant detectors and build new ones with better performance and innovative features. There are two new promising categories of resonant detectors: DUAL and spherical [10][11]. We will focus our attention on spherical ones, that are approaching operation [12]. We expect from them an increase of sensitivity and omnidirectionality.

Spherical detectors

A spherical antenna is an unique and interesting instrument to detect GW. First of all, it is omni-directional, capable to detect GW coming from any directions in the sky and having any polarization state. Moreover, a single spherical detector is sufficient to determine all five parameters of an incident GW: the amplitude of the two states of polarization, the direction of the source and the scalar component if present. In a spherical detector the five quadrupolar components that characterize a GW can be put in relation, one to one, to the five quadrupolar spherical modes of the sphere that can be excited by the GW. The problem of reading the amplitude of these modes from a sphere has been solved; one possible solution is the so-called TIGA configuration: it is possible to fasten 6 transducers on chosen faces of a truncated icosahedron, and to linearly combine their outputs by obtaining five mode channels corresponding to the five quadrupolar modes. A further advantages of a spherical antenna is the mass, about 20 times larger than that of a cylindrical detector working at the same frequency.

Recently, detectors based on hollow spheres were proposed [13]. A hollow sphere has the same geometrical properties of a solid one, it has a slightly smaller mass, but it can be more easily fabricated in large dimensions. A hollow sphere of $M \approx 2 \cdot 10^5 kg$ having a main resonant frequency around 200 Hz can reach a sensitivity $h \approx 10^{-22}$ for short bursts of GW.

Due to omnidirectionality, a spherical detector would be an interesting apparatus to work in parallel to a large interferometer, providing information concerning some parameters of the GW that cannot be observed by a single interferometer. Furthermore, two such detectors can be used for cross correlation measurements of stochastic background.

Directionality and sky coverage

The ability of a detector to cover the largest possible area of the sky is an important parameter as we are looking for rare phenomena. A sphere makes it possible for us to gain on the overall coverage of the sky with respect to other kinds of detectors, and to observe with the best sensitivity every source. The coverage of the sky in the case of a resonant bar or an interferometer depends on its position on the Earth and its azimuth with respect to North. For an interferometer this position is fixed once for all, instead for a bar it can be changed with minor effort. The resonant detectors working in the world were aligned in such a way to be parallel and therefore they have at any time the same response for signals coming from any source in the sky. In Fig. 6 is reported the percentage of time coverage for each area of the sky, considering a sensitivity at least 80% of the maximum, for three different detectors: EXPLORER (but it is similar for every other bar detector as they are all parallel), the VIRGO interferometer and a spherical detector. A spherical detector explores every area of the sky with the best sensitivity 24 hours a day, the other detectors observe with the best sensitivity only a small part of the sky for a time not larger then 60% of total operating time. In this analysis no hypothesis on the state of polarization of the wave has been done. If one takes into account the polarization, the advantage of a spherical detector, with respect to the other types, becomes larger.

In conclusion, we can imagine that in the future the GW search will greatly exploit spherical detectors that promise to be reliable instruments, complementary to the advanced large interferometers.

ACKNOWLEDGMENTS

I would like to thank my colleagues of ROG group and in particular Massimo Bassan, Eugenio Coccia and Lina Quintieri for useful discussions. All data shown here were produced by the ROG Collaboration.

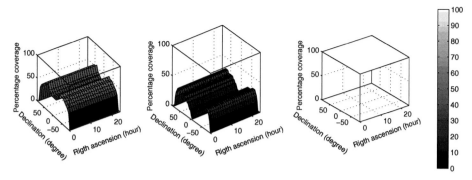

FIGURE 6. Percentage of observation time for each area of the sky for different detectors- starting from left the resonant bar EXPLORER - the interferometer VIRGO and a spherical detector installed in any place of the world

REFERENCES

1. Astone, P., Babusci, D., Baggio, L., Bassan, M., Blair, D. G., Bonaldi, M., Bonifazi, P., Busby, D., Carelli, P., Cerdonio, M., Coccia, E., Conti, L., Cosmelli, C., D'Antonio, S., Fafone, V., Falferi, P., Fortini, P., Frasca, S., Giordano, G., Hamilton, W. O., Heng, I. S., Ivanov, E. N., Johnson, W. W., Marini, A., Mauceli, E., McHugh, M. P., Mezzena, R., Minenkov, Y., Modena, I., Modestino, G., Moleti, A., Ortolan, A., Pallottino, G. V., Pizzella, G., Prodi, G. A., Quintieri, L., Rocchi, A., Rocco, E., Ronga, F., Salemi, F., Santostasi, G., Taffarello, L., Terenzi, R., Tobar, M. E., Torrioli, G., Vedovato, G., Vinante, A., Visco, M., Vitale, S., , and Zendri, J. P., *Phys. Rev. D*, **68**, 022001 (2003).
2. Bassan, M., Minenkov, Y., and Simonetti, R., "Advances in linear transducers for resonant gravitational wave antennas," in *Gravitational Waves Sources and Detectors*, edited by I. Ciufolini and F. Fidecaro, World Scientific, Singapore, 1997, pp. 225–228.
3. Astone, P., Bassan, M., Bonifazi, P., Carelli, P., Coccia, E., Fafone, V., D'Antonio, S., Frasca, S., Marini, A., Mauceli, E., Mazzitelli, G., Minenkov, Y., Modena, I., Modestino, G., Moleti, A., Pallottino, G. V., Papa, M. A., Pizzella, G., Ronga, F., Terenzi, R., Visco, M., and Votano, L., *Phys. Rev. Lett.*, **84**, 14–17 (2000).
4. Astone, P., Babusci, D., Bassan, M., Bonifazi, P., Carelli, P., Coccia, E., D'Antonio, S., Fafone, V., Giordano, G., Marini, A., Mazzitelli, G., Minenkov, Y., Modena, I., Modestino, G., Moleti, A., Pallottino, G. V., Pizzella, G., Quintieri, L., Rocchi, A., Ronga, F., Terenzi, R., and Visco, M., *Phys. Lett. B*, **540**, 179–184 (2002).
5. Astone, P., Bassan, M., Bonifazi, P., Carelli, P., Coccia, E., Cosmelli, C., D'Antonio, S., Fafone, V., Frasca, S., Minenkov, Y., Modena, I., Modestino, G., Moleti, A., Pallottino, G. V., Papa, M. A., Pizzella, G., Quintieri, L., Ronga, F., Terenzi, R., and Visco, M., *Phys. Rev. D*, **65**, 022001 (2002).
6. Astone, P., Babusci, D., Bassan, M., Borkowski, K. M., Coccia, E., D'Antonio, S., Fafone, V., Giordano, G., Jaranowski, P., Krolak, A., Marini, A., Minenkov, Y., Modena, I., Modestino, G., Moleti, A., Pallottino, G. V., Pietka, M., Pizzella, G., Quintieri, L., Rocchi, A., Ronga, F., Terenzi, R., and Visco, M., *Class. Quantum Grav.*, **20**, S665–S676 (2003).
7. Astone, P., Bassan, M., Bonifazi, P., Carelli, P., Cosmelli, C., Coccia, E., Fafone, V., Frasca, S., Marini, A., Minenkov, Y., Modena, I., Modestino, G., Moleti, A., Pallottino, G. V., Papa, M. A., Pizzella, G., Ronga, F., Terenzi, R., Visco, M., and Votano, L., *Astron. Astrophys.*, **351**, 811–814 (1999).
8. P. Astone, D. Babusci, M. Bassan, P. Bonifazi, P. Carelli, G. Cavallari, E. Coccia, C. Cosmelli, S. D'Antonio, V. Fafone, G. Federici, S. Frasca, G. Giordano, A. Marini, Y. Minenkov, I. Modena, G. Modestino, A. Moleti, G.V. Pallottino, G. Pizzella, L. Quintieri, A. Rocchi, F. Ronga, R. Terenzi, R. Torrioli, and M. Visco, *Class. Quantum Grav.*, **19**, 5449–5463 (2002).

9. Astone, P., Babusci, D., Bassan, M., Bonifazi, P., Carelli, P., Cavallari, G., Coccia, E., Cosmelli, C., D'Antonio, S., Fafone, V., Frasca, S., Giordano, G., Marini, A., Minenkov, Y., Modena, I., Modestino, G., Moleti, A., Pallottino, G. V., Pizzella, G., Quintieri, L., Rocchi, A., Ronga, F., Terenzi, R., Torrioli, G., and Visco, M., *Class. Quantum Grav.*, **20**, S785–S788 (2003).
10. Briant, T., Cerdonio, M., Conti, L., Heidmann, A., Lobo, A., and Pinard, M., *Phys. Rev. Lett.*, **67**, 102005 (2003).
11. Johnson, W. W., and Sphephen, M. M., *Phys. Rev. Lett.*, **70**, 2367–2370 (1993).
12. de Waard, A., Gottardi, L., Bassan, M., Coccia, E., Fafone, V., Flokstra, J., Karbalai-Sadegh, A., Minenkov, Y., Moleti, A., Pallottino, G. V., Podt, M., Pors, B. J., Reincke, W., Rocchi, A., Shumack, A., Srinivas, S., Visco, M., and Frossati, G., *Class. Quantum Grav.*, **21**, S465–S471 (2004).
13. Coccia, E., Fafone, V., Frossati, G., Lobo J. A., and Ortega, J., *Phys. Rev. D*, **57**, 2051–2060 (1998).

WORKSHOP—ASTROPHYSICS AND COSMOLOGY

Minisuperspace, WKB and Quantum States of General Relativistic Extended Objects

S. Ansoldi

Dipartimento di Matematica e Informatica, Università di Udine, and I.N.F.N. Sezione di Trieste via delle Scienze, 206 - I-33100 Udine (UD) - ITALY

Abstract. The dynamics of relativistic thin shells is a recurrent topic in the literature about the classical theory of gravitating systems and the still ongoing attempts to obtain a coherent description of their quantum behavior. Certainly, a good reason to make this system a preferred one for many models is the clear, synthetic description given by the Israel junction conditions. On using some results from an effective Lagrangian approach to the dynamics of spherically symmetric shells, we show a general way to obtain WKB states for the system; a simple example is also analyzed.

The study of the dynamics of an (infinitesimally)[1] thin surface layer separating two domains of spacetime is an interesting problem in General Relativity. The system can be described in a very concise and geometrically flavored way by using the Israel junction conditions [1, 2, 3]. Starting from these, many authors have then tackled the problem of finding some hints about the properties of the *still undiscovered* quantum theory of gravitational phenomena using shells as convenient models. In this context, just as examples of what can be found in the literature, we quote the seminal works of Berezin [5] and Visser [6], that date back to the early nineties, or the more recent [7, 8] and references therein.

What we are going to discuss shortly in the present contribution is set in this last perspective and suggests a semiclassical approach to define WKB quantum states for spherically symmetric shells. This method has already been used in [4].

Let us then consider a spherical shell (we refer the reader to [3] for very concise/clear background material and for definitions). For our purpose the relevant result is equation (4) in [3], i.e. the junction conditions[2] $K_{ij}^- - K_{ij}^+ \propto S_{ij} - g_{ij}S/2$. K_{ij} is the extrinsic curvature of the shell and can take different values on the two sides (+ and − spacetime regions) of it. S_{ij} is the stress-energy tensor describing the energy/matter content of the shell (S is its trace). For a spherical shell these equations reduce to the single condition

$$\varepsilon_-(\dot{R}^2 + f_-(R))^{1/2} - \varepsilon_+(\dot{R}^2 + f_+(R))^{1/2} = M(R)/R, \tag{1}$$

where $f_\pm(r)$ are the metric functions in the static coordinate systems adapted to the

---

[1] Far from being only idealizations of more realistic situations, shells have been extensively used to build concrete models of many astrophysical and cosmological scenarios (for a detailed bibliography, see the references in [4]).

[2] Conventions are as in [9] and the definition of the (quantities relevant for the concept of) extrinsic curvature can be found in [3, 9].

CP751, *General Relativity and Gravitational Physics, 16th SIGRAV Conference*, edited by G. Esposito et al.
© 2005 American Institute of Physics 0-7354-0236-1/05/$22.50

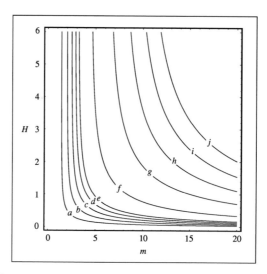

FIGURE 1. Plot of some WKB levels for the example discussed in the text. The curves a, b, c, d, e, f, g, h, i, j correspond to the quantum numbers $1, 2, 3, 4, 5, 10, 20, 30, 40, 50$, respectively.

spherical symmetry of the 4-dimensional spacetime regions joined across the shell; $\varepsilon_\pm = +1, -1$ are signs and R and $M(R)$ are the radius (a function of the proper time τ of a co-moving observer[3]) and the matter content (what remains of S_{ij}) of the spherical shell, respectively. As shown for example in [10, 11, 12] the above equation can be obtained from an effective Lagrangian[4] $L_{EFF}(R, \dot{R})$, as a first integral of the second order Euler–Lagrange equation. From $L_{EFF}(R, \dot{R})$ the momentum conjugate to the single degree of freedom R can be obtained as usual, $P(R, \dot{R}) = \partial L_{EFF}(R, \dot{R})/(\partial \dot{R})$. We are not interested in the explicit form of P here, we just point out that it is a function of R and $\dot{R}$ highly *non-linear* in $\dot{R}$. This non-linearity spoils the natural and simple interpretation of momentum that we know from classical analytical mechanics. Nevertheless, we can still solve (1) for the classically allowed trajectories of the shell, by using a *standard* analogy with the motion of a *unitary mass particle with zero energy in an effective potential* [16, 17, 18]. This gives $\dot{R}$ as a function of R and, substituting this expression for $\dot{R}$ in $P(R, \dot{R})$, we obtain the *conjugate momentum on a solution of the classical equations of motion*. In what follows we are going to indicate the momentum *evaluated on a classical trajectory* as $P(R; \mathscr{S})$: this emphasizes that it is a function of R, that it depends on the set $\mathscr{S}$ of the other parameters of the problem, but, of course, it is not a function of $\dot{R}$.

By integrating the expression for $P(R; \mathscr{S})$ on a classically allowed trajectory with turning points R_{MIN} and R_{MAX}, we can compute the value of the classical action for that trajectory. It is a function of the set of parameters characterizing the matter content and of the geometry, $\mathscr{S}$. WKB quantum states of the system can now be defined as states for

---

[3] As usual an overdot, " $\dot{}$ ", denotes a total derivative with respect to τ.

[4] For a more general and deeper discussion see [13] and references therein or also, in addition, [14, 15].

which the above action is a multiple of the quantum

$$S(\mathscr{S}) = 2 \int_{R_{\text{MIN}}}^{R_{\text{MAX}}} P(R; \mathscr{S}) dR \sim n, \quad n = 0, 1, 2, 3, \ldots \tag{2}$$

In quantum gravity we expect to have a theory that selects some geometries from the set of all possible ones, consistently with quantum dynamics. In our discussion we have limited the treatment to a *minisuperspace approximation*, but, indeed, we see that the quantization condition (2) does select only some of the possible geometries, those in which the parameters of the model are related by the quantum number n as in (2). We can see this more explicitly in the following simple model: a dust shell ($M(R) = m$) separating two domains of anti-de Sitter spacetime with the same cosmological constant ($f_-(r) = f_+(r) = f(r) = 1 + r^2/H^2$): it is possible to prove that a non-trivial junction of the two spacetimes can be performed, although we are not going to discuss this aspect here nor are we going to describe the resulting global spacetime structure. In this case the set of parameters is $\mathscr{S} = \{m, H\}$ and the quantization condition (2) becomes $S(m, H) \sim n, n = 0, 1, 2, 3, \ldots$. Some levels are plotted in figure 1 and clearly show that, given one of the parameters (say m), only a discrete set of values for the other H is allowed. Thus, the quantization condition restricts the possible values that can be given to the parameters that characterize the spacetime geometry and/or the matter content of the shell. This is consistent with the general picture of a quantum gravitational scenario.

Other applications of this general approach are presently under study. In particular, generalizations to higher dimensions [19] could be relevant, for example, in the context of brane cosmological models.

REFERENCES

1. Israel, W., *Nuovo Cimento B*, **44**, 1 (1966).
2. Israel, W., *Nuovo Cimento B*, **48**, 463 (1967).
3. Barrabes, C., and Israel, W., *Phys. Rev. D*, **43**, 1129–1142 (1991).
4. Ansoldi, S., *Class. Quantum Grav.*, **19**, 6321–6344 (2002).
5. Berezin, V. A., *Phys. Lett. B*, **241**, 194–200 (1990).
6. Visser, M., *Phys. Rev. D*, **43**, 402–409 (1991).
7. Berezin, V., *J. Mod. Phys. A*, **17**, 979–988 (2002).
8. Corichi, A., Cruz-Pacheco, G., Minzoni, A., Padilla, P., Rosenbaum, M., Ryan Jr., M. P., Smyth, N. F., and Vukasinac, T., *Phys. Rev. D*, **65**, 064006(1–13) (2002).
9. Misner, C. W., Thorne, K. S., and Wheeler, J. A., *"Gravitation"*, W. H. Freeman and Company, 1973.
10. Farhi, E., Guth, A. H., and Guven, J., *Nucl. Phys. B*, **339**, 417–490 (1990).
11. Ansoldi, S., Aurilia, A., Balbinot, R., and Spallucci, E., *Class. Quantum Grav.*, **14**, 2727–2755 (1997).
12. Ansoldi, S., *Graduation Thesis (in Italian)*, University of Trieste (1994).
13. Hajicek, P., and Kijowski, J., *Phys. Rev. D*, **62**, 044025(1–5) (2000).
14. Gladush, V. D., *J. Math. Phys.*, **42**, 2590–2610 (2001).
15. Sindoni, L., *Graduation Thesis (in Italian)*, University of Trieste (2004).
16. Berezin, V. A., Kuzmin, V. A., and Tkachev, I. I., *Phys. Rev. D*, **36**, 2919–2944 (1987).
17. Blau, S. K., Guendelman, E. I., and Guth, A. H., *Phys. Rev. D*, **35**, 1747–1766 (1987).
18. Aurilia, A., Palmer, M., and Spallucci, E., *Phys. Rev. D*, **40**, 2511–2518 (1989).
19. Ansoldi, S., Guendelman, E. I., and Ishihara, H. (In preparation).

Proper-time regulators and RG flow in QEG

Alfio Bonanno* and Martin Reuter†

*INAF - Osservatorio Astrofisico di Catania, Via S. Sofia 78, I-95123 Catania, Italy
INFN, Via S. Sofia 64, I-95123 Catania, Italy
†Institut für Physik, Universität Mainz
Staudingerweg 7, D-55099 Mainz, Germany

Abstract. A proper time renormalization group equation for Quantum Einstein Gravity is studied in the Einstein–Hilbert truncation and its predictions are compared to those of the conceptually different exact renormalization group equation of the effective average action. A smooth infrared regulator of a special type is known to give rise to extremely precise critical exponents in scalar theories. In particular the proper time equation, too, predicts the existence of a non-Gaussian fixed point as is necessary for the conjectured nonperturbative renormalizability of Quantum Einstein Gravity.

By definition, Quantum Einstein Gravity (QEG) is the conjectured quantum field theory of the spacetime metric whose bare action is (infinitesimally close to) an ultraviolet (UV) attractive non-Gaussian fixed point of the renormalization group (RG) flow. As outlined by S. Weinberg long ago [1] this theory would be nonperturbatively renormalizable or "asymptotically safe" and, most probably, predictive and internally consistent even at the shortest sub-Planckian distances. Most of the existing RG calculations were performed within the framework of the exact RG equation pertaining to the effective average action [2], applied to a truncated space of action functionals (theory space). It turned out that the UV fixed point necessary for asymptotic safety does indeed exist within the Einstein–Hilbert [3, 4, 5] and the R^2-truncation [6], and detailed analyses of the reliability of those approximations [4, 6] revealed that the fixed point found is unlikely to be a truncation artifact. But clearly it is necessary to collect further evidence for its existence at the exact level, both within the effective average action formalism (by generalizing the truncation) and with conceptually independent methods.

In the present study we are going to analyze the non-Gaussian fixed point of QEG using the "proper time renormalization group" which has been extensively used for other theories recently [7, 8]. Our approach implements a certain RG improvement of one-loop perturbation-theory, employing an infrared (IR)-regularized proper time representation of the corresponding one-loop determinants. We shall derive a flow equation for the scale dependent gravitational action from it, whose structure is quite different from that of the exact RG equation of the average action, for every possible choice of its built-in cutoff operator $\mathscr{R}_k$. Apart from the conceptual independence, the main motivation for the present analysis are the remarkably precise critical exponents which have been obtained for the Wilson–Fisher fixed point with this method [7]. In fact, we shall be mainly interested in the universal properties of the non-Gaussian fixed point of QEG, (g_*, λ_*), in particular its critical exponents θ' and θ'' and the product $g_* \lambda_*$.

CP751, *General Relativity and Gravitational Physics, 16th SIGRAV Conference*, edited by G. Esposito et al.

The ultimate goal of the RG approach is the calculation of the functional integral $Z = \int \mathscr{D}\gamma_{\mu\nu} \exp(-S[\gamma_{\mu\nu}])$ over all metrics $\gamma_{\mu\nu}$ where S is an arbitrary bare action, invariant under general coordinate transformations. In the Wilsonian approach one is interested in a determination of the "blocked action", i.e. an effective action generated by integrating out only the degrees of freedom above the scale k. The "proper time RG equation" for the functional $\widehat{S}_k$ is then a non-perturbative functional flow equation for the "blocked" effective action at the scale k, and it reads as (see [9] for details)

$$\partial_t \widehat{S}_k[g, \bar{g}] = -\frac{1}{2} \mathrm{Tr} \int_0^\infty \frac{ds}{s} \, \partial_t f_k^m(s) [\exp(-s\widehat{S}_k^{(2)}) - 2\exp(-sS_{\mathrm{gh}}^{(2)})], \tag{1}$$

where $\widehat{S}_k$ is the blocked action, and $S_{\mathrm{gh}}^{(2)}$ is the second functional derivative of the ghost action contribution.

For actual calculations we shall use the following one-parameter family of smooth cutoffs $f_k \equiv f_k^m$ which has been extensively used in the literature, in particular in high precision calculation of critical exponents:

$$f_k^m(s) = \frac{\Gamma(m+1, sk^2) - \Gamma(m+1, s\Lambda^2)}{\Gamma(m+1)}. \tag{2}$$

Here m is an arbitrary real, positive parameter which controls the shape of the f_k^m in the interpolating regions, and $\Gamma(\alpha, x) = \int_x^\infty dt \, t^{\alpha-1} e^{-t}$ denotes the incomplete Gamma-function. We shall also consider the "rescaled" smearing functions $\bar{f}_k^m(s) = f_k^m(ms)$ that have also been used for the calculation of critical exponents and first-order phase transition [10].

The RG equation (1) describes a flow on the infinite-dimensional space of functionals $\widehat{S}[g, \bar{g}]$. It is possible to obtain nonperturbative solutions (RG trajectories) by truncating this theory space. In the following we shall consider the Einstein–Hilbert approximation [11] whose truncated theory space is two-dimensional. The corresponding ansatz for $\widehat{S}_k$ is parametrized by a k-dependent Newton constant G_k and cosmological constant $\bar{\lambda}_k$. It is important to rewrite the projected flow equations in terms of the dimensionless Newton constant $g(k) \equiv k^{d-2} G_k \equiv k^{d-2} Z_{Nk}^{-1} \bar{G}$ and the dimensionless cosmological constant $\lambda(k) \equiv k^{-2} \bar{\lambda}_k$. This leads to the following system of equations:

$$\partial_t g = \beta_g(g, \lambda) \equiv [d - 2 + \eta_N] g, \tag{3a}$$

$$\partial_t \lambda = \beta_\lambda(g, \lambda), \tag{3b}$$

where anomalous dimension $\eta_N \equiv -\partial_t \ln Z_{Nk}$ and β-function of λ are described in [9]. It turns out that the β-functions are much simpler than their counterparts from the average action. In particular they do not contain terms (proportional to η_N) stemming from the differentiation of the $\mathscr{Z}$-factors (see ref. [11]), and they allow for an explicit evaluation of the threshold functions. We have calculated the non-trivial fixed point $(g_*, \lambda_*) \neq 0$ in $d = 4$, implied by the simultaneous vanishing of β_g and β_λ in (3), and the critical exponents θ', θ'' [4, 5] associated to the stability matrix for the non-trivial fixed point. The results are depicted in Tab.1 for various values of the cut-off parameter m. The

TABLE 1. Fixed-point values and critical exponents for various values of the cutoff parameter m. Note the plateau behavior of the universal quantities θ', θ'', and $\lambda_* g_*$ as the value of m increases. λ_*^m and g_*^m are the fixed-point values for the $\bar{f}_k^m$ cutoff. In this latter case the universal quantities are the same within the numerical error.

m	g_*	λ_*	$\lambda_* g_*$	θ'	θ''	g_*^m	λ_*^m
3/2	0.763	0.192	0.147	2.000	1.658	0.509	0.289
2	1.663	0.118	0.138	1.834	1.230	0.583	0.236
3	1.890	0.066	0.125	1.769	1.081	0.623	0.199
4	2.589	0.046	0.119	1.750	1.001	0.647	0.183
5	3.281	0.035	0.115	1.742	0.959	0.656	0.175
6	3.970	0.028	0.113	1.737	0.934	0.661	0.170
10	6.718	0.016	0.108	1.729	0.886	0.671	0.161
40	27.271	0.0038	0.103	1.722	0.840	0.681	0.152

most important result is that the proper time RG equation, too, predicts a non-Gaussian fixed point. It exists for any value of m. The nonuniversal coordinates λ_* and g_* show a significant m-dependence, and it is impressive to see how this m-dependence cancels out in the product $\lambda_* g_*$ which is universal in an exact calculation [5]. For every value of m, the stability matrix $(-\partial_i \beta_j)$, $i, j \in \{g, \lambda\}$, has a pair of complex conjugate eigenvalues $\theta' \pm i\theta''$. Remarkably, those critical exponents and the product $\lambda_* g_*$ tend to constant values as m increases; they form a "plateau". The universal quantities θ', θ'', and $\lambda_* g_*$ show excellent stability properties for this class of regulators, and these values are in complete agreement with the values found in the framework of the exact flow equation for the effective average action for gravity [4, 5]. In fact, the differences between the proper time and the average action RG equations are of the same order of magnitude as the residual scale dependence which is present in each of the two approaches and which results from the truncated theory space.

We have then shown that the universal quantities determined by the non-Gaussian fixed point are in very good agreement with the results from the average action [4, 5, 6], and in particular, that they show a very weak dependence on the regulator. Our result can be taken as a further, conceptually different indication that the fixed point is rather "robust" and should survive also in a more general truncation.

REFERENCES

1. Weinberg, S., *In General Relativity, an Einstein Centenary Survey*, Hawking, S.W., Israel, W. (Eds.): Cambridge University Press (1979).
2. Wetterich, C., *Phys. Lett.*, **B301**, 90–94 (1993).
3. Souma, W., *Prog. Theor. Phys.*, **102**, 181–195 (1999).
4. Lauscher, O., and Reuter, M., *Phys. Rev. D*, **65**, 025013 (2002).
5. Reuter, M., and Saueressig, F., *Phys. Rev. D*, **65**, 065016 (2002).
6. Lauscher, O., and Reuter, M., *Class. Quantum Grav.*, **19**, 483–492 (2002).
7. Bonanno, A., and Zappala, D., *Phys. Lett.*, **B504**, 181–187 (2001).
8. Litim, D. F., and Pawlowski, J. M., *Phys. Lett.*, **B546**, 279–286 (2002).
9. Bonanno, A., and Reuter, M., hep-th/0410191 (2004).
10. Bonanno, A., and Lacagnina, G., *Nucl. Phys.*, **B693**, 36–50 (2004).
11. Reuter, M., *Phys. Rev. D*, **57**, 971–985 (1998).

Three-Dimensional Chern–Simons and BF Theories

Andrzej Borowiec* and Mauro Francaviglia[†]

*Institute of Theoretical Physics, University of Wrocław, pl. Maxa Borna 9
PL-50-204 WROCŁAW, POLAND
[†]Dipartimento di Matematica, Università di Torino, Via C. Alberto 10
10123 TORINO, ITALY

Abstract. Our aim in this note is to clarify a relationship between covariant Chern–Simons 3-dimensional theory and Schwartz type topological field theory known also as BF theory.

INTRODUCTION.

Chern–Simons theory gives an interesting example of topological field theory. Its Lagrangian 3-form lives on a principal G-bundle and after pulling back to space-time (base) manifold provides, in general, a family of local, non-covariant Lagrangian densities [3, 4]. [1] Because of this, it is also more difficult to analyze, in this case, Nöether conserved quantities [2]. A more standard approach to the problem of symmetries and conservation laws has been applied in the so-called covariant formalism [3]. It exploits the transgression 3-form as a global and covariant Chern–Simons Lagrangian with two dynamical gauge fields. This formalism has been used for the calculation of Nöther currents and their identically vanishing parts - superpotentials. Augmented variational principle and relative conservation laws have been recently proposed in [6]. Our aim in the present note, which can be viewed as an appendix to [3], is to explain a link between covariant Chern–Simons theory and the so-called BF theories [1, 5, 7].

CHANGE OF VARIABLES

Let us consider a principal bundle $P(M, G)$ over a three-dimensional manifold M with a (semisimple) structure group G. Let ω_i $(i = 0, 1)$ be two principal connection 1-forms with the corresponding curvature 2-forms

$$\Omega_i = d\omega_i + \omega_i^2 = d\omega_i + \frac{1}{2}[\omega_i, \omega_i]. \tag{1}$$

Denote by $\alpha = \omega_1 - \omega_0$ a tensorial 1-form.

---

[1] However, the corresponding Euler–Lagrange equations of motion have well-defined global meaning.

CP751, *General Relativity and Gravitational Physics, 16th SIGRAV Conference*, edited by G. Esposito et al.
© 2005 American Institute of Physics 0-7354-0236-1/05/$22.50

The transgression 3-form is given by the well-known formula

$$Q(\omega_1, \omega_0) = -Q(\omega_0, \omega_1) = tr\left(2\Omega_0 \wedge \alpha + D_0\alpha \wedge \alpha + \frac{2}{3}\alpha^3\right)$$

$$= tr\left(2\Omega_1 \wedge \alpha - D_1\alpha \wedge \alpha + \frac{2}{3}\alpha^3\right), \tag{2}$$

where $D_i\alpha = d\alpha + [\omega_i, \alpha]$ denotes the covariant derivative of α with respect to the connection ω_i. Thus, $Q(\omega_1, \omega_0)$ is a tensorial (covariant) object which well defines the corresponding global 3-form on M. It undergoes a non-covariant splitting as a difference of two Chern–Simons Lagrangians

$$Q(\omega_1, \omega_0) = CS(\omega_1) - CS(\omega_0) + dtr(\omega_0 \wedge \omega_1), \tag{3}$$

where

$$CS(\omega) = tr\left(\Omega \wedge \omega - \frac{1}{3}\omega^3\right) \tag{4}$$

stands for non-covariant Chern–Simons Lagrangian.

Note that in the case of two connections one has

$$2\Omega_0 + D_0\alpha = 2\Omega_1 - D_1\alpha =$$
$$\Omega_0 + \Omega_1 + \frac{1}{2}(D_0\alpha - D_1\alpha) = \Omega_0 + \Omega_1 - \alpha^2. \tag{5}$$

The last equality makes it possible for us to rewrite

$$Q(\omega_1, \omega_0) = 2tr\left(\bar{\Omega}\alpha + \frac{1}{12}\alpha^3\right), \tag{6}$$

where $\bar{\omega} = \frac{1}{2}(\omega_1 + \omega_0)$ is a new (average) connection and $\bar{\Omega} = d\bar{\omega} + \bar{\omega}^2$. Of course, one has $\omega_1 = \bar{\omega} + \frac{1}{2}\alpha$, $\omega_0 = \bar{\omega} - \frac{1}{2}\alpha$.

Thus, the Lagrangian $Q(\omega_1, \omega_0)$ can be treated in three different (but equivalent) ways:

- with two (flat) connections ω_0, ω_1 as dynamical variables; see (3). In this case $\Omega_0 = 0$ and $\Omega_1 = 0$ are equations of motion. This point of view was presented in [3].
- with a (flat) connection ω_1 and tensorial 1-form α as dynamical variables; see (2). In this case equations of motion are $\Omega_1 = 0$ and $D_1\alpha = \alpha^2$.
- with an "average" (non-flat) connection $\bar{\omega} = \frac{1}{2}(\omega1 + \omega2)$ and tensorial 1-form α as independent dynamical variables; see (6). In this case $\Omega = -\frac{1}{4}\alpha^2$ and $D\alpha = 0$ are equations of motion. This is the so-called BF theory with a cosmological constant $\Lambda = 1$ (see e.g references [1, 5, 8, 9]).

More generally, one can define a new connection $\omega_t = t\omega_1 + (1-t)\omega_0 = \omega_0 + t\alpha$ as a convex combination of two connections with parameter $0 \leq t \leq 1$. The inverse transformation is $\omega_0 = \omega_t - t\alpha$, $\omega_1 = \omega_t + (1-t)\alpha$. In this case

$$\Omega_t = t\Omega_1 + (1-t)\Omega_0 - t(1-t)\alpha^2 .$$

Now the equation (5) can be replaced by the more general one

$$2\Omega_1 - D_1\alpha = 2\Omega_0 + D_0\alpha = \qquad (7)$$
$$= 2\Omega_t - 2t(1-t)\alpha^2 - (2t-1)D_t\alpha,$$

(notice that $t\omega_1 + (t-1)\omega_0 = (2t-1)\omega_t + 2t(1-t)\alpha$).
In these new variables (ω_t, α) we obtain

$$Q(\omega_1, \omega_0) = 2tr\left(\Omega_t \wedge \alpha - \left(t - \frac{1}{2}\right)D_t\alpha \wedge \alpha + (\frac{1}{3} - t + t^2)\alpha^3\right). \qquad (8)$$

The corresponding equations of motion are

$$\Omega_t = -t(1-t)\alpha^2 \quad , \quad D_t\alpha = (2t-1)\alpha^2. \qquad (9)$$

Thus, the choices $t = \frac{1}{2}, 0, 1$ lead to the simplest formulae.

It is interesting to point out that the superpotential related to the (infinitesimal) gauge transformation χ remains independent of the choice of variables (ω_t, α) (compare formula (19) in [3]), i.e.

$$U(\chi) = tr(\alpha\chi).$$

Instead, an explicit expression for the superpotential related to (infinitesimal) diffeomorphism transformation driven by a vector field ξ does depend on the variables (ω_t, α) and is equal to (compare formulae (23,25) in [3])

$$U(\xi) = tr[\alpha(2\omega_t(\xi) + (1-2t)\alpha(\xi))] .$$

REFERENCES

1. D. Birmingham, M. Blau, M. Rakowski and G. Thompson, *Phys. Rep.*, **209**, 129 (1991).
2. A. Borowiec, M. Ferraris and M. Francaviglia, *J. Phys. A: Math. Gen.*, **36**, 8823-8833 (1998).
3. A. Borowiec, M. Ferraris and M. Francaviglia, *J. Phys. A: Math. Gen.*, **31**, 2589-2598 (2003).
4. A. Borowiec, M. Ferraris, M. Francaviglia and M. Palese, *Universitatis Iagellonicae Acta Mathematica*, **XLI**, 319-331 (2003).
5. A .S. Cattaneo, P. Cotta-Ramusino, J. Fröhlich and M. Martellini, *J. Math. Phys.*, **36**, 6137-6160 (1995).
6. L. Fatibene, M. Ferraris, and M. Francaviglia, "Augmented Variational Principles and Relative Conservation Laws in Classical Field Theory" *math-phys/0411029* (2004).
7. G .T. Horowitz, *Commun. Math. Phys.*, **125**, 417 (1989).
8. M. Montesinos, *Class. Quantum Grav.*, **20**, 3569-3575 (2003).
9. G. Sardanashvily, *Int. J. Geom. Meth. Mod. Phys.*, **1**, 4 (2004).

Singular PP waves, Junction Conditions and BPS States

Fabrizio Canfora, Gaetano Vilasi

Università di Salerno, Dipartimento di Fisica "E.R.Caianiello", Via S. Allende, 84081 Baronissi
(Salerno) Italy
Istituto Nazionale di Fisica Nucleare, GC di Salerno, Italy.

Abstract. A simple model to study the collision of PP waves via the Israel junction conditions is proposed. The junction conditions are interpreted as topological conservation laws, and the relation with BPS states is shortly described.

INTRODUCTION

The PP waves, first introduced by Peres [23] [24], in the last years received much attention from the theorists since supergravity and superstring theory on a PP-wave background have very interesting properties. Moreover, asymptotically flat PP waves have a longitudinal polarization [23] [24] [9] [10], and are very naturally coupled to cosmic strings and γ-ray bursts. Thus, it is not unlikely that, once experimental devices will detect gravitational waves, it will be possible to distinguish the experimental signature of a PP wave from the one of standard gravitational waves [8].

In this contribution, the matching between two PP waves moving in opposite spatial directions will be analyzed. This analysis is interesting since it makes it possible to build a space-time of highly non-trivial physical content in a very natural way.

The paper is organized as follows: in the first section some useful properties of PP waves are shortly described. Then, the Israel matching conditions are introduced. In the third section the matching between two PP waves travelling in opposite spatial direction is analyzed by using the Israel method and the corresponding physical consequences on the dynamics of cosmic strings and Σ are drawn. The extension to space-time of arbitrary dimension is discussed and some interesting features related to the *brane-world scenarios* (see, for example, [15] [3] [4] [26] and references therein) are briefly described. Particular emphasis has been given to the five-dimensional case in connection with the role of BPS states in gravity[1], and the relation between the Dirac quantization condition and the mass quantization of extremal charged black holes.

---

[1] Such as the Majumdar–Papapetrou solutions which, as it will be explained below, are naturally related to PP waves.

CP751, *General Relativity and Gravitational Physics, 16th SIGRAV Conference*, edited by G. Esposito et al.
© 2005 American Institute of Physics 0-7354-0236-1/05/$22.50

THE METRIC

PP waves provide interesting non-trivial backgrounds for superstring theory (see, for example, [5] [1] [19] [16] [27]). However, the PP-waves that string theorists usually consider are non-singular: this means that the harmonic function which parametrizes such waves is assumed to be quadratic. Instead, we will consider asymptotically flat PP waves whose harmonic function has a finite number of $\delta-$like singularities. Such PP-wave space-times, as we will now explain, have a finite number of string-like singularities.

A PP wave propagating in the positive z-direction reads as

$$g^+ = 2dudv + H^+(x, y, u)du^2 - dx^2 - dy^2, \tag{1}$$

where $z = (u - v)/2$, and the vacuum field equations imply

$$\left(\partial_x^2 + \partial_y^2\right) H^+ = 0. \tag{2}$$

Because of Eq. (2), coordinates (u, v, x, y) are harmonic. However, in order to have spatially asymptotically flat gravitational fields (a finite number of) $\delta-$like singularities in the x-y plane (that is, by taking into account the third spatial dimension, string-like singularities) have to be allowed [8], otherwise the metric (2) would not approach the flat metric. Thus, H^+ fulfils the following equation:

$$\left(\partial_x^2 + \partial_y^2\right) H^+ = \sum_i \mu_i^+ \delta^{(2)}\left(x_i(u), y_i(u)\right), \tag{3}$$

where the positions of singularities, in principle, could change with u and μ_i^+ represents the energy per unit length of the $i-$th cosmic string. Thus, cosmic strings are naturally coupled to asymptotically flat PP waves. An asymptotically flat PP wave propagating in the negative z-direction has the following form:

$$g^- = 2dudv + H^-(x, y, v)dv^2 - dx^2 - dy^2, \tag{4}$$

$$\left(\partial_x^2 + \partial_y^2\right) H^- = \sum_i \mu_i^- \delta^{(2)}\left(x_i(v), y_i(v)\right). \tag{5}$$

To construct a space-time containing PP waves and cosmic strings we will match the two above gravitational fields on the three-dimensional[2] time-like hypersurface $2z = u - v = 0$ by using the elegant method of matching conditions introduced by Israel in 1966 [17].

THE ISRAEL MATCHING CONDITIONS

Starting with a right propagating g^+ and a left propagating g^- PP-wave solutions of the Einstein equations, we can match them to obtain a global (piecewise C^2) solution

---

[2] Of course, in general N-dimensional space-time, such a timelike hypersurface would be $(N-1)$-dimensional.

provided the following conditions are fulfilled [17]:

$$\lim_{x \to \Sigma^+} g^+(x) = \lim_{x \to \Sigma^-} g^-(x), \tag{6}$$

$$2 \langle K_{ab} - h_{ab}K \rangle = -\kappa^2 S_{ab}, \tag{7}$$

$$S_{ab} = 2 \frac{\delta L^{\Sigma}_{mat}}{\delta h^{ab}} - h_{ab} L^{\Sigma}_{mat}, \tag{8}$$

where $\langle X \rangle = [X(\Sigma^+) - X(\Sigma^-)]/2$ denotes the jump of X between the two sides ($\Sigma^{\pm}$) of the hypersurface and S_{ab} is the energy-momentum tensor of the hypersurface [17].

We will match the two colliding waves on the timelike hypersurface $z = 0$, z being the propagation direction of the two waves. The Israel conditions cannot be solved in vacuum, that is for $S_{ab} = 0$, however a very simple S_{ab} is able to do the job. Let us use the induced local coordinates on the hypersurface

$$(t, x, y) \equiv (1, 2, 3),$$

where $2t = u + v$. The first condition, when $S_{ab} = 0$, admits only the trivial solution. Indeed, by continuity, one should have

$$\lim_{u \to v^+} H^+ = \lim_{v \to u^+} H^+ = 0,$$

so that H^+ has to depend also on v while H^- has to depend also on u, since

$$\lim_{u \to v^+} H^+ = 0 \Rightarrow \lim_{u \to v^+} \frac{H^+}{(u-v)^{\beta}} = 1, \quad \beta > 0;$$

a similar argument also holds for H^-. However, according to our ansatz, H^+ does not depend on v and H^- does not depend on u.

It is useful to slightly modify two metrics on the two sides of Σ by replacing $H^{\pm}$ with $f(u-v)H^{\pm}$, where $f(u-v)$ is an even function such that

$$\lim_{x \to 0} f(x) = 0, \quad \lim_{x \to +\infty} f(x) = 1. \tag{9}$$

For example, one could consider

$$f(x) = \exp\left[-\frac{L^2}{(u-v)^2} \right].$$

In this way, to within few L lengths in the z-direction away from Σ, one recovers the gravitational field of two non-interacting PP waves, while on Σ (once conditions (10) and (11) have been fulfilled) condition (6) is satisfied because of Eq. (9), so that Σ can be thought of as the interaction region. By letting L approach zero one can see that the Israel matching conditions cannot be satisfied although, to compensate this "mismatching", it is enough to consider an energy-momentum tensor S_{ab} whose only non-vanishing component is S_{tt}, the one with two indices along the time direction. Such an energy-momentum tensor is the one of a tensionless membrane[3] placed at $z = 0$.

Thus, cosmic strings naturally end up on membranes and, at the same time, membranes allow for topological interactions between cosmic strings. This is very much like what happens in *superstring theory* [25], in which superstrings end up on *D-branes*.

Let us consider the remaining Israel condition: an explicit evaluation of the left-hand side of Eq. (7) gives

$$\langle K_{tx} - h_{tx}K \rangle = 0 \Rightarrow \partial_x H^+ = \partial_x H^-, \tag{10}$$

$$\langle K_{ty} - h_{ty}K \rangle = 0 \Rightarrow \partial_y H^+ = \partial_y H^-. \tag{11}$$

We will give a nice interpretation of the above equation in a moment. It is well known [29] that cosmic strings give rise to conical singularities and the deficit angle Δ is related in a simple way to the energy per unit length μ of the cosmic string:

$$\Delta = 8\pi G \mu.$$

Thus, the two deficit angles are

$$\Delta^{\pm} = 8\pi G \sum_i \mu_i^{\pm}.$$

Conditions (10) and (11) imply that the two-dimensional gradients $\vec{\nabla} H^{\pm}$ coincide on Σ. Let γ_t be a closed curve on Σ encircling some of the singularities in the $(x$-$y)$-cone at a given t. Because of Eqs. (3) and (5) the two-dimensional Gauss law implies that

$$\oint_{\gamma_t} \vec{\nabla} H^+ \cdot \vec{n} = \oint_{\gamma_t} \vec{\nabla} H^- \cdot \vec{n} = \sum_i^{N_\gamma} \mu_i^{\pm}, \tag{12}$$

$\vec{n}$ being the external normal to γ_t and N_γ the number of singularities encircled by γ_t. By taking into account Eqs. (10) and (11), we conclude that

$$\Delta|_{\gamma_t}^+ = \Delta|_{\gamma_t}^-, \tag{13}$$

$\Delta|_{\gamma_t}^{\pm}$ being the total deficit angle encircled by γ_t; therefore, the deficit angles, the natural topological charges occurring in this model, are conserved across the membrane. Moreover, at any given time t, the two kinds of cosmic strings (the ones sourcing PP waves travelling in the positive z-direction and the others sourcing PP waves travelling in the negative z-direction) have to glue on Σ at the same points, otherwise it would be possible to construct a γ_t able to violate Eq. (12)[4]. The previous arguments can be easily generalized in every dimension. In the general case also the Israel matching condition cannot be fulfilled in vacuum but the introduction of a (mem)brane at $z = 0$ does the job.

---

[3] So that it does not suffer from the typical dynamical instability of domain walls [29].
[4] This could be done by considering a γ_t small enough to encircle the intersection with Σ of only one kind of cosmic strings, for example, one on the left of Σ.

Of course, the deficit angle is conserved across Σ. In the next section, in the spirit of the brane-world scenarios [15] [3] [4] [26], we will analyze in some detail the physical meaning of the deficit angle in the five-dimensional case.

THE FIVE-DIMENSIONAL CASE

The four-dimensional geometry of Σ in the case in which the space-time in Eq. (1) is five-dimensional reads as

$$g|_\Sigma = (1+H)dt^2 - dr^2 - r^2 d\Omega^2, \tag{14}$$

where r, θ and φ are spherical coordinates on Σ and H is a solution of the Poisson Equation that, on Σ, takes the form

$$^3\Delta H = \sum_i^q \mu_i \delta^{(3)} \left(r_i(t), \theta_i(t), \varphi_i(t)\right). \tag{15}$$

This gravitational field corresponds to the weak-field limit of q spherically symmetric asymptotically flat black holes. The weak-field limit can be obtained by looking at the g_{00} component of the metric. In our case, from the relation

$$g_{00} \sim 1 + 2\Phi, \tag{16}$$

we obtain for the Newtonian potential

$$\Phi = \frac{1}{2}H. \tag{17}$$

Thus, because of Eq. (15), the metric describes q spherically symmetric black holes of masses μ_i. Since these q spherically symmetric black holes are almost in equilibrium (that is, the relative locations of the q point-like sources in Eq. (15) can be changed slowly without doing any work), the four-dimensional metric in Eq. (14) coincides asymptotically (that is, far away from the point-like sources) with the Majumdar–Papapetrou solution [18] [22]. In fact, the Majumdar–Papapetrou solution represents the gravitational field of q spherically symmetric extremal charged black holes in equilibrium because the gravitational attraction is balanced by the electrostatic repulsion [6]. Let us introduce in the space-time (1), which in this section is assumed to be five-dimensional, the vector potential A and two-form field

$$F = dA = du \wedge dH, \tag{18}$$

with

$$A = H_{,i}dx^i,$$

then the associated energy-momentum tensor has only one non-vanishing component T_{uu}, and can be represented as massless dust

$$T_{\mu\nu} = \rho U_\mu U_\nu,$$

ρ being the energy density of the five-dimensional electromagnetic field. Non-vacuum five-dimensional Einstein equations read as

$$^3\Delta H = \rho,$$

so that when ρ approaches the finite sum of δ-functions on the rhs of the Poisson Equation (that is, when we consider the field generated by q charged cosmic strings which are in one-to-one correspondence with δ-functions on the rhs of Eq. (15)), we get back, on Σ, the gravitational field (14). When restricted on the membrane Σ, the two-form field F is a purely electric field

$$F|_\Sigma = dt \wedge dH,$$

and represents the Coulomb field of q charged particles. To compute the charges, say $\tilde{\mu}_i$, let us write down the curved five-dimensional Maxwell equations

$$\nabla_\mu F^{\mu\alpha} = j^\alpha,$$

j^α being the charge density five vector. The only non-trivial equation is

$$\nabla_\mu F^{\mu\upsilon} = \sum_i^q \tilde{\mu}_i \delta^{(3)} \left(r_i(u), \theta_i(u), \varphi_i(u)\right).$$

Since the first member of the above equation is nothing but $^3\Delta H$, we immediately see that, because of Poisson Equation and Eq. (15), the charges are equal to the masses.

Finally, let us speculate upon an interesting feature of this model which is related to the generalized Dirac quantization conditions in the brane-world scenarios and, more generally, in superstring and D-brane theory and to its BPS nature.

Let us recall that the BPS bound in the case of "almost" Majumdar–Papapetrou solution is given by

$$M \geq \frac{1}{G}\sqrt{Q^2 + \Pi^2}, \tag{19}$$

where G, Q and Π are the Newton constant, the electric charge and the magnetic charge, respectively. In the purely four-dimensional case, quantization conditions can only be achieved by going outside the framework of classical supergravity with Abelian gauge fields [21]. In fact, we can get the desired relation by interpreting four-dimensional Majumdar–Papapetrou solutions as a slice of a five-dimensional spacetime with superconducting cosmic strings, a sort of five-dimensional analogue of the Abrikosov–Nielsen–Olesen vortex solutions [2] [20] (which carry either electric or magnetic charges[5]). For superconducting cosmic strings the electric Q and magnetic Π charges satisfy the standard quantization condition (see, for example, [29])

$$Q\Pi = 2\pi n.$$

---

[5] Let us note that the five-dimensional electromagnetic field in Eq. (18) has either an electric or a magnetic component.

Thus, from Eq. (19), we get

$$M \geq \frac{\Pi}{G}\sqrt{1 + \frac{4\pi^2 n^2}{\Pi^4}}. \tag{20}$$

In this model, the masses of extremal Reissner–Nordstrom black holes are quantized. As it might be expected on holographic ground (since Σ can be seen as the boundary of a bulk in which there are propagating PP waves), in the perturbative spectrum of string theory on PP-wave background the nth quantized mode (in the light cone gauge) of the string has energy [5]

$$\omega_n = \sqrt{\mu^2 + \frac{n^2}{(\alpha' p^+)^2}}, \tag{21}$$

α' being the string tension and μ the "strength" of the PP wave.

The resemblance of Eqs. (20) and (21) could be a consequence of the fact that, in many cases in which holography holds, the strong-coupling regime on the boundary (the BPS-extreme black-hole solutions on Σ) corresponds to the weak-coupling regime of the bulk theory. However, only the embedding of the above model in a genuine superstring-brane theoretic model would provide sound physical basis to the previous considerations.

ACKNOWLEDGMENTS

The authors would like to thank Professors M. Bianchi, K. Panigrahi, A Sagnotti and O. Zapata for enlightening discussions and suggestions.

REFERENCES

1. M. Alishahiha, A. Kumar, Phys. Lett. B **542**, 130 (2002).
2. A. A. Abrikosov, Sov. Phys. JETP **5**, 1174 (1957).
3. N. Arkani-Hamed, S. Dimopoulos, G. Dvali, Phys. Lett. B **429**, 263 (1998).
4. I. Antoniadis, N. Arkani-Hamed, S. Dimopoulos, G. Dvali, Phys. Lett. B **436**, 257 (1998).
5. D. Bernstein, J. M. Maldacena, H. Nastase, JHEP **0204**, 013 (2002).
6. S. Chandrasekar, *The mathematical theory of black holes* (Clarendon Press, Oxford, 1983).
7. C. G. Callan, J. M. Maldacena, Nucl. Phys. B **513**, 198 (1998).
8. F. Canfora, G.Vilasi, Phys. Lett. B **585**, 193 (2004).
9. F. Canfora, G. Vilasi, P. Vitale, Phys. Lett. B **545**, 373 (2002).
10. F. Canfora, G. Vilasi, P. Vitale, Int. J. Mod. Phys. B **18**, 527 (2004).
11. G. W. Gibbons, C. M. Hull, Phys. Lett. B **109**, 190 (1982).
12. G. W. Gibbons, S. W. Hawking, G. T. Horowitz, M. J. Perry, *Commun. Math. Phys.* **88**, 295 (1983).
13. G. W. Gibbons, M. J. Perry, Nucl. Phys. B **248**, 629 (1984).
14. J. B. Griffiths, *Colliding Plane Waves in General Relativity* (Clarendon Press, Oxford, 1991).
15. P. Horava, E. Witten, Nucl. Phys. B **460**, 506 (1996).
16. N. Itzhaki, I. R. Klebanov, S. Mukhi, JHEP **0203**, 048 (2002).
17. W. Israel, Nuovo Cim. B **44**, 1 (1966).
18. S. D. Majumdar, Phys. Rev. **72**, 390 (1947).
19. R. R. Metsaev, Nucl. Phys. B **625**, 70 (2002).
20. H. B. Nielsen, P. Olesen, Nucl. Phys. B **61**, 45 (1973).
21. G. W. Gibbons in *Duality and Supersymmetric Theories* (Cambridge University Press, 1999, D. I. Olive, P. C. West eds).

22. A. Papapetrou, Proc. Roy. Irish Acad. A **51**, 191 (1947).
23. A. Peres, Phys. Rev. Lett. **3**, 571 (1959).
24. A. Peres, Phys. Rev. **118**, 1105 (1960).
25. J. Polchinski, Phys. Rev. Lett. **75**, 4724 (1995).
26. L. Randall, R. Sundrum, Phys. Rev. Lett. **83**, 4690 (1999).
27. J. G. Russo, A. A. Tseytlin, JHEP **0204**, 021 (2002).
28. M. Sakellariadou, *hep-ph*/0212365 (2002).
29. A. Vilenkin, E. P. S. Shellard, *Cosmic Strings and Other Topological Defects* (Cambridge University Press, 2000).

Holography and BMS field theory

G. Arcioni* and C. Dappiaggi[1†]

*The Racah Institute of Physics, The Hebrew University,
Jerusalem 91904, Israel.
†Dipartimento di Fisica Nucleare e Teorica,
Università degli Studi di Pavia, and INFN, Sezione di Pavia,
via A. Bassi 6, I-27100 Pavia, Italy

Abstract. We study the key ingredients of a candidate holographic correspondence in asymptotically flat spacetimes; in particular, we develop the kinematical and the classical dynamical data of a BMS invariant field theory living at null infinity.

Since 't Hooft foundational paper [1], the holographic principle has played a key role in improving our understanding on the nature of gravitational degrees of freedom in a quantum field theory over a curved background. Originally the principle was proposed in order to solve the apparent information paradox of black holes by means of a theory living on a lower dimensional hypersurface (usually the boundary) with respect to bulk spacetime where all the physical information of the manifold is encoded. Moreover, motivated by the Bekenstein entropy formula, the density of data on the "holographic screen" should not exceed the Planck density; this implies that there is a high redundancy in the way we usually count degrees of freedom in a quantum field theory, since if we excite more than $\frac{A}{4}$ degrees of freedom, we end up with a black hole.

A way to proceed within this framework is based on the reconstruction of the bulk starting directly from boundary data describing how they are generated, their dynamics and mainly how they can reproduce classical spacetime geometry. The main example is the AdS/CFT correspondence [2] (see [3] for a recent review) where, in the low energy limit, a supergravity theory living on $AdS_d \times M^{10-d}$ is a SU(N) conformal gauge field theory living on the boundary of AdS. The whole approach is based on the assumption of the equivalence of partition sum of gravity and gauge theory once asymptotically AdS boundary conditions are imposed on the bulk spacetime. Thus, it seems rather natural to investigate whether we could find a similar holographic description once we choose a different class of manifolds and thus of boundary conditions and in particular we will refer to asymptotically flat (AF) spacetimes, a scenario where this problem has been addressed only in the last few years.

Different approaches have been proposed: in a recent one [4] [5], a Minkowski background is considered and it is divided in AdS and dS slices; the idea is to apply separately both AdS/CFT and dS/CFT correspondence and then patch together the results. This scenario is interesting but it is limited up to now only to the flat case and it

---

[1] contributing author

is unclear how to extend it to a generic asymptotically flat spacetime[2].

In [7], [8], instead, we have investigated the holographic principle in four-dimensional AF spacetimes through the study of the asymptotic symmetry group at null infinity $\Im^\pm \sim \mathbb{R} \times S^2$, i.e. the Bondi-Metzner-Sachs group (BMS). In the Penrose intrinsic construction of $\Im^\pm$ and in the so-called Bondi reference frame $(u = t - r, r, z, \bar{z})$, the BMS is the diffeomorphism group preserving the degenerate boundary metric $ds^2 = 0 \cdot du^2 + d\Omega^2$, i.e.

$$z \to z' = \frac{az+b}{cz+d} \quad ad - bc = 1, \qquad u \to u' = K(z, \bar{z})(u + \alpha(z, \bar{z})),$$

where α is an arbitrary square integrable function over S^2 and K is a fixed multiplicative function [7]. Thus, the BMS group has the structure of the semidirect product

$$BMS_4 = SL(2, \mathbb{C}) \ltimes L^2(S^2). \tag{1}$$

The underlying strategy we have followed has been both to point out the differences between the flat and the AdS scenario and to study the key ingredients of a field theory living on $\Im^\pm$ and invariant under (1). The first step (described in detail in [7]) has been to construct, by means of pure group theoretical techniques, the kinematical data of the boundary theory, i.e. the full particle spectrum of a BMS field theory. Following the Wigner approach [9] and working in a momentum representation, we can introduce a *BMS invariant field* as a map $\psi^\lambda : L^2(S^2) \to \mathscr{H}^\lambda$ where $\mathscr{H}^\lambda$ is a suitable target Hilbert space; ψ^λ transforms under the action of $g = (\Lambda, p(z, \bar{z})) \in BMS_4$ through

$$U(g)\psi^\lambda(p') = e^{i \int_{S^2} \frac{dz d\bar{z}}{1 + |z|^2} p(z, \bar{z}) p'(z, \bar{z})} D^\lambda(\Lambda) \psi^\lambda(\Lambda^{-1} p'), \tag{2}$$

where $D^\lambda(\Lambda)$ is a unitary $SL(2, \mathbb{C})$ representation. At the same time we can introduce the *induced wave function*, which transforms under a unitary and irreducible representation of BMS_4 constructed from the set of BMS little groups $L = H \ltimes L^2(S^2)$, where H is any compact subgroup of $SL(2, \mathbb{C})$, i.e.

$$\tilde{\psi}^j : \mathscr{O} \sim \frac{SL(2, \mathbb{C})}{H} \to \mathscr{H}^j, \tag{3}$$

where $\mathscr{H}^j$ is a suitable target Hilbert space. Thus, the particle spectrum of a BMS invariant theory is characterized by fields labelled by each different little group H and by a mass identified through the following projection:

$$\pi(p(z, \bar{z})) = \pi \left(\sum_{l=0}^\infty \sum_{m=-l}^l p_{lm} Y_{lm}(z, \bar{z}) \right) = (p_{00}, ..., p_{11}) = p_\mu \implies m^2 = \eta^{\mu\nu} p_\mu p_\nu. \tag{4}$$

---

[2] Another recent paper [6] suggests relating holography in a Ricci flat spacetime to a Goursat problem, i.e. a characteristic problem in a Lorentzian setting.

The key components of the overall picture can be summarized in the following table:

BMS field	labels	orbit $\mathscr{O}$
SU(2)	$m^2 > 0$	$\frac{SL(2,\mathbb{C})}{SU(2)} \sim \mathbb{R}^3$
SO(2)	$m^2 = 0$	$\frac{SL(2,\mathbb{C})}{SO(2)} \sim \mathbb{R}^3 \times S^2$

In the Wigner approach, the equations of motion for each BMS particle arise as a set of constraints to impose on (2) in order to reduce its form to that of (3). In detail, each field has to satisfy an orbit constraint reducing $L^2(S^2)$ to each $\mathscr{O}(H)$, a mass equation enforcing (4), and eventually, since $\mathscr{H}^\lambda$ is bigger than $\mathscr{H}^j$, an orthoprojection reducing the components in excess in ψ^λ. To give an example, let us consider the BMS $SU(2)$ scalar field in the covariant form $\psi : L^2(S^2) \to \mathbb{R}$ and in the induced one $\tilde{\psi} : \frac{SL(2,\mathbb{C})}{SU(2)} \to \mathbb{R}$; the equations of motion are (BMS Klein-Gordon equation):

$$[p(z,\bar{z}) - \pi(p)]\psi(p) = 0 \text{ and } [\eta^{\mu\nu}\pi(p)_\mu \pi(p)_\nu - m^2]\psi(p) = 0.$$

Starting from the above data, we have studied in [8] the classical dynamics of each BMS field and in particular we have addressed the question of whether the equations of motion might be derived from a variational principle. Working in a Hamiltonian framework, we have constructed for each field the covariant phase space, i.e. the set of dynamically allowed configurations and, through symplectic techniques, the set of possible energy functions. Surprisingly, we have found a 1:1 correspondence between the covariant phase space either of BMS $SU(2)$ fields and Poincaré SU(2) fields, or of BMS $SO(2)$ fields and Poincaré $E(2)$ fields. From an holographic point of view this implies that the boundary theory, already at a classical level, encodes all the information from the bulk (Poincaré invariant) theory; nonetheless this is not sufficient to claim about an holographic correspondence since the latter is fully manifest at a quantum level and we still lack a proper understanding of the quantum BMS field theory, not to mention a way to reconstruct bulk data from boundary ones. On the other hand, we also suspect that the BMS field theory is related to the IR sectors of pure gravity, and this subject is currently under investigation [10].

REFERENCES

1. G. 't Hooft, arXiv:gr-qc/9310026.
2. O. Aharony, S. S. Gubser, J. M. Maldacena, H. Ooguri and Y. Oz, Phys. Rept. **323** (2000) 183
3. J. de Boer, L. Maoz and A. Naqvi, arXiv:hep-th/0407212.
4. J. de Boer and S. N. Solodukhin, Nucl. Phys. B **665** (2003) 545 [arXiv:hep-th/0303006].
5. S. N. Solodukhin, arXiv:hep-th/0405252.
6. E. Alvarez, J. Conde and L. Hernandez, Nucl. Phys. B **689** (2004) 257 [arXiv:hep-th/0401220].
7. G. Arcioni and C. Dappiaggi, Nucl. Phys. B **674** (2003) 553 [arXiv:hep-th/0306142].
8. G. Arcioni and C. Dappiaggi, arXiv:hep-th/0312196
9. A.O. Barut, R. Raczka: *"Theory of group representation and applications"* World Scientific (1986).
10. G. Arcioni and C. Dappiaggi, work in progress

SOME PROPERTIES OF THE DE BROGLIE GRAVITATIONAL WAVES

ANTONIO FEOLI* and SREERAM VALLURI†

*Dipartimento d'Ingegneria, Università del Sannio,
Corso Garibaldi n. 107, Palazzo Bosco Lucarelli,
82100 - Benevento - Italy.
†Department of Physics and Astronomy, Dept of Applied Mathematics
University of Western Ontario, London, ON, Canada N6A 3K7

Abstract. We study some interesting properties of the de Broglie gravitational waves such as the energy-momentum tensor and the effect of the wave on a sphere of test particles. This kind of waves are responsible for a longitudinal shift of test particles which does not occur with the standard transverse gravitational waves.

In 1924 de Broglie proposed his relation $P^\nu = \hbar K^\nu$ between the classical constant momentum $P^\nu = mcu^\nu = m\gamma(c,\vec{v}) = (E/c,\vec{p})$ of a free particle and the wave number $K^\nu \equiv (\omega/c,\vec{k})$ of the associated plane wave. He believed that the wave field is not only a mathematical tool to compute probabilities but also a *real* "pilot wave" $\phi = R(t,x,y,z)exp\left[iP_\mu x^\mu/\hbar\right]$ which leads the particle in spacetime.

Our aim is to show that it is possible to obtain de Broglie waves from the linearized Einstein field equations and to study some properties of this special kind of gravitational waves. We start from a metric tensor

$$g_{\mu\nu} = \eta_{\mu\nu} + h_{\mu\nu}(x), \quad |h_{\mu\nu}| << 1, \tag{1}$$

The linearized vacuum Einstein field equations are

$$\nabla_\alpha\nabla^\alpha h_{\mu\nu} = 0 \ and \ \nabla_\mu\left(h_\nu^\mu - \frac{1}{2}\delta_\nu^\mu h\right) = \nabla_\mu\sigma_\nu^\mu = 0. \tag{2}$$

The classical solution is a plane wave with $K_\nu K^\nu = 0$. On the contrary, we found [1] a solution in the form of a wave packet of cylindrical symmetry around the propagation direction:

$$h_{\mu\nu} = e_{\mu\nu}(K_0,K_1)AJ_o(\sqrt{y^2+z^2}/\lambda)\cos\left(K^0ct - K^1x\right), \tag{3}$$

such that $K^\mu K_\mu = \lambda^{-2}$, which represents a gravitational wave propagating along the x-axis with a phase velocity $V_{ph} = c\sqrt{1+(\lambda k)^{-2}}$. The nonvanishing components of the polarization tensor, that turn out to be depending on the wave number, are

$$e_{00} = e_{11} = \frac{e_{10}}{2}\left(\frac{K_1}{K_0} + \frac{K_0}{K_1}\right), \tag{4}$$

CP751, *General Relativity and Gravitational Physics, 16th SIGRAV Conference*, edited by G. Esposito et al.
© 2005 American Institute of Physics 0-7354-0236-1/05/$22.50

$$e_{22} = e_{33} = \frac{e_{10}}{2}\left(\frac{K_1}{K_0} - \frac{K_0}{K_1}\right). \tag{5}$$

The new dispersion relation indicates the existence of an effective mass; thus, on considering $\lambda = \hbar/mc$, the solution (3) can be interpreted as the de Broglie wave associated with a particle of rest mass m moving with constant velocity v along the x - axis in such way that $K^0 ct - K^1 x = (Et - px)/\hbar$. As in the standard QM, we can add waves with a slightly different velocity to obtain [1,2] a de Broglie wave packet with a group velocity such that $V_{ph} \cdot V_G = c^2$. An interesting limiting case is when $K_1 = 0$, i.e. when we have an oscillation in time but not a propagation. In that case we find $e_{10} = 0$ and $e_{22} = e_{33} = -e_{00} = -e_{11}$, and we argue that e_{10} must be proportional to v. If we choose

$$e_{01} = e_{10} = -\frac{v}{c(1 - v^2/c^2)}, \tag{6}$$

we can write

$$\sigma_{\mu\nu} = h_{\mu\nu} - \frac{1}{2}\eta_{\mu\nu}h = AJ_0 \cos\left(\frac{Et - px}{\hbar}\right) V_\mu V_\nu, \tag{7}$$

where $V_\mu = \gamma(-v/c, 1, 0, 0)$ has the same nonvanishing components of u_μ in the inverse order. Then, from (4) and (5), the other nonvanishing components of the polarization tensor become

$$e_{00} = e_{11} = \frac{c^2 + v^2}{2(c^2 - v^2)}, \quad e_{22} = e_{33} = -\frac{1}{2}, \tag{8}$$

but the metric holds only for velocities such that $|h_{\mu\nu}| \ll 1$. A different choice of e_{10} (done in previous papers [1,2,3]) leads to a metric that holds for any value of v.

We have found it impossible [3] to gauge away simultaneously all components of the solution (3), (4) and (5), so our waves are not an apparent field. This is confirmed also by calculating the associated energy-momentum tensor, that in our case is

$$t_{(1)}^{\mu\nu} = \frac{1}{4}\left[2\nabla^\mu \sigma^{\alpha\beta}\nabla^\nu \sigma_{\alpha\beta} - \nabla^\mu \sigma \nabla^\nu \sigma - \eta^{\mu\nu}\left(\nabla^\rho \sigma^{\alpha\beta}\nabla_\rho \sigma_{\alpha\beta} - \frac{1}{2}\nabla_\rho \sigma \nabla^\rho \sigma\right)\right]. \tag{9}$$

Here the subscript indicates that $t_{(1)}^{\mu\nu}$ is only a first approximation to the exact expression for the energy-momentum tensor. We obtain

$$t_{00} = t_{11} = -A^2 e_{22}^2 \left\{J_0^2\left(K_0^2 + K_1^2\right)\sin^2 K^\alpha x_\alpha + J_1^2 \lambda^{-2}\cos^2 K^\alpha x_\alpha,\right\} \tag{10}$$

$$t_{10} = 2A^2 J_0^2 e_{22}^2 K_1 K_0 \sin^2 K^\alpha x_\alpha, \tag{11}$$

$$t_{22} = -A^2 e_{22}^2 \left\{\left[(\partial_2 J_0)^2 - (\partial_3 J_0)^2\right]\cos^2 K^\alpha x_\alpha + J_0^2 \lambda^{-2}\sin^2 K^\alpha x_\alpha\right\}, \tag{12}$$

$$t_{33} = -A^2 e_{22}^2 \left\{-\left[(\partial_2 J_0)^2 - (\partial_3 J_0)^2\right]\cos^2 K^\alpha x_\alpha + J_0^2 \lambda^{-2}\sin^2 K^\alpha x_\alpha\right\}. \tag{13}$$

The other nonvanishing components of $t_{\mu\nu}$ can be calculated without any problem. The trace is not vanishing in our case and it reduces to the form $t_\mu^\mu = 2A^2 e_{22}^2 J_0^2 \lambda^{-2}\sin^2 K^\alpha x_\alpha$. Finally note that $t_{10}/t_\mu^\mu = \lambda^2 K_1 K_0 = -e_{10}/2e_{22} = e_{10}$.

As the de Broglie wave passes, moving along the x-axis, it deforms what was a sphere of test particles as measured in the proper frame of the central particle. This can be shown by calculating the tidal acceleration between two particles, using the formula

$$\frac{D^2\eta^\mu}{c^2 d\tau^2} = -R^\mu_{\nu\gamma\delta}\beta^\nu\beta^\delta\eta^\gamma,\tag{14}$$

where τ is the proper time along the worldline of the two masses, β^μ is their four-velocity and η^μ their relative distance. We can use two approximations: the velocity of test particles is very small compared with the speed of light, so $\beta^\mu \simeq (1,0,0,0)$ and the displacement δ^μ of the two masses resulting from the wave is also very small. Thus, if the initial position of the first particle is at the origin and the position of the second one is at $\xi^\mu = (0,x,y,z)$, we have $\eta^\mu = (0,x+\delta^1,y+\delta^2,z+\delta^3)$ and we can consider $x^i >> \delta^i$. In this framework the acceleration of geodesic deviation is $\ddot{\eta}^k \simeq -c^2 R^k_{0j0}\xi^j$, where a dot indicates a differentiation with respect to time t and $i,j = 1,2,3$. For simplicity we write explicitly the nonvanishing shifts only for the cases:

a) The mass m_1 in $\xi^j = (\ell,0,0)$

$$\eta_1 = \frac{A\ell}{2K_0^2}J_0(0)\left[(K_0^2+K_1^2)e_{00} - 2K_0K_1e_{01}\right]\cos\left(K_0ct + K_1\ell\right) + \ell.\tag{15}$$

b) The mass m_2 in $\xi^j = (0,\ell,0)$

$$\eta_1 = \frac{A\ell}{2\lambda K_0^2}\left(K_0e_{10} - K_1e_{00}\right)J_1(\ell/\lambda)\sin\left(K_0ct\right),\tag{16}$$

$$\eta_2 = \frac{A\ell}{2K_0^2}\left\{e_{22}K_0^2J_0(\ell/\lambda) + e_{00}\left(\frac{J_2(\ell/\lambda) - J_0(\ell/\lambda)}{2\lambda^2}\right)\right\}\cos\left(K_0ct\right) + \ell.\tag{17}$$

c) The mass m_3 in $\xi^j = (0,0,\ell)$

$$(\eta_1)_{m_3} = (\eta_1)_{m_2} \quad (\eta_2)_{m_3} = 0 \quad (\eta_3)_{m_3} = (\eta_2)_{m_2}.\tag{18}$$

Calculations of the geodesic deviation tell us that there is a longitudinal displacement η_1 of the particles. This effect is a characteristic feature of these waves and differs from the standard gravitational waves. Einstein and Rosen [4] have shown that massless longitudinal waves, propagating at the speed of light, are unphysical, but the de Broglie gravitational waves, having an effective mass, give the possibility to observe longitudinal effects. This effective mass can be interpreted as the mass of the graviton itself or, as suggested in previous papers [1,2], as the mass of whatever quantum particle associated with the corresponding de Broglie wave.

REFERENCES

1. A. Feoli and G. Scarpetta, Found. Phys. Lett. **11** (1998) 395.
2. A. Feoli, Europhys. Lett. **58** (2002) 169.
3. A. Feoli and S. Valluri, Int. Jour. Mod. Phys. D **13** (2004) 907.
4. A. Einstein and N.Rosen, J. Franklin Inst. **223** (1937) 43.

Simplicial Aspects of String Dualities

Mauro Carfora, Claudio Dappiaggi, <u>Valeria Gili</u> [1]

Dipartimento di Fisica Nucleare e Teorica,
Università degli Studi di Pavia,
and
Istituto Nazionale di Fisica Nucleare, Sezione di Pavia,
via A. Bassi 6, I-27100 Pavia, Italy

Abstract. We show how the study of randomly triangulated surfaces merges with the study of open/closed string dualities. In particular, we discuss the Conformal Field Theory which arises in the open string sector and its implications.

String dualities provide a powerful tool to study UV properties of a Quantum Field Theory by means of IR techniques proper of String Theory. This approach had led to the formulation of AdS/CFT correspondence but, at mathematical level, it has been explained only in a topological setting by Gopakumar and Vafa in [1] and, more recently, by Gaiotto and Rastelli in [2]. In such a framework, a paradigmatical result has been established by Gopakumar in [3, 4]. Starting from a Schwinger parameterization of the free gauge field correlators of an $\mathcal{N} = 4$ SYM $SU(N)$ gauge theory, he exploits an analogy with electrical networks which allows him to sum over internal loop momenta to obtain a skeleton graph with a number of vertices equal to the number of holes of the free open theory and which shows the basic connectivity of the original correlator. By performing a change of variables into the Schwinger parameter spaces, he is able to fill the holes and obtain a closed AdS tree diagram. In this context it is important to stress that planar graphs with different connectivities give rise to different skeleton diagrams, and all these different skeleton contributions need to be summed over to obtain the closed string dual of a single open free field diagram. Moreover, it is important to recognize that all these structures are in one-to-one correspondence with the moduli space of a sphere with n holes, moduli space which arises as a natural structure in the large N limit framework, proper of gauge/gravity correspondence.

Motivated by the ubiquitous role that simplicial methods play in the above result, we have recently introduced a geometrical framework [5] in which it is possible to implement new examples of open/closed string dualities. Our approach is based on a careful use of uniformization theory for triangulated surfaces carrying curvature degrees of freedom.

In order to show how this uniformization arises, let us consider the dual polytope associated with a Random Regge Triangulation [6] $|T_l| \to M$ of a Riemannian manifold M. Using the properties of Jenkins-Strebel quadratic differentials [5] it is possible to

---

[1] contributing author

CP751, *General Relativity and Gravitational Physics, 16th SIGRAV Conference,* edited by G. Esposito et al.
© 2005 American Institute of Physics 0-7354-0236-1/05/$22.50

decorate the neighborhood of each curvature supporting vertex with a punctured disk uniformized by a conical metric

$$ds^2_{(k)} \doteq \frac{[L(k)]^2}{4\pi^2}|\zeta(k)|^{-2(\frac{\varepsilon(k)}{2\pi})}|d\zeta(k)|^2.$$

Alternatively, we can blow up every such a cone into a corresponding finite cylindrical end, by introducing a finite annulus $\Delta^*_{\varepsilon(k)} \doteq \left\{\zeta(k) \in \mathbb{C}| \quad \exp-\frac{2\pi}{2\pi-\varepsilon(k)} \leq |\zeta(k)| \leq 1\right\}$ endowed with the cylindrical metric

$$|\phi(k)| \doteq \frac{[L(k)]^2}{4\pi^2}|\zeta(k)|^{-2}|d\zeta(k)|^2.$$

It is important to stress the different role that the deficit angle plays in such two unformizations. In the "closed" uniformization the deficit angle $\varepsilon(k)$ plays the usual role of localized curvature degrees of freedom and, together with the perimeter of the polytopal cells, provide the geometrical information of the underlying triangulation. Conversely, in the "open" uniformization, the deficit angle associated with the k-th polytope cell defines the geometric moduli of the k-th cylindrical end. As a matter of fact each annulus can be mapped into a cylinder of circumpherence $L(k)$ and height $\frac{L(k)}{2\pi-\varepsilon(k)}$, thus $\frac{1}{2\pi-\varepsilon(k)}$ is the geometrical moduli of the cylinder. This shows how the uniformization process works quite differently from the one used in Kontsevich-Witten models, in which the whole punctured disk is uniformized with a cylindrical metric. In this case the disk can be mapped into a semi-infinite cylinder, no role is played by the deficit angle and the model is topological; conversely, in our case, we are able to deal with a non topological theory.

In the closed sector both the coupling of the geometry of the triangulation with D bosonic fields and the quantization of the theory can be performed under the paradigm of critical field theory. However, in order to discuss Polyakov string theory directly over the dual open Riemann surface so defined, we have to deal with a Boundary Conformal Field Theory (BCFT) defined over each cylindrical end. The unwrapping of the cones into finite cylinders suggests compactifying each field defined on the k-th cylindrical end along a circle of radius $\frac{R(k)}{L(k)}$:

$$X^\alpha(k) \xrightarrow{\vartheta(k)\to\vartheta(k)+2\pi} X^\alpha(k) + 2\pi v^\alpha(k)\frac{R^\alpha(k)}{L(k)} \qquad v(k) \in \mathbb{Z}.$$

Under these assumptions, it is possible to quantize the theory and to compute the quantum amplitude over each cylindrical end: writing it as an amplitude between an initial and final state, we can extract suitable boundary states which arise as a generalization of the states introduced by Langlands in [7]. As they stand, these boundary states preserve neither the conformal symmetry nor the $U(1)_L \times U(1)_R$ symmetry generated by the cylindrical geometry. It is then necessary to impose on them suitable gluing conditions relating the holomorphic and anti-holomorphic generators on the boundary. These restrictions generate the usual families of Neumann and Dirichlet boundary states.

Within this framework, the next step in the quantization of the theory is to define the correct interaction of the N_0 copies of the cylindrical CFT on the ribbon graph associated with the underlying Regge Polytope. This can be achieved via the introduction over each strip of the graph of Boundary Insertion Operators (BIO) $\psi_{\lambda(p,q)}^{\lambda(p)\lambda(q)}$ which act as a coordinate dependent homomorphism from $V_{\lambda(p)} \star V_{\lambda(p,q)}$ and $V_{\lambda(q)}$, mediating in such a way the change of boundary conditions. Here $V_{\lambda(\bullet)}$ denotes the Verma module generated by the action of the Virasoro generators over the $\lambda(\bullet)$ highest weight, and $\star$ denotes the fusion of the two representations.

In the limit in which the theory is rational (*i.e.* when the compactification radius is an integer multiple of the self-dual radius $R_{s.d.} = L(k)/\sqrt{2}$) the compactified boson theory is the same as an $SU(2)_{k=1}$ WZW model, thus it is possible to identify the BIO as primary operators with well defined conformal dimension and correlators. Moreover, on considering the coordinates of three points in the neighborhood of a generic vertex of the ribbon graph, we can write the OPEs describing the insertion of such operators in each vertex. Considering four adjacent boundary components, it is then possible to show that the OPE coefficients $C_{j(r,p)j(q,r)j(p,q)}^{j_p j_r j_q}$ are provided by the fusion matrices $F_{j_r j(p,q)} \begin{bmatrix} j_p & j_q \\ j_{(r,p)} & j_{(q,r)} \end{bmatrix}$, which in WZW models coincide with the $6j$-symbols of the quantum group $SU(2)_{e^{\frac{\pi}{3}i}}$:

$$C_{j(r,p)j(q,r)j(p,q)}^{j_p j_r j_q} = \left\{ \begin{matrix} j_{(r,p)} & j_p & j_r \\ j_q & j_{(q,r)} & j_{(p,q)} \end{matrix} \right\}_{Q=e^{\frac{\pi}{3}i}}.$$

From these data, through edge-vertex factorization we can characterize the general structure of the partition function for this model [8] as a sum over all possible $SU(2)$ primary quantum numbers describing the propagation of the Virasoro modes along the N_0 cylinders $\{\Delta_{\varepsilon(k)}^*\}$.

The overall picture which emerges is that of N_0 cylindrical ends glued through their inner boundaries to the ribbon graph, while their outer boundaries lay on D-branes. Each D-brane acts naturally as a source for gauge fields: it allows us to introduce open string degrees of freedom whose information is traded through the cylinder to the ribbon graph, whose edges thus acquire naturally a gauge coloring. This provides a new kinematical set-up for discussing gauge/gravity correspondence [9].

REFERENCES

1. Gopakumar, R., and Vafa, C., *Adv. Theor. Math. Phys.*, **3**, 1415 (1999), [hep-th/9811131].
2. Gaiotto, D., and Rastelli, L. (2003), [hep-th/0312196].
3. Gopakumar, R., *Phys. Rev.*, **D70**, 025009 (2004), [hep-th/0308184].
4. Gopakumar, R., *Phys. Rev.*, **D70**, 025010 (2004), [hep-th/0402063].
5. Carfora, M., Dappiaggi, C., and Marzuoli, A., *Class. Quant. Grav.*, **19**, 5195 (2002), [gr-qc/0206077].
6. Carfora, M., and Marzuoli, A., *Adv. Theor. Math. Phys.*, **6**, 357–401 (2003), [math-ph/0107028].
7. Langlands, R. P., Lewis, M.-A., and Saint-Aubin, Y. (1999), [hep-th/9904088].
8. Arcioni, G., Carfora, M., Dappiaggi, C., and Marzuoli, A., *Jour. Geom. Phys.*, **52**, 137 (2004), [hep-th/0209031].
9. Carfora, M., Dappiaggi, C., and Gili, V. (2004), in preparation.

Static Axially Symmetric Sources of the Gravitational Field

Daniele Malafarina*, Giulio Magli† and Luis Herrera**

Dipartimento di Matematica, Politecnico di Milano, Italy
e-mail: malafarina@mate.polimi.it
†*Dipartimento di Matematica, Politecnico di Milano, Italy*
e-mail: magli@mate.polimi.it
**Escuela de Fisica, Facultad de Ciencias, Universidad Central de Venezuela, Caracas, Venezuela*
e-mail: laherrera@telcel.net.ve

Abstract. The gamma metric, a static axially symmetric vacuum solution of Einstein field equations, is reviewed, the nature of its singularity and the shape of its sources are studied. This space-time is matched with an interior metric satisfying physically viable conditions. The model obtained, which can represent a dense astrophysical object, is used to investigate the boundary between black hole (i.e. Schwarzschild) and naked singularity as a function of the shape of the source.

INTRODUCTION

Analytical description of astrophysical objects within the framework of General Relativity is a complicated and largely undeveloped field. As a matter of fact, only a few models exist describing gravitating objects and most of them are restricted to the unphysical scenario of exact spherical symmetry.

The interest in finding sources of the gravitational field is twofold, since the problem deals not only with the astrophysical issue of describing realistic objects in space but also with the, equally important, theoretical issue of gravitational collapse and the Cosmic Censor Conjecture.

In the present paper we present a solution of the Einstein field equations in the presence of matter which satisfies all reasonable physical conditions and matches smoothly to a static axially symmetric vacuum space-time known as the gamma metric.

THE GAMMA METRIC

The most general static axially symmetric gravitational field in vacuum [1] depends only on two unknown functions, one of which must satisfy the Laplace equation in a fictitious flat two-dimensional space setting up a one-to-one correspondence between solutions of the Laplace equation and static axially symmetric space-times.

The gamma metric [2] is a particularly interesting space-time belonging to this class which corresponds to the solution of the Laplace equation for a thin rod source of constant density γ uniformly distributed along the z-axis. Written in Erez–Rosen coordinates

CP751, *General Relativity and Gravitational Physics, 16th SIGRAV Conference*, edited by G. Esposito et al.

$(t, r, \vartheta, \varphi)$ the metric takes the form

$$ds^2 = -\Delta^\gamma dt^2 + \Delta^{\gamma^2-\gamma-1}\Sigma^{1-\gamma^2}dr^2 + r^2\Delta^{\gamma^2-\gamma}\Sigma^{1-\gamma^2}d\vartheta^2 + r^2\sin^2\vartheta\Delta^{1-\gamma}d\varphi^2, \quad (1)$$

where

$$\Delta = \left(1 - \frac{2m}{r}\right), \Sigma = \left(1 - \frac{2m}{r} + \frac{m^2}{r^2}\sin^2\vartheta\right), \quad (2)$$

and the coordinate range is $-\infty \leq t \leq +\infty$, $r > 2m$, $0 \leq \vartheta \leq \pi$, $0 \leq \varphi \leq 2\pi$.
It is easy to check that, for $\gamma = 1$ the metric reduces to the Schwarzschild solution and that for $\gamma = 0$ it becomes the flat Minkowski space-time. Performing the multipole expansion of the metric we obtain the total mass $M = m\gamma$ and the quadrupole moment $Q = \frac{1}{3}m^3\gamma(1 - \gamma^2)$ which represent the shape of the source [3]. By comparison with the classical Newtonian potential in spherical coordinates we can say that values of $\gamma > 1$ account for oblate objects while $\gamma < 1$ do for prolate.
This result is in agreement with the shape of the sources obtained by comparing the area of a surface of revolution at fixed value of $r > 2m$ with the corresponding area for Schwarzschild [4]. According to the Israel theorem the gamma metric has a naked singular horizon as r approaches the critical value $2m$ whenever $\gamma \neq 1$. This singular behavior can be easily demonstrated by evaluation of the Kretschmann scalar. Furthermore, since $\Delta \to 0$ as $r \to 2m$ the corresponding area for the horizon vanishes [5].

SOURCES OF THE GAMMA METRIC

In order to find an interior model for the gamma metric we followed a procedure proposed by Hernandez and used by Stewart et al. [6] based upon the deformation of known Schwarzschild sources. Therefore we considered $\gamma = 1 + \varepsilon$, with ε small:

$$g_{00} = -\Delta - \varepsilon\Delta\ln\Delta \quad (3)$$
$$g_{11} = \Delta^{-1} + \varepsilon\Delta^{-1}(\ln\Delta - 2\ln\Sigma) \quad (4)$$
$$g_{22} = r^2 + \varepsilon r^2(\ln\Delta - 2\ln\Sigma) \quad (5)$$
$$g_{33} = r^2\sin^2\vartheta - \varepsilon r^2\sin^2\vartheta\ln\Delta. \quad (6)$$

For $\varepsilon = 0$ a very simple interior solution matching the Schwarzschild field at $r = r_b$ and satisfying all conditions can be obtained by putting $m = \mu(r)$ with

$$\mu(r) = \frac{4}{3}\pi E_0 r^3\left(1 - \frac{3}{4}\frac{r}{r_b}\right). \quad (7)$$

This simple choice is however not enough to construct a viable source for the gamma metric. In fact putting $m = \mu(r)$ in (3)-(6) does not satisfy the energy conditions for any given value of $\varepsilon \neq 0$. We therefore consider a more general ansatz for the interior metric taking $m = \mu(r)$, $\ln\Delta = F(r)$ and $\ln\Sigma = -G(r, \vartheta)$ being $F(r)$, $G(r, \vartheta)$ and $\mu(r)$ chosen in order to satisfy the physical conditions. In other words

$$F(r_b) = \ln\left(1 - \frac{2m}{r_b}\right), \quad G(r_b, \vartheta) = -2\ln\left(1 - \frac{2m}{r_b} + \frac{m^2}{r_b^2}\sin^2\vartheta\right) \quad (8)$$

$$F'(r_b) = \frac{2m}{r_b^2}\left(1 - \frac{2m}{r_b}\right)^{-1}, \quad G'(r_b, \vartheta) = -\frac{4m}{r_b^2}\frac{1 - \frac{m}{r_b}\sin^2\vartheta}{1 - \frac{2m}{r_b} + \frac{m^2}{r_b^2}\sin^2\vartheta} \tag{9}$$

$$\lim_{r\to 0}\frac{G(r,\vartheta)}{r^2} = 0, \quad \lim_{r\to 0}\left(2\frac{F'(r)}{r} + F''(r)\right) = c \geq 0 \tag{10}$$

$$G''(r_b, \vartheta) \leq \left(1 - \frac{2m}{r_b}\right)^{-1}\left(\frac{4m}{r_b^3} - \frac{1}{r_b^2}G_{\vartheta\vartheta}(r_b, \vartheta) - \frac{1}{r_b}\left(1 - \frac{m}{r_b}\right)G'(r_b, \vartheta)\right) \tag{11}$$

$$\left(1 - \frac{2\mu(r)}{r}\right)\left(\frac{2}{r}F'(r) - \frac{1}{2}G''(r,\vartheta)\right) \geq \frac{1}{2r^2}\left(G_{\vartheta\vartheta}(r,\vartheta) + \frac{\cos\vartheta}{\sin\vartheta}G_\vartheta(r,\vartheta)\right) \tag{12}$$

Weak-energy conditions can be satisfied only for $\varepsilon > 0$ (i.e. for oblate objects). It is possible to choose $F(r)$ and $G(r,\vartheta)$ so that all conditions are satisfied [4] and in figure the energy density and the radial energy condition are shown for an object with $M = 2M_\odot$ and $r_b = 10 Km$.

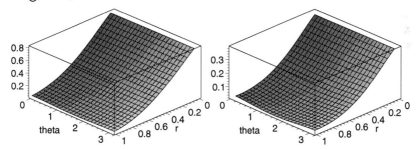

FIGURE 1. *Plots for the energy density T_0^0 (left) and for $T_0^0 - T_1^1$ (right), in the dimensionless unit obtained multiplying by r_b^2, for a fixed value of $\varepsilon = \frac{1}{10}$ as a function of ϑ and the radial variable $\frac{r}{r_b}$.*

REFERENCES

1. H. Weyl, *Ann. Physik* **54**, 117 (1917); *ibid* **159**, 185 (1919); J. L. Synge, *Relativity, the general theory*, North–Holland Publ. Co, Amsterdam, 1960.
2. R. Bach and H. Weyl, *Math. Z.* **13**, 134 (1922); G. Darmois *Les equations de la Gravitation Einsteinienne*, Gauthier-Villars, Paris, 1927; G. Erez and N. Rosen, *Bull. Res. Council Israel* **8F**, 47 (1959); D. M. Zipoy *J. Math. Phys.* **7**, 1137 (1966); R. Gautreau and J. L. Anderson, *Phys.Lett.* A **25** , 291 (1967); B. H. Voorhees, *Phys. Rev. D* **2**, 2119 (1970); F. Espósito and L. Witten , *Phys. Lett.* **58B**, 357 (1975); D. Papadopoulos, B. Stewart and L. Witten, *Phys. Rev. D* **24**, 320 (1981); L. Herrera, F. Paiva and N. O. Santos. *J. Math. Phys.* **40**, 4064 (1999).
3. L. Herrera and J. L. Hernández-Pastora, *J. Math. Phys.* **41**, 7544 (2000).
4. L.Herrera, G. Magli and D. Malafarina, arXiv:gr-qc/0407037 (2004).
5. K. S. Virbhadra, arXiv:gr-qc/9606004 v1 (1996); H. Kodama and W. Hikida, gr-qc/0304064 (2003).
6. W. C. Jr. Hernández, *Phys. Rev.* **153**, 1359 (1967); B. Stewart, D. Papadopoulos, L. Witten, R. Berezdivin and L. Herrera, *Gen. Rel. Grav.* **14**, 97 (1982).

Bergmann–Bianchi identities in field theories[1]

M. Francaviglia, M. Palese and E. Winterroth

Department of Mathematics, University of Torino, via C. Alberto 10, 10123 Torino, Italy

Abstract.
We relate the generalized Bergmann–Bianchi identities for Lagrangian field theories on gauge-natural bundles with the kernel of the associated gauge-natural Jacobi morphism. A Hamiltonian form is then canonically associated with a given gauge-natural variational problem.

2000 MSC: 58A20,58A32,58E30,58E40,58J10,58J70.
Key words: gauge-natural bundles, Bergmann–Bianchi identities, Jacobi morphism.

THE BERGMANN–BIANCHI MORPHISM

Our general framework is the calculus of variations on finite order *gauge-natural bundles* [3, 8]. Such geometric structures have been widely recognized to suitably describe so-called gauge-natural field theories, *i.e.* physical theories in which right-invariant infinitesimal automorphisms of the structure bundle P uniquely define the transformation laws of the fields themselves (see *e.g.* [4] and references quoted therein). We shall in particular consider *finite order variational sequences on gauge-natural bundles*, whereby fundamental objects of calculus of variations such as Lagrangians, Euler–Lagrange and Jacobi morphisms are conveniently represented as quotient morphisms (see *e.g.* [6, 9]). For basic notions and fixing notation we refer *e.g.* to [1, 3, 4, 5, 6, 8, 9, 11].

Recall that generalized Bergmann–Bianchi identities for field theories are necessary and (locally) sufficient conditions for the Noether conserved current to be not only closed but also the divergence of a skew-symmetric (tensor) density, along solutions of the Euler–Lagrange equations [1]. In the general theory of relativity these identities coincide with the contracted Bianchi identities for the curvature tensor of the pseudo-Riemannian metric.

Let Y_ζ be a gauge-natural bundle and let λ be a gauge-natural Lagrangian [4, 8] on some s-th order prolongation $J_s Y_\zeta$. Let $\mathscr{A}^{(r,k)}$ be the vector bundle of right-invariant principal infinitesimal automorphisms of the underlying principal structure bundle P. In the following we shall consider variation vector fields which are vertical parts of gauge-natural lifts of a given $\bar{\Xi} \in \mathscr{A}^{(r,k)}$. Let $\mathscr{C}_{2s}^*[\mathscr{A}^{(r,k)}] \simeq J_{2s+1}\mathscr{A}^{(r,k)} \times_{J_{2s}\mathscr{A}^{(r,k)}} V J_{2s}\mathscr{A}^{(r,k)}$. By a slight abuse of notation we denote by $\mathfrak{G}(\bar{\Xi})_V$ the vertical part – with respect to the contact structure induced by the projections $J_{s+1}Y_\zeta \to J_s Y_\zeta$ – of (jet prolongation of)

---

[1] Work partially supported by MIUR (PRIN 2003) and University of Torino

the gauge-natural lift $\mathfrak{G}(\bar{\Xi})$ [3, 4, 5]. We set

$$\omega(\lambda,\mathfrak{G}(\bar{\Xi})_V) \equiv \pounds_{\bar{\Xi}}\rfloor \mathscr{E}_n(\lambda) : J_{2s}Y_\zeta \to \mathscr{C}_{2s}^*[\mathscr{A}^{(r,k)}]\otimes\mathscr{C}_0^*[\mathscr{A}^{(r,k)}]\wedge(\overset{n}{\wedge}T^*X),$$

where $\pounds_{\bar{\Xi}}$ is the Lie derivative operator acting on sections of the gauge-natural bundle [5], $\rfloor$ is the interior product and $\mathscr{E}_n(\lambda)$ is the Euler-Lagrange operator associated with λ [6]. The morphism $\omega(\lambda,\mathfrak{G}(\bar{\Xi})_V)$ so defined is a generalized Lagrangian associated with the field equations of the original Lagrangian λ and it has been considered in applications *e.g.* in General Relativity. By the linearity of $\pounds$ we can regard $\omega(\lambda,\mathfrak{G}(\bar{\Xi})_V)$ as an extended morphism defined on $J_{2s}Y_\zeta \underset{X}{\times} VJ_{2s}\mathscr{A}^{(r,k)}$. We have $D_H\omega(\lambda,\mathfrak{G}(\bar{\Xi})_V)=0$, where D_H is the horizontal exterior differential on the above mentioned space; thus, as a consequence of a global decomposition formula for vertical morphisms due to Kolář [7], we can state the following [11].

Lemma 1 *Let $\omega(\lambda,\mathfrak{G}(\bar{\Xi})_V)$ be as above. On the domain of $\omega(\lambda,\mathfrak{G}(\bar{\Xi})_V)$ we have (up to pull-backs):*

$$\omega(\lambda,\mathfrak{G}(\bar{\Xi})_V) = \beta(\lambda,\mathfrak{G}(\bar{\Xi})_V) + F_{\omega(\lambda,\mathfrak{G}(\bar{\Xi})_V)},$$

where

$$\beta(\lambda,\mathfrak{G}(\bar{\Xi})_V):=E_{\omega(\lambda,\mathfrak{G}(\bar{\Xi})_V)},$$

$E_{\omega(\lambda,\mathfrak{G}(\bar{\Xi})_V)}$ *being the generalized Euler–Lagrange morphism associated with* $\omega(\lambda,\mathfrak{G}(\bar{\Xi})_V)$*; and, locally, $F_{\omega(\lambda,\mathfrak{G}(\bar{\Xi})_V)} = D_H M_{\omega(\lambda,\mathfrak{G}(\bar{\Xi})_V)}$.*

Definition 1 We call the global morphism $\beta(\lambda,\mathfrak{G}(\bar{\Xi})_V)$ the *generalized Bergmann–Bianchi morphism* associated with the Lagrangian λ and the variation vector field $\bar{\Xi}$.

$\square$

Bergmann–Bianchi identities and Hamiltonian structure

Let $\delta_{\mathfrak{G}}^2\lambda$ be the second variational vertical derivative of λ with respect to $\mathfrak{G}(\bar{\Xi})_V$ and let $\mathscr{J}(\lambda,\mathfrak{G}(\bar{\Xi})_V):=E_{\mathfrak{G}(\bar{\Xi})_V\rfloor\chi(\lambda,\mathfrak{G}(\bar{\Xi})_V)} - \chi(\lambda,\mathfrak{G}(\bar{\Xi})_V)$ being the generalized Helmholtz morphism conveniently represented in [6] - be the generalized gauge-natural Jacobi morphism [2, 11]. Let $\mathfrak{K}$ be the *kernel* of $\mathscr{J}(\lambda,\mathfrak{G}(\bar{\Xi})_V)$. We have the following characterization of the Bergmann-Bianchi identities for gauge-natural theories [11]:

Theorem 1 *The generalized Bergmann–Bianchi morphism is globally vanishing if and only if $\delta_{\mathfrak{G}}^2\lambda \equiv \mathscr{J}(\lambda,\mathfrak{G}(\bar{\Xi})_V) = 0$, i.e. if and only if $\mathfrak{G}(\bar{\Xi})_V \in \mathfrak{K}$.*

From now on we shall write $\omega(\lambda,\mathfrak{K})$ to denote $\omega(\lambda,\mathfrak{G}(\bar{\Xi})_V)$ when $\mathfrak{G}(\bar{\Xi})_V$ belongs to $\mathfrak{K}$. Analogously for β and other morphisms.

Let $\mathscr{L}_{j_s\bar{\Xi}}$ be the variational Lie derivative operator [6] acting on generalized variational morphisms. Then the following holds:

Proposition 1 *For each $\bar{\Xi} \in \mathscr{A}^{(r,k)}$ such that $\bar{\Xi}_V \in \mathfrak{K}$, we have*

$$\mathscr{L}_{j_s\bar{\Xi}_H}\omega(\lambda,\mathfrak{K}) = -D_H(-j_s\pounds_{\bar{\Xi}}\rfloor p_{D_V\omega(\lambda,\mathfrak{K})}).$$

PROOF. We have $\mathscr{L}_{j_s\bar{\Xi}_V}\omega(\lambda,\mathfrak{K}) = \mathscr{L}_{j_s\bar{\Xi}_V}\mathscr{L}_{j_s\bar{\Xi}}\lambda = \mathscr{L}_{j_s[\bar{\Xi}_V,\bar{\Xi}_H]}\lambda$. On the other hand it is also easy to verify that $\mathscr{L}_{j_s\bar{\Xi}_H}\omega(\lambda,\mathfrak{K}) = \mathscr{L}_{j_s[\bar{\Xi}_H,\bar{\Xi}_V]}\lambda = -\mathscr{L}_{j_s\bar{\Xi}_V}\omega(\lambda,\mathfrak{K})$. Thus, from Theorem 1 above, since

$$\mathscr{L}_{j_s\bar{\Xi}_V}\omega(\lambda,\mathfrak{K}) = -\pounds_{\bar{\Xi}}\rfloor\mathscr{E}_n(\omega(\lambda,\mathfrak{K})) + D_H(-j_s\pounds_{\bar{\Xi}}\rfloor p_{D_V\omega(\lambda,\mathfrak{K})}) =$$
$$= \beta(\lambda,\mathfrak{K}) + D_H(-j_s\pounds_{\bar{\Xi}}\rfloor p_{D_V\omega(\lambda,\mathfrak{K})}),$$

we get the assertion. $\qquad\square$

The new generalized Lagrangian $\omega(\lambda,\mathfrak{K})$ is gauge-natural invariant too, *i.e.* $\mathscr{L}_{j_s\bar{\Xi}}\omega(\lambda,\mathfrak{K}) = 0$ holds. Even more, we can state the following:

Proposition 2 *Let* $\bar{\Xi}_V \in \mathfrak{K}$. *We have*

$$\mathscr{L}_{j_s\bar{\Xi}_H}\omega(\lambda,\mathfrak{K}) = 0.$$

Corollary 1 *Let* $\bar{\Xi}_V \in \mathfrak{K}$. *We have the* covariant "strong" *conservation law:*

$$D_H(-j_s\pounds_{\bar{\Xi}}\rfloor p_{D_V\omega(\lambda,\mathfrak{K})}) = 0.$$

This result can be compared with [2].

Definition 2 We define the covariantly conserved current

$$\mathscr{H}(\lambda,\mathfrak{K}) = -j_s\pounds\rfloor p_{D_V\omega(\lambda,\mathfrak{K})},$$

to be a Hamiltonian form for $\omega(\lambda,\mathfrak{K})$ (in the sense of [10]). $\qquad\square$

REFERENCES

1. P.G. BERGMANN: Conservation Laws in General relativity as the Generators of Coordinate Transformations, *Phys. Rev.* **112** (1) (1958) 287–289.
2. B. CASCIARO, M. FRANCAVIGLIA: A new variational characterization of Jacobi fields along geodesics. Ann. Mat. Pura Appl. (4) 172 (1997), 219–228.
3. D.J. ECK: Gauge-natural bundles and generalized gauge theories, *Mem. Amer. Math. Soc.* **247** (1981) 1–48.
4. L. FATIBENE, M. FRANCAVIGLIA: *Natural and gauge natural formalism for classical field theories. A geometric perspective including spinors and gauge theories*; Kluwer Academic Publishers, Dordrecht, 2003.
5. L. FATIBENE, M. FRANCAVIGLIA, M. PALESE: Conservation laws and variational sequences in gauge-natural theories, *Math. Proc. Camb. Phil. Soc.* **130** (2001) 555–569.
6. M. FRANCAVIGLIA, M. PALESE, R. VITOLO: Symmetries in Finite Order Variational Sequences, *Czech. Math. J.* **52(127)** (2002) 197–213.
7. I. KOLÁŘ: A Geometrical Version of the Higher Order Hamilton Formalism in Fibred Manifolds, *J. Geom. Phys.*, **1** (2) (1984) 127–137.
8. I. KOLÁŘ, P.W. MICHOR, J. SLOVÁK: *Natural Operations in Differential Geometry*, (Springer–Verlag, N.Y., 1993).
9. D. KRUPKA: Variational Sequences on Finite Order Jet Spaces, *Proc. Diff. Geom. and its Appl.* (Brno, 1989); J. Janyška, D. Krupka eds.; World Scientific (Singapore, 1990) 236–254.
10. L. MANGIAROTTI, G. SARDANASHVILY: *Connections in Classical and Quantum Field Theory*, (World Scientific, Singapore, 2000).
11. M. PALESE, E. WINTERROTH: Global Generalized Bianchi Identities for Invariant Variational Problems on Gauge-natural Bundles, to appear in *Arch. Math. (Brno)*.

Penrose limit and duality between string and gauge theories

Alessandro Tanzini

S.I.S.S.A. / I.S.A.S., via Beirut 2/4, 34100, Trieste, Italy

Abstract. We give a brief introduction to the Penrose limit and its use in the AdS/CFT correspondence. Related developments on the relationship between gauge theories and integrable systems are discussed also for non–conformal theories.

In General Relativity there is a remarkably simple argument, due to Penrose [1], which shows that any space–time has a plane wave as a limit. The Penrose limit has been recently generalized to string theory [2], and has attracted a fair amount of activity since then. The basic motivation is that this limiting procedure yields new supersymmetric *curved* backgrounds with non-trivial fluxes, on which the string spectrum can be computed *exactly* in the (inverse) string tension α'. In this note we will focus on the maximally supersymmetric plane wave in type IIB string theory, which can be obtained as a Penrose limit of the $AdS_5 \times S^5$ space [3]. Berenstein, Maldacena and Nastase proposed a very concrete description of string theory in this background in terms of a particular sector of the $\mathcal{N} = 4$ Super Yang-Mills theory [4]. This is the first example in which the world–sheet of a well defined string theory is reproduced starting from gauge theory amplitudes, and provides at the moment the best theoretical laboratory to study and test the gauge/string duality. Let us start by recalling the $AdS_5 \times S^5$ solution of IIB supergravity in global coordinates

$$
\begin{aligned}
ds^2 &= R^2 \left[-\cosh^2 r\, dt^2 + dr^2 + \sinh^2 r\, d\Omega_3^2 + \cos^2\theta\, d\psi^2 + d\theta^2 + \sin^2\theta\, d\Omega_3'^2 \right], \\
F_5 &= \frac{1}{R} \left(dV_{AdS_5} + dV_{S^5} \right) , \quad \phi = \text{constant} ,
\end{aligned}
\tag{1}
$$

where ϕ is the dilaton field and F_5 the Ramond–Ramond five form. The solution (1) preserves the maximal number of supersymmetries and is believed to be exact to all orders in α'. The Penrose limit consists in focusing on the neighborhood of a null geodesic and rescaling coordinates and metric in order to blow up this neighborhood to the whole space. In supergravity also the other forms have to be rescaled properly in the limit: the homogeneity of the action under these rescalings ensures that the space obtained in the Penrose limit is still an exact solution [2]. For the $AdS_5 \times S^5$ space, there are essentially two different Penrose limits, according to the different choices of the null geodesic [3]: if this lies entirely in the AdS space, one gets a limiting flat space. By choosing instead the geodesic of a particle rotating around the sphere, *i.e.* $t = \psi$ and

CP751, *General Relativity and Gravitational Physics, 16th SIGRAV Conference*, edited by G. Esposito et al.
© 2005 American Institute of Physics 0-7354-0236-1/05/$22.50

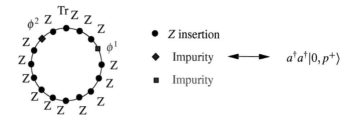

$r = \theta = 0$, and going to the light–cone coordinates

$$x^+ = \frac{t+\psi}{2\mu} , \quad x^- = \mu R^2 \frac{t-\psi}{2} , \quad \hat{r} = Rr , \quad y = R\theta , \tag{2}$$

one gets in the Penrose limit $R \to \infty$ with $x^\pm, \hat{r}, y$ fixed

$$ds^2 = -4dx^+dx^- - \mu^2(x_I)^2dx^{+\,2} + dx^{I\,2} , $$
$$F_{+1234} = F_{+5678} = 2\mu , \quad \phi = \text{constant.} \tag{3}$$

The key features of the above plane-wave solution are that *(i)* it is a curved background $(R_{++} \sim \mu)$ with non-trivial Ramond–Ramond flux, and *(ii)* it preserves all 32 supersymmetries of type IIB string theory. This last feature results from a general property of the Penrose limit, according to which the number of (super)symmetries cannot decrease in the limit [3]. The physical interpretation of the Penrose limit is a *truncation* of the string spectrum to the states describing small oscillations around a point-like string spinning along the five–sphere. The geometry (3) seen by these states is greatly simplified, so that their spectrum can be computed *exactly*. In fact the world–sheet action in the light–cone gauge turns out to be free [5]. The light–cone Hamiltonian at fixed light–cone momentum p^+ is a sum of eight bosonic and eight fermionic harmonic oscillators, all with the same frequency

$$\frac{H}{\mu} = \sum_{n=-\infty}^{+\infty} \omega_n \left(a_n^{I\dagger} a_n^I + b_n^{\alpha\dagger} b_n^\alpha \right) , \quad \omega_n = \sqrt{1 + \frac{n^2}{(\mu\alpha'p^+)^2}} , \tag{4}$$

with $I = 1,\dots,8$, $\alpha = 1,\dots,8$ [5]. The basic idea of [4] was to regard some gauge-invariant operators of the $\mathcal{N} = 4$ Super Yang–Mills theory containing a large number J of fields as a discretised version of the physical type IIB string on the plane-wave background. The BMN operators are single trace operators formed by a long chain of one of the elementary scalar fields of $\mathcal{N} = 4$, with the insertion of a few other fields and covariant derivatives (called *impurities*), each of them corresponding to a different excitation of the string (see figure).

The anomalous dimensions of these operators are expected to coincide with the mass of the corresponding string state. On the gauge theory side, the Penrose limit corresponds to take the 't Hooft coupling λ to infinity while keeping the ratio $\lambda' = \lambda/J^2$ fixed. It is thus different from the usual 't Hooft expansion. The remarkable feature of the spectrum (4) is that, when using the AdS/CFT dictionary $\lambda' = 1/\mu p^+ \alpha'$, one gets an *analytic* function of the (effective) gauge coupling λ'. This makes it possible to directly compare the string spectrum with the perturbative expansion of the gauge theory, finding agreement [4]! Notice that this comparison is based on the extrapolation of the perturbative gauge theory results to the strong coupling regime $\lambda \to \infty$, λ' fixed. A definite proof of the validity of this extrapolation is still lacking, and would certainly improve our understanding of the duality in the large J limit.

A lot of work has been done in order to include the string interaction in the duality. In [6], we constructed a supersymmetric 3-string interaction vertex on the plane-wave background. At the leading order in λ', the resulting amplitudes are in agreement with that extracted from three-point correlators of BMN operators. However, the extension of this correspondence at the subleading order remains an open problem (for a review, see [7]). The main obstacle is that in the Penrose limit the boundary of the AdS space is washed away (recall that the null geodesic (2) sits at the center of AdS space), and this makes it rather problematic to define a clear holographic principle. See however [8] for some promising progress.

The study of the AdS/CFT correspondence in the plane-wave limit has posed the difficult problem to compute the anomalous dimensions of composite operators of the $\mathcal{N} = 4$ SYM containing a large number of fields $J \to \infty$. A very useful result in this context [9] is that, in the large-N limit, the matrix of one-loop anomalous dimensions can be mapped into the Hamiltonian of an integrable system. This intriguing remark applies not only to the BMN operators, but also to a wider class of operators containing a large number of impurities $J_i \sim J$. In particular, for the (closed) subsector of composite operators of two scalar fields, one gets the Hamiltonian of an XXX Heisenberg spin chain [9]. In [10] we have analysed the analogous sector for the $\mathcal{N} = 2$ SYM theory, finding an XXZ spin-chain. From the dual string theory point of view, the presence of the anisotropy in one of the spin directions should be related to the non-trivial fluxes which break the supersymmetry from $\mathcal{N} = 4$ to $\mathcal{N} = 2$. One interesting lesson from the $\mathcal{N} = 2$ case is that the breaking of conformal invariance by virtue of the non-vanishing beta function $\beta \sim -g^3$ does *not* affect the renormalization-group equations for these operators at one-loop order. This is a general property of gauge theories which is not related to supersymmetry. These features make it particularly interesting to investigate whether some relation can be found between the integrability of some subsectors of gauge theories in the large N limit and the existence of a dual string theory description for them.

REFERENCES

1. R. Penrose, in *Differential Geometry and Relativity*, ed. M. Chaen and M. Flato, Dordrecht (1976).
2. R. Gueven, Phys. Lett. B **482**, 255 (2000).
3. M. Blau, J. Figueroa–O'Farrill, and G. Papadopoulos, Class. Quantum Grav. **19**, 4753 (2002).
4. D. Berenstein, J. M. Maldacena, and H. Nastase, JHEP **04**, 013 (2002).
5. R. R. Metsaev, Nucl. Phys. B **625**, 70 (2002).
6. P. Di Vecchia, J. L. Petersen, M. Petrini, R. Russo and A. Tanzini, Class. Quantum Grav. **21**, 2221 (2004).
7. R. Russo and A. Tanzini, Class. Quantum Grav. **21**, S1265 (2004).
8. S. Dobashi and T. Yoneya, hep-th/0406225; S. Lee and R. Russo, hep-th/0409261.
9. J. A. Minahan and K. Zarembo, JHEP **03**, 013 (2003).
10. P. Di Vecchia and A. Tanzini, J. Geom. Phys., to appear, hep-th/0401155.

Asymptotics of Quasinormal Modes for Schwarzschild-de Sitter Black Holes

L. Vanzo [*] and S. Zerbini[1*]

*Dipartimento di Fisica,
Università degli Studi di Trento,
INFN, Gruppo Collegato Di Trento
via Sommarive 14, 38050 Povo, Italy*

Abstract. The asymptotics of Quasinormal Modes for Schwarzschild-de Sitter Black Holes is investigated in the large angular momentum limit by means of the analytic dilatation method. The case of Nariai space-time is exactly treated.

Quasinormal modes (QNMs) are small perturbations associated with static (eventually stationary) solutions of the Einstein equation with spherical symmetry [1, 2]. The motivations for the interest in QNMs are several. Phenomenological motivations: relevance in the search for gravitational waves; physical motivations: at fundamental level, they may shed light on the degrees of freedom in quantum gravity. General motivations: in physical applications and static space-times, after time factorization $e^{i\omega t}$, one is dealing with an elliptic second-order self-adjoint differential operator L on some spatial N-dimensional manifold. For a compact manifold, the related spectrum is discrete, and the spectral data are the eigenvalues λ_i of a suitable self-adjoint operator L. The heat-kernel asymptotics plays a decisive role

$$Tre^{-tL} = \sum_{\lambda_i} e^{-t\lambda_i} \simeq \sum_r A_r t^{r-N/2}.$$

The Seeley-De Witt spectral coefficients A_r are computable and give information about one-loop divergences in QFT and asymptotic evaluation of the one-loop effective action (see for example [3]).

In black hole physics one is dealing with non compact manifolds and here one has to deal with a continuous spectrum. As a result, the spectral data are given by the so-called "resonance set", a set of complex eigenvalues associated with a suitable operator L. Physically they represent QNM frequencies in classical relativity.

Our aim is to investigate with a mathematically well-established method, i.e. the analytic dilatation method, the asymptotics for QNMs associated with a generic D-dimensional black hole. Working in D dimensions may be justified by the phenomenological interest in extra spatial dimensions which has recently appeared in the literature, triggered by string theory.

---

[1] contributing author

CP751, *General Relativity and Gravitational Physics, 16th SIGRAV Conference*, edited by G. Esposito et al.

We recall that the master equation for gravitational perturbations in the Regge–Wheeler coordinate x reads as

$$L_l \phi_l(x) = \left[-\frac{d^2}{dx^2} + V_l(r(x)) \right] \phi_l(x) = \omega^2 \phi_l(x) ,$$

where the potential V_l depends on the index l of spherical harmonics and on the tensorial nature of the perturbation. In $D = 4$

$$V_l(r) = \frac{A(r)}{r^2} \left[l(l+1) + W_{l,s}(r) \right] .$$

The related Sturm–Liouville problem for QNMs reads as: $L\phi_l = \omega^2 \phi_l$ in $(-\infty, +\infty)$ plus the in-going and out-going boundary conditions, respectively:

$$\phi(x) \to e^{i\omega x} \quad \text{as } x \to -\infty, \quad \phi(x) \to e^{-i\omega x} \quad \text{as } x \to \infty.$$

Such boundary conditions render the operator L not self-adjoint. Thus, one has, in general, complex ω^2.

Now the key point is that the QNM problem can be considered as a scattering problem. We may give a new definition of QNMs: QNMs are the scattering resonances associated with the Master operator L_l. Recall that the resonance frequencies are the poles of the meromorphic continuation for complex ω of the related scattering matrix. The analytic dilatation method (see, for example, [4] and references therein) provides an elegant and powerful method for investigating the resonances. The method leads to the following equivalent Sturm–Liouville problem:

$$\left[-\frac{d^2}{dx^2} + i V_l(\sqrt{i}x) \right] \Psi_l(x) = i\omega^2 \Psi_l(x).$$

Here one can work in a suitable Hilbert space with $\Psi_l(x)$ vanishing at infinity. For large l, our main result for the asymptotics is [4]

$$\omega_{l,n} \simeq \frac{\sqrt{A(r_0)}}{r_0} \left[\pm \left(l + \frac{1}{2} \right) - i \left(n + \frac{1}{2} \right) \sqrt{ \left| \frac{r_0 A''(r_0)}{2} - A(r_0) \right| } \right] ,$$

where r_0 is the value of r at which the local maximum of the potential $V_l(r)$ is reached. The multiplicity is present and is the one of $(D-2)$-dimensional spherical harmonics.

As an example, let us consider a Schwarzschild-de Sitter BH in D-dimensions

$$A(r) = 1 - \left(\frac{r_S}{r} \right)^{D-3} - \frac{r^2}{L^2}, \quad r_S^{D-3} = c_D M .$$

On denoting by M_N the BH critical mass $M_N = \frac{2}{c_D(D-3)} \left(\frac{D-3}{D-1} \right)^{(D-1)/2} L^{D-3}$, one has

$$\omega_{l,n} \simeq \frac{1}{L} \left[\left(\left(\frac{M}{M_N} \right)^{\frac{2}{D-3}} - 1 \right) \right]^{1/2} \left[\pm \left(l + \frac{D-3}{2} \right) - i \left(n + \frac{1}{2} \right)(D-3) \right] .$$

For $D = 4$

$$\omega_{l,n} \simeq \frac{\left(1 - 9\Lambda M^2\right)^{1/2}}{3\sqrt{3}M}\left[\pm(l+\frac{1}{2}) - i(n+\frac{1}{2})\right].$$

The pure Schwarzschild case may be obtained in the limit of large L or vanishing cosmological constant. Recall that these expressions are valid for very large l, but for $n = 0, 1, 2, 3, ...$ Other methods, numerical and analytic, give results for low and fixed l and very large n (see for example [5] and references quoted therein).

As an exact computation, we consider the D-dimensional Nariai space (extremal limit of S-dS BH). The metric reads as

$$ds^2 = -\left(1 - \frac{(D-1)y^2}{L^2}\right)dt^2 + \frac{dy^2}{\left(1 - \frac{(D-1)y^2}{L^2}\right)} + r_N^2 dS_{D-2}^2,$$

where $r_N^2 = \frac{D-3}{D-1}L^2$. The topology is $dS_2 \times S_{D-2}$. Since dS_2 is not simply connected, this space-time has two horizons at $y_H = \pm\frac{L}{\sqrt{D-1}}$, $|y| < |y_H|$. Both horizons have the same Hawking temperature $T_H = \frac{\sqrt{D-1}}{2\pi L}$.

The radial master equation becomes

$$\left[-\frac{d^2}{dx^2} + \frac{U_l}{\cosh^2 2\pi T_H x}\right]\Psi_l(x) = \omega^2 \Psi_l(x),$$

where $U_l = \frac{\lambda_l^2}{r_N^2}$, and $\lambda_l^2 = l(l+D-3)$ are eigenvalues related to the $(D-2)$-dimensional spherical harmonics. For the one-dimensional Schroedinger equation with a Pöschl-Teller potential, the exact QNM frequencies can be evaluated as poles in the associated analytically continued scattering matrix. The result is

$$\omega_{n,l} = 2\pi T_H\left(\pm\sqrt{\frac{\lambda_l^2}{D-3} - \frac{1}{4}} - i(n+\frac{1}{2})\right).$$

In the large l limit and making use of a suitable identification [4], the approximated expression coincides with the one obtained previously.

REFERENCES

1. H. P. Nollert, Class. Quant. Grav. **16**, R159 (1999);
2. K.D. Kokkotas and B.G. Schmidt, Living Rev. Rel. **2**, 2 (1999).
3. A.A. Bytsenko, G. Cognola, L. Vanzo and S. Zerbini, Phys. Reports **256**, 1 (1996).
4. L. Vanzo and S. Zerbini, Phys. Rev. **D 70**, 044030 (2004).
5. N. Andersson, Class. Qauntum Grav. **10**, L61 (1993); H.P. Nollert, Phys. Rev. **47**, 5253 (1993); L. Motl, Adv. Theor. Math. Phys. **6**, 1135 (2003); L. Motl and A. Neitzke, Adv. Theor. Math. Phys. **7**, 307 (2003).

WORKSHOP—CLASSICAL AND QUANTUM GRAVITY

Weak lensing and cosmological investigation

Viviana Acquaviva[†]

SISSA/ISAS, Via Beirut 2-4, 34014 Trieste, Italy
INFN, Sezione di Trieste, Via Valerio 2, 34127 Trieste, Italy

Abstract. In the last few years the scientific community has been dealing with the challenging issue of identifying the dark energy component. We regard weak gravitational lensing as a brand new, and extremely important, tool for cosmological investigation in this field. In fact, the features imprinted on the Cosmic Microwave Background radiation by the lensing from the intervening distribution of matter represent a pretty unbiased estimator, and can thus be used for putting constraints on different dark energy models. This is true in particular for the magnetic-type B-modes of CMB polarization, whose unlensed spectrum at large multipoles ($l \simeq 1000$) is very small even in presence of an amount of gravitational waves as large as currently allowed by the experiments: therefore, on these scales the lensing phenomenon is the only responsible for the observed power, and this signal turns out to be a faithful tracer of the dark energy dynamics.

We first recall the formal apparatus of the weak lensing in extended theories of gravity, introducing the physical observables suitable to cast the bridge between lensing and cosmology, and then evaluate the amplitude of the expected effect in the particular case of a Non-Minimally-Coupled model, featuring a quadratic coupling between quintessence and Ricci scalar.

GENERAL FEATURES OF LENSING ON THE CMB

In ordinary ΛCDM models, the effect of the lensing on the CMB from Large-Scale Structure has been well exploited by several authors (see ie [1]). The typical size of the lenses is those of the largest bound structures, the clusters of galaxies: it is usually a few Mpc and it states the relevant physical scale for the analysis of the temperature multipoles. On scales which are much larger, corresponding to multipoles up to one hundred, the lensing effect is randomized and therefore negligible, while for $200 < l < 2000$ the signal is coherent and it appears as a smoothing of the peak-and-troughs structure.

However, we are even more interested in the effect on the polarization spectra: in fact, lensing has the remarkable property of transferring power from the electric E-modes to the magnetic B-ones, and this injection of power peaks at $l \simeq 1000$, at a scale much smaller than that of gravitational waves and reionization effects, which are the other mechanisms imprinting a signal on the B-mode spectrum. Thus, the bump of power at that scale is straightforwardly related to the lensing dynamics, and we expect to be able to use its properties in order to discriminate between different dark energy models [2].

In the weak lensing regime, the mapping of the unlensed signal to the lensed one can be given in terms of the symmetric, two-dimensional *distortion tensor* ψ_{ij}; its diagonal components are related to the observed isotropic magnification of an image, and are encoded in a scalar quantity called *convergence*. On the other hand, the off-diagonal terms describe the anisotropic distortion, and are expressed as the components

CP751, *General Relativity and Gravitational Physics, 16th SIGRAV Conference*, edited by G. Esposito et al.
© 2005 American Institute of Physics 0-7354-0236-1/05/$22.50

of a two-dimensional vector, the *shear*. Again as a consequence of the weak lensing approximation, these two quantities have identical statistical properties, and the power spectrum of each of them can be used as the relevant lensing-related observable. In order to link this quantity to the underlying cosmology, we need an equation of motion, which is simply the null geodesic equation for the photons: this will allow to obtain, for example, the relation between the convergence power spectrum and that of the gravitational potential.

The next step is connecting the gravitational potential spectrum to the observed matter density distribution, and this will be made by means of the Poisson equation.

COSMOLOGICAL SETTING

The formulation of the dark energy hypothesis has been driven by the observation of cosmic acceleration [3, 4]. The most credited explanations for this phenomenon belong to one of the following categories:

- the presence of a spacetime-independent vacuum energy, often referred to as the Cosmological Constant;
- a dynamical scalar field, commonly known as Quintessence;
- a modification in the gravitational sector of the standard theory of General Relativity, i.e., in the Einstein–Hilbert Lagrangian.

Therefore, the analysis has to be carried out in a fairly general framework: we compute our equation of motion from the following action [5]:

$$S = \int d^4x \sqrt{-g} \left[\frac{1}{2\kappa} f(\phi, R) - \frac{1}{2} \omega(\phi) \phi^{;\mu} \phi_{;\mu} - V(\phi) + \mathscr{L}_{\text{fluid}} \right], \tag{1}$$

where g is the determinant of the background metric, R is the Ricci scalar, ω generalizes the kinetic term, and $\mathscr{L}_{\text{fluid}}$ includes contributions from the matter and radiation cosmological components; $\kappa = 8\pi G_*$ plays the role of the "bare" gravitational constant, and the usual gravity term $R/16\pi G$ has been replaced by the generic function $f/2\kappa$.

As background cosmology we will adopt a FRW scenario, making the simplifying assumption of spatially flat geometry, and make use of linear perturbation theory.

NEW FEATURES IN GENERALIZED THEORIES OF GRAVITY

The effects arising on the lensing signal and resulting from the dark energy or modified gravity dynamics are present both at the background and at the perturbative level. The main effect on the background dynamics is the modification of distance measurements, since we let the strength of the gravitational field be time-varying. For the perturbed quantities there are two main different phenomena: first, the gravitational potential Φ is modified from its interactions with the fluctuations in the scalar field $\delta\phi$; second, there will be one more gravitational potential, Ψ, signaling the presence of an extra degree of freedom, usually referred to as anisotropic stress.

We can be more specific and give analytical results for a class of models known as Non-Minimally-Coupled, where the function $f(\phi, R)$ is of the form

$$f/\kappa = FR \quad \text{and} \quad F(\phi) = \frac{1}{8\pi G} + \xi (\phi^2 - \phi_0^2),$$ (2)

where ϕ_0 is the present value of the quintessence field and ξ is a coupling parameter. In these models the most relevant effect is that of the variable gravitational constant, thus we won't take into account the interactions with the field fluctuations. The resulting correction to the convergence power spectrum is

$$\delta P_\kappa(l) = -128 G \xi \phi_0^2 \int_0^{\chi_\infty} d\chi \, f(\chi) \int_0^{\chi_\infty} d\chi' f(\chi') \left(\frac{\phi^2}{\phi_0^2} - 1 \right)$$
$$\times \int dk \, k^6 \, j_l(k\chi) j_l(k, \chi') \langle \Phi(k, \chi) \Phi(k, \chi') \rangle.$$ (3)

In this expression there is a "hidden" projection effect, encoded in the geometric part containing $f(\chi)$, which will be responsible for an alteration in the position of the peaks; there is then an amplitude term, $(\phi^2/\phi_0^2 - 1)$, which can be safely regarded as slowly dependent of the redshift and is of order unity in all models under study, allowing to gain an estimate of the size of the effect as

$$\delta P_\kappa / P_\kappa \simeq -128 G \xi \, \phi_0^2.$$ (4)

Thus, since in typical Extended Quintessence scenarios [6] the product $G \phi_0^2$ is close to one, the correction in the convergence power spectrum is sizeable even with values of the coupling parameter as small as 10^{-3}, and this could yield significant constraints on Brans–Dicke theories, in a complementary fashion to those coming from Solar System experiments [7].

ACKNOWLEDGMENTS

I am grateful to Carlo Baccigalupi and Francesca Perrotta for a stimulating and fruitful collaboration.

REFERENCES

1. W. Hu, Phys. Rev. D **66**, 083515 (2002).
2. V. Acquaviva, C. Baccigalupi and F. Perrotta, Phys. Rev. D **70**, 023515 (2004).
3. S. Perlmutter et al., Astrophys. J. **517**, 565 (1999).
4. A.G. Riess et al., Astron. J **116**, 1009 (1998).
5. J. Hwang, Astrophys. J. **375**, 443 (1991).
6. S. Matarrese, C. Baccigalupi and F. Perrotta, Phys. Rev. D **70**, 061301 (2004).
7. B. Bertotti, L. Iess and P. Tortora, Nature **425**, 374 (2003).

Pulsar magnetism and dynamo actions

Alfio Bonanno*, V. Urpin† and G. Belvedere**

*INAF - Osservatorio Astrofisico di Catania, Via S. Sofia 78, I-95123 Catania, Italy
†A.F. Ioffe Institute of Physics and Technology, St. Petersburg, Russia
Isaac Newton Institute of Chile, Branch in St. Petersburg, 194021 St. Petersburg, Russia
**Dipartimento di Fisica ed Astronomia, Via S. Sofia 78, I-95123, Catania, Italy

Abstract. We have investigated the turbulent mean-field dynamo action in protoneutron stars that are subject to convective and neutron-finger instabilities during the early evolutionary phase. By solving the mean-field induction equation we have evaluated the critical spin period at which dynamo action is possible and we have found that a mean-field dynamo will be operating in most of the protoneutron stars.

The nature of pulsar magnetism is a subject of debate for decades. In a simple magnetic dipole braking model, the polar field strength inferred from observational data can reach $\sim 5 \times 10^{13}$G. One possibility is that the magnetic field of a progenitor star is amplified by many orders of magnitude because of the conservation of magnetic flux during the collapse stage. However, the "fossil field" hypothesis, despite being seemingly attractive and plausible, meets a number of problems. For instance, the progenitor star should possess a sufficiently strong magnetic field that does not agree with observational data (see discussion in [1]).

Recently, it has been shown [2] that turbulence can drive both small and large scale dynamos in PNS. In this study, we consider the mean-field dynamo action in PNS in more detail. We extend our study to the case of non-axisymmetric fields that are of particular interest in pulsars. The main goal of this paper is to show that the mean-field dynamo can operate in a wide range of parameters of PNS and generate the field of a strength comparable to that of "standard" pulsars.

To investigate the efficiency of a mean-field dynamo action, we model the PNS as a sphere of radius R with two substantially different turbulent zones separated at $R_c < R$. The inner part ($r < R_c$) corresponds to the convectively unstable region, while the outer one ($R_c < r < R$) to the neutron-finger unstable region. The boundary between the two regions moves inward on a timescale comparable to the cooling timescale (i.e. $\sim 1 - 10$ s) that is much longer than the turnover time for both unstable zones. The energy of turbulence is generally non-stationary as well, developing rapidly soon after the collapse, reaching a quasi-stationary regime after a few seconds, and then progressively disappearing. However, changes take place on the cooling time scale and, therefore, all parameters of turbulence can be treated in quasi-steady approximation. In this case, the mean-field induction equation for a turbulent PNS can be written as

$$\frac{\partial \vec{B}}{\partial t} = \nabla \times (\vec{v} \times \vec{B} + \alpha \vec{B}) - \nabla \times (\eta \nabla \times \vec{B}) , \tag{1}$$

CP751, *General Relativity and Gravitational Physics, 16th SIGRAV Conference*, edited by G. Esposito et al.
© 2005 American Institute of Physics 0-7354-0236-1/05/$22.50

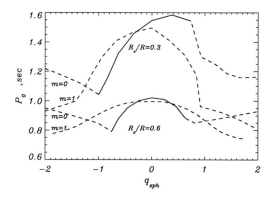

FIGURE 1. Critical period as a function of the differential rotation parameter q_{sph} for $F(r) = 1$. The two pairs of curves refer to different values of R_c, with the solid parts corresponding to a stationary dynamo and the dashed parts to an oscillatory dynamo (travelling waves for $m = 1$). The number m corresponds to different values of the azimuthal wavenumber in a polynomial expansion for the magnetic field.

where η is the turbulent magnetic diffusivity, α is a pseudo-scalar measuring the efficiency of the dynamo (the "α-parameter"), and $\vec{v}$ is the velocity of a mean fluid motion. Boundary conditions for the magnetic field need to be specified at the stellar surface, where we impose vacuum boundary conditions, and at the center of the star, where we impose the vanishing of the toroidal magnetic field.

The PNS is assumed to be differentially rotating, that is often observed in numerical simulations [3]. In Fig. 1 we plot the critical period, P_0, below which the dynamo action becomes possible for the case of a shell-type rotation (2). The critical period is shown as a function of the parameter $q_{sph} = \Omega(r = R) - \Omega(r = 0)/\Omega(r = R)$ that characterizes differential rotation. Progressively varying the differential rotation parameter, q_{sph}, we have solved the induction equation (1) to determine the critical value α_0 corresponding to the marginal stability of the dynamo. The seed magnetic field will grow if $\alpha_{nf0} > \alpha_0$ and decay if $\alpha_{nf0} < \alpha_0$, α_{nf0} being the value of the turbulence helicity in the neutron finger region. Since $\alpha_{nf0} \approx \Omega L$, the critical value α_0 effectively selects a critical value for the spin period, $P_0 \equiv 2\pi L/\alpha_0$, such that magnetic field generation via a mean-field dynamo action will be possible only if the stellar spin period is shorter than the critical one. The different types of dynamo can be distinguished according to whether the generated field exhibits periodic oscillations (oscillatory dynamo, dashed lines in Fig. 2) or not (stationary dynamo, solid lines).

As shown in Fig. 1, a stationary dynamo dominates the axisymmetric magnetic field generation process for $|q| \lesssim 1$, while an oscillatory dynamo is more efficient for $1 \lesssim |q|$. These two regimes correspond to α^2-dynamo and $\alpha\Omega$-dynamo, respectively. Given the large differential rotation required for a generation of the axisymmetric oscillatory magnetic field, it may be difficult to achieve in practice. Hence, the α^2-dynamo appears to be the most likely source of axisymmetric magnetic field generation via dynamo processes in PNS. The situation is, however, different for the generation of a non-axisymmetric magnetic field that is of interest for PNS since the observed pulsars have non-axisymmetric fields. The fields with azimuthal wavenumber $m = 1$ are travelling waves in the azimuthal direction.

Our conclusions are as follows: the PNS is subject to two substantially different instabilities, with a convective instability operating in the inner region of the star and a neutron-finger instability being more efficient in the outer region. The turbulent motions are more rapid in the convective zone, where the Rossby number is large but the α-parameter, characterizing the mean-field dynamo action, is likely small. In the neutron-finger unstable region, on the other hand, the turbulent turnover time is considerably longer, the Rossby number small, and the α-parameter can be sufficiently large to drive a mean-field dynamo.

The α-parameter depends on the stellar rotation being larger for rapidly rotating stars. Our simulations show that even relatively slowly rotating PNS can be subject to a dynamo action, with the α^2-dynamo being the most efficient mechanism of generation for both axisymmetric and non-axisymmetric fields if differential rotation is not extremely strong. The calculated critical value of the spin period that determines the onset of dynamo in PNS is $P_0 \sim 1$ s for a wide range of models.

This value is essentially larger even than the characteristic spin period of very young pulsars as inferred from observations ($\sim 50 - 100$ ms) but likely PNS can rotate even faster. As a result, a turbulent mean-field dynamo can be effective in the early stages of the life of most PNS. The critical period is not very different for axisymmetric and non-axisymmetric fields and likely both these magnetic configurations can be generated by the PNS dynamo. The generation of a non-axisymmetric component is the key point for magnetism of pulsars, since they have a substantial non-axisymmetric field. It should be stressed that, as the neutron-finger unstable region shrinks towards the surface, the difference between axisymmetric and non-axisymmetric critical periods disappears, thus providing an opportunity to generate a non-axisymmetric field in a very general framework.

The final strength of the generated small-scale field turns out to be the same for both unstable zones. For the largest turbulent scale, $\ell_T = L \sim 1 - 3$ km, we obtain $B_{eq} \sim 3 \times 10^{13}$ G. Using this estimate of B_{eq}, we can conclude that the strength of a large-scale poloidal field generated by the mean-field dynamo is in a good agreement with the observed magnetic fields of the majority of pulsars. For example, the generated poloidal field is $\sim (1 - 2) \times 10^{13}$ G if the star rotates with the period ~ 100 ms. Note that extremely rapidly rotating PNS ($P \sim 1$ ms) may possess a very strong poloidal field $\sim (3 - 6) \times 10^{14}$ G comparable to that of magnetars.

REFERENCES

1. Thompson C., Duncan R. 1993, ApJ, 408, 194
2. Bonanno A., Rezzolla L., Urpin V. 2003. A&A, 410, L33
3. Dimmelmeier H., Font J. A., Müller E. 2002, A&A, 393, 523

Newman–Penrose quantities as valuable tools in astrophysical relativity

Marco Bruni*, Andrea Nerozzi* and Frances White*

*Institute of Cosmology and Gravitation, University of Portsmouth, Portsmouth PO1 2EG, UK.

Abstract. In this talk I will briefly outline work in progress in two different contexts in astrophysical relativity, i.e. the study of rotating star spacetimes and the problem of reliably extracting gravitational wave templates in numerical relativity. In both cases the use of Weyl scalars and curvature invariants helps to clarify important issues.

INTRODUCTION

The Weyl scalars - the components of the Weyl tensor over a null tetrad - were known and used in relativity before the introduction of the Newman–Penrose (NP) formalism, but within the latter they acquired a new relevance. Here I will summarise the use of Weyl scalars as tools in two different contexts in astrophysical relativity. First, I will briefly summarize recent work [1] aimed at assessing the validity of the Hartle–Thorne (HT) slow-rotation approximation for describing stationary axisymmetric rotating Neutron Stars (NS), introducing work in progress [2] to extend the analysis. In this context the Weyl scalars are used to construct invariant measures of the deviation of the exterior spacetime from Petrov type D, in view of a possible development of a Teukolsky-like perturbative formalism for rotating NS.

In the second part I will outline how the Weyl scalars may be used in numerical relativity in order to construct a wave extraction formalism for simulations dealing with spacetimes that will settle to a perturbed black hole (BH) at late times. In this case the Teukolsky BH perturbation formalism [3] is in principle applicable, but it is difficult to extract a BH background spacetime, i.e. the gravitational mass and angular momentum, from a given simulation. Introducing the notion of a quasi-Kinnersly frame (also used in [1]), in [4, 5] a method was proposed that bypasses this difficulty, by not requiring a background, and allows direct wave extraction. I will present here work in progress [6] where the method is directly applicable.

THE SPACETIME OF ROTATING STARS

Using a variety of models for NS of different masses and equations of state, and comparing with full general relativistic numerical models, it was shown in [1] that the HT approximation to the metric of a rotating relativistic star is very good for most astrophysical applications, even at the rotation rates of the fastest known milli-second

CP751, *General Relativity and Gravitational Physics, 16^(th) SIGRAV Conference*, edited by G. Esposito et al.

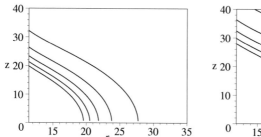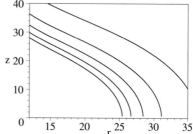

FIGURE 1. Contour plots of $(1 - |S|) \times 10^4$ for decreasing values (from bottom left) $5, 4, 3, 2, 1$, for a 1.4 $M_\odot$ star, for a representative equation of state. Left panel: $\varepsilon = 0.39258$; right panel: $\varepsilon = 0.54440$.

pulsar. For instance, the innermost stable circular orbit is predicted with an accuracy of $\sim 1\%$. It was also shown in [1] that, although the spacetime of these stars is of Petrov type I (general; it would be type D for a spherical star), the deviation from type D is small, at least on the equatorial plane. The HT metric is obtained as a perturbative solution to second order in $\varepsilon = \Omega/\Omega^*$, where Ω is the star's angular velocity and $\Omega^* = (M/R^3)^{1/2}$ is a "Keplerian" rotational scale. The deviation from type D is measured by $1 - |S|$, where $S = 27J^2/I^3$ is an invariant curvature scalar, with $S = 1$ for type D. Fig. 1 [2] shows that $1 - |S|$ is small also out of the equatorial plane. In fact, $1 - |S|$ decreases more rapidly to zero out of the equatorial plane, as one would have expected. Thus, in this sense $1 - |S|$ in the equatorial plane is a good upper limit for the deviation from type D.

WAVE EXTRACTION

In [4, 5] methods were introduced to identify, for a general numerical relativity implementation, what was dubbed the Quasi-Kinnersley frame, an equivalence class of tetrads that reduce, in the limit where the spacetime approaches type D, to the Kinnersly tetrad used in [3] for the BH background. In this tetrad the Weyl scalar Ψ_4 carries information on outgoing gravitational radiation, and Ψ_0 on ingoing radiation. Work is in progress to identify one specific and physically significant tetrad from the equivalence class, appropriate for a generic numerical relativity code using an arbitrary ADM slicing. However, the method of [4, 5] is already applicable when using a null slicing, in particular to the Bondi metric used in [7], where non-linear oscillations of a BH were analysed. In the case of the Bondi metric the gravitational wave signal can be extracted using the news function $\gamma_{,v}$ (v is retarded time), directly related to the outgoing energy. In the linear regime, one expects $\Psi_4 = -\gamma_{,vv}$. Thus, this case is ideally suited to test our method, since we can compare [6] the news function obtained directly from the Bondi metric with that obtained via Ψ_4 in the Quasi-Kinnersley frame. Fig. 2 shows the comparison of Ψ_4 with $\gamma_{,vv}$, where in our axisymmetric case $2\gamma = h_{\theta\theta}^{TT}$, i.e. the other polarization $h_{\theta\phi}^{TT}$ vanishes. Clearly the agreement is excellent at late times, as predicted, with an error

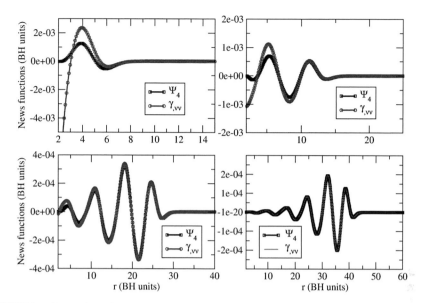

FIGURE 2. Comparison (from [6]) of Ψ_4 and $\gamma_{,vv}$ at retarded times $v_1 = 1$, $v_2 = 20$, $v_3 = 50$, $v_4 = 80$.

$\Delta = |\Psi_4 + \gamma_{vv}| \sim 10^{-6}$ at $v_4 = 80$.

CONCLUSIONS

Here we have shown how the method obtained in [4, 5] works, reproducing very well results obtained in [7], when applied in the context of a code using the characteristic approach [6]. Work is in progress to find a method to choose one particular Quasi-Kinnersly tetrad out of the general class in order to properly extract gravitational wave templates from any numerical relativity code using an arbitrary ADM slicing.

For a rotating NS, we have shown that the deviation of the spacetime from Petrov type D is always very small, with $1 - |S|$ rapidly decreasing out of the equatorial plan and with increasing distance from the star.

REFERENCES

1. E. Berti, F. White, A. Maniopoulou and M. Bruni, *Mon. Not. R. Astr. Soc.* , submitted, (2004). [gr-qc/0405146]
2. F. White, PhD thesis, University of Portsmouth, *in preparation*.
3. S. A. Teukolsky, *Astrophys. J.* **185**, 635 (1973).
4. A. Nerozzi, C. Beetle, M. Bruni, L. M. Burko and D. Pollney, *Phys. Rev. D*, submitted (2004). [gr-qc/0407013]
5. C. Beetle, M. Bruni, L. M. Burko and A. Nerozzi, *Phys. Rev. D*, submitted (2004). [gr-qc/0407012]
6. A. Nerozzi, PhD thesis, University of Portsmouth, (2004).
7. P. Papadopoulos, *Phys. Rev. D* **65**, 084016 (2002).

Neutrino mixing and cosmological constant

M. Blasone[*†], A. Capolupo[*], S. Capozziello[*], S. Carloni[*] and G. Vitiello[*†]

[*]Dipartimento di Fisica "E.R. Caianiello" and INFN, Università di Salerno, I-84100 Salerno, Italy
[†]Unità INFM, Salerno, Italy

Abstract. We report on the recent result according to which the non-perturbative vacuum structure associated with neutrino mixing leads to a non-vanishing contribution to the value of the cosmological constant. Its value is estimated by using the natural cut-off appearing in the quantum field theory formalism for neutrino mixing.

In this paper we show that the vacuum energy induced by neutrino mixing may contribute to the value of the cosmological constant [1].

It is known that the vacuum for neutrinos with definite mass is not invariant under the field mixing transformation, and in the infinite-volume limit it is unitarily inequivalent to the vacuum for neutrinos with definite flavor [2]-[10]. This phenomenon is crucial in order to obtain a non-vanishing contribution to the cosmological constant [1]; it also affects the oscillation formula which turns out to be different from the usual Pontecorvo formula [11]. In the following, for simplicity we restrict ourselves to the two-flavor mixing and we use Dirac neutrino fields.

It was shown in Ref. [6] that, in Quantum Field Theory (QFT), it is possible to construct flavor states for neutrino fields. In the infinite-volume limit, these states are orthogonal to the mass eigenstates, i.e. we have two inequivalent vacua related to each other by the mixing generator $G_\theta(t)$: $|0(t)\rangle_f \equiv G_\theta^{-1}(t)|0\rangle_m$. Here, θ is the mixing angle, t is the time variable, $|0(t)\rangle_f$ and $|0\rangle_m$ are the flavor and mass vacua, respectively. $G_\theta(t)$ is given by

$$G_\theta(t) = \exp\left[\theta \int d^3\mathbf{x} \left(v_1^\dagger(x)v_2(x) - v_2^\dagger(x)v_1(x)\right)\right]. \tag{1}$$

A Bogolubov transformation is involved in connecting the flavor annihilation operators to the mass annihilation operators. We consider in particular the Bogolubov coefficient $V_\mathbf{k}$ which is related to the condensate content of the flavor vacuum [2]:

$$_f\langle 0|\alpha_{\mathbf{k},j}^{r\dagger}\alpha_{\mathbf{k},j}^r|0\rangle_f = {}_f\langle 0|\beta_{\mathbf{k},j}^{r\dagger}\beta_{\mathbf{k},j}^r|0\rangle_f = |V_\mathbf{k}|^2 \sin^2\theta, \qquad j = 1,2, \tag{2}$$

where $\alpha_{\mathbf{k},j}^r$, $\beta_{\mathbf{k},j}^r$ are the annihilation operators for neutrino fields v_1, v_2 with definite masses, m_1, m_2. $|V_\mathbf{k}|^2$ vanishes for $m_1 = m_2$, it has a maximum at $|\mathbf{k}| = \sqrt{m_1 m_2}$. For $|\mathbf{k}| \gg \sqrt{m_1 m_2}$, it goes like $|V_\mathbf{k}|^2 \simeq (m_2 - m_1)^2/(4|\mathbf{k}|^2)$.

We now use this formalism to derive a contribution to the value of the cosmological constant Λ. The link between the vacuum energy density $\langle \rho_{vac}\rangle$ and Λ is provided by the usual relation $\langle \rho_{vac}\rangle = \Lambda/4\pi G$, where G is the gravitational constant.

CP751, General Relativity and Gravitational Physics, 16th SIGRAV Conference, edited by G. Esposito et al.
© 2005 American Institute of Physics 0-7354-0236-1/05/$22.50

To compute Λ we can use the $(0,0)$ component of the energy-momentum tensor in flat space-time T_{00}^{Flat}. Indeed, one can see that the temporal component of the spinorial derivative in the FRW metric is just the standard time derivative [1]: $D_0 = \partial_0$. Thus, $\mathcal{T}_{00} = \mathcal{T}_{00}^{Flat}$. We then obtain

$$\mathcal{T}_{00}(x) = \frac{i}{2} \sum_{\sigma=e,\mu} : \left(\bar{v}_\sigma(x) \gamma_0 \overset{\leftrightarrow}{\partial}_0 v_\sigma(x) \right) := \frac{i}{2} \sum_{j=1,2} : \left(\bar{v}_j(x) \gamma_0 \overset{\leftrightarrow}{\partial}_0 v_j(x) \right) : \qquad (3)$$

where $: ... :$ denotes the customary normal ordering with respect to the mass vacuum in flat space-time. In terms of the annihilation and creation operators of fields v_1 and v_2, the energy-momentum tensor $T_{00} = \int d^3x \, \mathcal{T}_{00}(x)$ is given by

$$T_{00} = \sum_{r,j} \int d^3k \, \omega_{k,j} \left(\alpha_{\mathbf{k},j}^{r\dagger} \alpha_{\mathbf{k},j}^r + \beta_{-\mathbf{k},j}^{r\dagger} \beta_{-\mathbf{k},j}^r \right). \qquad (4)$$

Note that T_{00} is time independent.

The expectation value of T_{00} in the flavor vacuum $|0\rangle_f$ gives the contribution $\langle \rho_{vac}^{mix} \rangle$ of neutrino mixing to the vacuum energy density:

$$_f\langle 0|T_{00}|0\rangle_f = \langle \rho_{vac}^{mix} \rangle \eta_{00} . \qquad (5)$$

Within the QFT formalism for neutrino mixing we have $_f\langle 0|T_{00}|0\rangle_f = {}_f\langle 0(t)|T_{00}|0(t)\rangle_f$ for any t. We then obtain

$$_f\langle 0|T_{00}|0\rangle_f = \sum_{r,j} \int d^3k \, \omega_{k,j} \left({}_f\langle 0|\alpha_{\mathbf{k},j}^{r\dagger} \alpha_{\mathbf{k},j}^r|0\rangle_f + {}_f\langle 0|\beta_{\mathbf{k},j}^{r\dagger} \beta_{\mathbf{k},j}^r|0\rangle_f \right),$$

and

$$_f\langle 0|T_{00}|0\rangle_f = 8\sin^2\theta \int d^3k \, (\omega_{k,1} + \omega_{k,2}) |V_\mathbf{k}|^2 = \langle \rho_{vac}^{mix} \rangle \eta_{00}, \qquad (6)$$

i.e.

$$\langle \rho_{vac}^{mix} \rangle = 32\pi^2 \sin^2\theta \int_0^K dk \, k^2 (\omega_{k,1} + \omega_{k,2}) |V_\mathbf{k}|^2, \qquad (7)$$

where the cut-off K has been introduced. Eq. (7) is our result: it shows that the cosmological constant gets a non-vanishing contribution induced from neutrino mixing only [1]. Note that such a contribution is indeed vanishing in the no-mixing limit ($\theta = 0$ and/or $m_1 = m_2$). Moreover, the contribution is absent in the traditional phenomenological (Pontecorvo) mixing treatment.

We may try to estimate $\langle \rho_{vac}^{mix} \rangle$ by fixing the cut-off. If we choose the cut-off proportional to the natural scale appearing in the mixing phenomenon, i.e. $k_0 \simeq \sqrt{m_1 m_2}$ [2], and if we use $K \sim k_0$, $m_1 = 7 \times 10^{-3} eV$, $m_2 = 5 \times 10^{-2} eV$, and $\sin^2\theta \simeq 0.3$ [12] in Eq. (7), we obtain $\langle \rho_{vac}^{mix} \rangle = 0.43 \times 10^{-47} GeV^4$ and $\Lambda \sim 10^{-56} cm^{-2}$, which is in agreement with the upper bound on Λ [13]. Another possible choice is to use the electro-weak scale cut-off: $K \approx 100 GeV$. We then have $\Lambda \sim 10^{-24} cm^{-2}$, which is, however, beyond the accepted upper bound.

In a recent paper [14], it was suggested that, in the context of hierarchical neutrino models, the cut-off scale can be taken as the sum of the two neutrino masses, $K = m_1 + m_2$, resulting in a contribution of the right order.

In conclusion, the QFT treatment of neutrino mixing leads to a non-vanishing contribution to the cosmological constant [1]. By choosing the cut-off as given by the natural scale of the neutrino mixing phenomenon, we obtain a value of Λ which is consistent with its accepted upper bound. Our result discloses a new possible, non-perturbative mechanism contributing to the cosmological constant value.

Partial financial support by MURST, INFN, INFM and ESF Program COSLAB is acknowledged.

REFERENCES

1. M. Blasone, A. Capolupo, S. Capozziello, S. Carloni and G. Vitiello, Phys. Lett. A **323** (2004) 182.
2. M. Blasone and G. Vitiello, Annals Phys. **244** (1995) 283 [Erratum-ibid. **249** (1995) 363].
3. K. C. Hannabuss and D. C. Latimer, J. Phys. A **33** (2000) 1369; J. Phys. A **36** (2003) L69.
4. M. Binger and C. R. Ji, Phys. Rev. D **60** (1999) 056005; C. R. Ji and Y. Mishchenko, Phys. Rev. D **64** (2001) 076004; Phys. Rev. D **65** (2002) 096015.
5. K. Fujii, C. Habe and T. Yabuki, Phys. Rev. D **59** (1999) 113003 [Erratum-ibid. D **60** (1999) 099903]; Phys. Rev. D **64** (2001) 013011; K. Fujii, C. Habe and M. Blasone, [hep-ph/0212076].
6. M. Blasone and G. Vitiello, Phys. Rev. **D60** (1999) 111302; M. Blasone, P. Jizba and G. Vitiello, Phys. Lett. **B 517** (2001) 471; M. Blasone, P. P. Pacheco and H. W. Tseung, Phys. Rev. D **67** (2003) 073011.
7. M. Blasone, P. A. Henning and G. Vitiello, Phys. Lett. B **451** (1999) 140.
8. M. Blasone, A. Capolupo, O. Romei and G. Vitiello, Phys. Rev. D **63** (2001) 125015; A. Capolupo, C.R. Ji, Y. Mishchenko and G. Vitiello, Phys. Lett. B **594** (2004) 135.
9. M. Blasone, A. Capolupo and G. Vitiello, Phys. Rev. **D66** (2002) 025033.
10. A. Capolupo, Ph.D. thesis, hep-th/0408228 (2004).
11. S. M. Bilenky and B. Pontecorvo, *Phys. Rep.* **41** (1978) 225.
12. G.L. Fogli, E. Lisi, A. Marrone, A. Melchiorri, A. Palazzo, P. Serra, J. Silk, hep-ph/0408045.
13. Ya. B. Zeldovich, I. D. Novikov, *Structure and evolution of the universe*, Moscow, Izdatel'stvo Nauka (1975).
14. G. Barenboim and N. E. Mavromatos, hep-ph/0406035.

Gravitational waves from neutron stars described by modern EOS

O. Benhar, V. Ferrari, L. Gualtieri[†]

Dipartimento di Fisica "G. Marconi", Universitá degli Studi di Roma, "La Sapienza", P.le A. Moro 2, 00185 Roma, Italy
INFN, Sezione Roma 1, P.le A. Moro 2, 00185 Roma, Italy

Abstract. The frequencies and damping times of neutron star (and quark star) oscillations have been computed using the most recent equations of state available in the literature. We find that some of the empirical relations that connect the frequencies and damping times of the modes to the mass and radius of the star, and that were previously derived in the literature, need to be modified.

Asteroseismology, i.e. the study of stellar properties through the analysis of proper oscillation frequencies, is a very useful tool. For instance, it has been succesfully applied to study the internal composition of the Sun.

The oscillation modes of compact stars, like neutron stars (NS) or quark stars, give a clear signature in the spectrum of gravitational waves that these stars may emit in several astrophysical processes. It should be stressed that, according to General Relativity, the modes of compact stars are not normal modes, because they are damped by gravitational wave (GW) emission; for this reason they are called "quasi-normal modes" (QNM), with characteristic frequencies and damping times, that carry information on the structure of the star and on the behavior of nuclear matter in the interior. The study of QNM is called *gravitational wave asteroseismology* [1].

Some years ago, Andersson and Kokkotas computed the frequencies and damping times of the most relevant oscillation modes [1] (that is, the modes that most likely would be excited by a perturbing event) of a non rotating NS for a number of equations of state (EOS) available at that time. They fitted the data with appropriate functions of the radius and the mass of the star, showing how these empirical relations could be used to put constraints on these parameters if the frequency of one or more modes could be identified in a detected gravitational signal. Knowing the mass and the radius, we would gain information on the behavior of matter in a NS core, at density that cannot be reproduced in a laboratory.

In recent years, a number of new EOS have been proposed to describe matter at supranuclear densities, some of them allowing for the formation of a core of strange baryons and/or deconfined quarks. In ref. [2], that we summarize here, we have verified whether, in the light of recent developments, the empirical relations derived in [1] are still appropriate or need to be updated.

We have considered a variety of EOS. For any of them we have obtained the equilibrium configurations for assigned values of the mass, and solved the equations of stellar perturbations by computing the frequencies and damping times of the QNM. Then, we

CP751, *General Relativity and Gravitational Physics, 16th SIGRAV Conference*, edited by G. Esposito et al.
© 2005 American Institute of Physics 0-7354-0236-1/05/$22.50

have fitted our data with suitable functions of M and R to see whether the fits agree with those of [1]. We have extended the results of [1] in two respects: we have considered more recent EOS, and we have studied a larger set of QNM.

A NS is believed to be composed mainly by three different layers of different composition: an outer crust, composed by heavy nuclei and free electrons; an inner crust, consisting of heavy nuclei, free electrons and neutrons; a core, given by leptons, nucleons, and, in some models, also hyperons or quarks. There is an overall agreement on the EOS describing the crust [3, 4], while the composition of the core is poorly known, due to the present limited understanding of hadronic interactions. We have modeled the crust as in [3, 4], and used various models for the matter in the core, which we summarize in Table 1.

TABLE 1.   EOS included in our study

APR2	Akmal, Pandharipande, Ravenhall [5, 6]
APRB200 , APRB120	APR2 [5, 6] + quark inner core [7], [8]
BBS1	Baldo, Burgio, Schultze without hyperons [9]
BBS2	Baldo, Burgio, Schultze with hyperons [9]
G240	Glendenning, mean field approximation [10]

In addition to the above models we have considered the possibility that a star entirely made of quarks (strange star) may form. The models denoted by SS1 and SS2 correspond to a quark star, described by the MIT bag model [7], with or without a crust.

In order to find the frequencies and damping times of the quasi-normal modes, we have solved the equations describing nonradial perturbations of a nonrotating star in general relativity [11, 12]. In [2] we have considered several oscillation modes. Here we only report the results on the fundamental mode (f-mode), which gives the major contribution to GW emission.

The data we derived can be fitted by the following expressions:

$$ v_f = a + b\sqrt{\frac{M}{R^3}}, \quad \tau_f = \frac{R^4}{cM^3}\left[c + d\frac{M}{R}\right]^{-1}, \tag{1} $$

with $a = 0.79 \pm 0.09$ kHz, $b = 33 \pm 2$ km kHz, $c = [8.7 \pm 0.2] \cdot 10^{-2}$ and $d = -0.271 \pm 0.009$. Frequencies are expressed in kHz, masses and radii in km, damping times in s. The data for the f-mode and the fits are shown in Fig. 1. In the left panel we plot v_f versus $\sqrt{M/R^3}$, for all considered stellar models. Our fit (1) is plotted as a thick solid line, and the fit computed by Andersson and Kokkotas in [1], which is based on the EOS considered in that paper, is plotted as a dashed line labelled as 'AK-fit'. In the right panel we plot the damping time τ_f versus the compactness M/R, our fit and the corresponding AK-fit. We can see that our new fit for v_f is sistematically lower than the AK fit by about 100 Hz. Conversely, our fit for the damping time is very similar to that found in [1]. The empirical relations derived above can be used, as described in [1, 13], to determine the mass and the radius of the star from the knowledge of the frequency and damping time of the modes.

By comparing our results with the sensitivity curves of existing gravitational detectors, we have shown in [2] that it is unlikely that the first generation of interferometric antennas will detect the GW emitted by an oscillating neutron star. However, new detectors are under investigation that should be much more sensitive at frequencies above 1-2

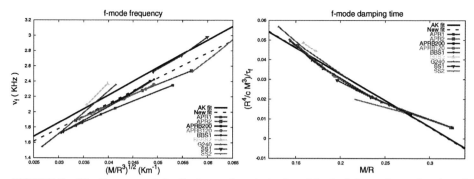

FIGURE 1. The frequency of the fundamental mode is plotted in the left panel as a function of the square root of the average density for the different EOS considered in this paper. The new fit is systematically lower (about 100 Hz) than that previously derived in the literature. The damping time of the fundamental mode is plotted in the right panel as a function of the compactness M/R.

kHz and that would be more appropriate to detect these signals. If the frequencies of the modes will be identified in a detected signal, the simultaneous knowledge of the mass of the emitting star will be crucial to understand its internal composition.

REFERENCES

1. N. Andersson, K.D. Kokkotas, MNRAS **299**, 1059 (1998).
2. O. Benhar, V. Ferrari, L. Gualtieri, astro-ph/0407529, submitted to Phys. Rev. D.
3. G. Baym, C.J. Pethick, P. Sutherland, Ap. J. **170**, 299 (1971).
4. C.J. Pethick, B.G. Ravenhall, C.P. Lorenz, Nucl. Phys. A **584**, 675 (1995).
5. A. Akmal, V. R. Pandharipande, Phys. Rev. C **56**, 2261 (1997).
6. A. Akmal, V. R. Pandharipande, D.G. Ravenhall, Phys. Rev. C **58**, 1804 (1998).
7. A. Chodos, R.L. Jaffe, K. Johnson, C.B. Tohm, V.F. Weisskopf, Phys. Rev. D **9**, 3471 (1974).
8. R. Rubino, thesis, Università "La Sapienza", Roma (unpublished);
   O. Benhar, R. Rubino, astro-ph/0410376.
9. M. Baldo, G.F. Burgio, H.J. Schulze, Phys. Rev. C **61**, 055801 (2000).
10. N.K. Glendenning, *Compact Stars* (Springer, New York, 2000)
11. L. Lindblom, S. Detweiler, Ap. J. Suppl. **53**, 73 (1983).
12. S. Chandrasekhar, V. Ferrari, Proc. R. Soc. Lond. A **434**, 449 (1991).
13. N. Andersson, K.D. Kokkotas, MNRAS **320**, 307 (1999).

Constraining the Equation of State of Neutron Stars with General Relativity

G. Lavagetto[*], L. Burderi[†], F. D'Antona[†], T. Di Salvo[*] and N. R. Robba[*]

[*]Dipartimento di Scienze Fisiche ed Astronomiche, Università di Palermo, via Archirafi n.36,
90123 Palermo, Italy.
[†]Osservatorio Astronomico di Roma, Via Frascati 33, 00040 Monteporzio Catone (Roma), Italy.

Abstract. When a radio pulsar breakes down by virtue of magnetodipole emission, its gravitational mass decreases accordingly. If the pulsar in hosted in a binary system, this mass loss will increase the orbital period of the system. We show that this relativistic effect can be indeed observable if the NS is fast and magnetized enough and that, if observed, it will help to put tight constraints on the equation of state of ultradense matter.

INTRODUCTION

Neutron stars (hereafter NS) are the densest objects we can observe directly: their density can exceed by an order of magnitude the nuclear density and therefore their binding energy is non-negligible [1]. Constraining the equation of state (EOS) of matter at supernuclear densities is of great interest. Anyway, only a few constraints that are sufficiently model-independent have been found from radio and X–ray observations. Moreover, these constraints did not allow to rule out most of the EOS proposed in the literature [1].

Millisecond radio pulsars (MSP) are some of the fastest spinning NS known to date, and their timing behavior is measurable with great precision thanks to their stability. Of particular interest are millisecond pulsars hosted in binary systems: in particular, timing studies of NS-NS binaries allowed us to test general relativity to an unprecedented precision (see for example [8]). Here we present a mechanism that, if observed in binary MSP, could make it possible to put strong constraints on the EOS of NS.

CONSTRAINING THE EOS WITH BINARY RADIO PULSAR TIMING

In a NS, the variation gravitational mass M_G per unit time depends on the variation of both the baryonic mass M_B and the angular momentum J of the star [6], i.e.

$$\dot{M}_G = \Phi \dot{M}_B + \frac{\omega}{c^2} \dot{J}, \tag{1}$$

where ω is the NS spin frequency, and Φ is the energy needed to bring a unit mass from infinity to the surface of the star. Although a pulsar does not lose matter ($\dot{M}_B = 0$, it loses

CP751, *General Relativity and Gravitational Physics, 16th SIGRAV Conference*, edited by G. Esposito et al.
© 2005 American Institute of Physics 0-7354-0236-1/05/$22.50

angular momentum via magnetodipole radiation, and therefore loses gravitational mass. This will have effects on the orbital evolution of the system. We obtain for the evolution of the orbital period P [4]

$$\frac{\dot{P}}{P} = -2\frac{\dot{M}_G}{M_c}\frac{q}{1+q} + \frac{\dot{P}_{GW}}{P}, \tag{2}$$

where $q = M_c/M_G$ and $\dot{P}_{GW}$ is the orbital period derivative due to the emission of gravitational waves, that is in general negative. As the gravitational mass of the NS decreases, the period of the binary system widens, yielding a positive contribution to the orbital period derivative, opposite to the contribution of gravitational waves emission. Since the effect of gravitational mass loss is $\propto P$, while the effect of gravitational waves emission is $\propto P^{-5/3}$ [2], the former effect will be dominant in systems with large enough orbital periods (say $P \geq 6$ h). To a good approximation we can write

$$\dot{M}_G = \frac{I}{c^2}\omega\dot{\omega}, \tag{3}$$

where I is the moment of inertia of the NS. In a binary MSP it is often possible to measure both the spin frequency and its derivative with high precision. The orbital period derivative depends then only on measured quantities $(\omega, \dot{\omega})$, on the masses of the two stars and on the moment of inertia of the neutron star (see equation 2). We can get information on the two masses from the mass function $f(M)$, that is measurable in binary MSP with very good precision. Using it, we can impose constraints on the momentum of inertia of the NS. Since the momentum of inertia depends strongly on the EOS of the NS [3], the detection of these effects will allow us to discriminate between various EOS on a firm observational basis. Suppose, for example, that a system with an orbital period $P = 8$ h, a spin period $P_s = 2$ ms, $\dot{P}_s = 3 \times 10^{-19}$ and a mass function of $0.005 M_\odot$ is observed, and that the orbital period derivative has been measured to be $+2.5 \times 10^{-14}$. In figure 1, we plot the values of the masses of the two stars that are compatible with the value of the orbital period derivative we obtained, under the hypothesis that the NS is governed by EOS A (the pure neutron EOS by Pandharipande [5]) or by EOS BBB (a realistic EOS by Baldo, Bombaci and Burgio, [7]) respectively. The mass function requires that the values of the two masses should be above the dashed line in figure 1. Therefore, this particular system would allow us to rule out EOS A on an almost model-independent basis. This could be a potentially very powerful method for constraining the EOS of NS with very precise measurements and with an almost model-independent method.

Which is, between the known millisecond binary radio pulsars, the best candidate for detecting such an effect? The most promising object we found is PSR J0218+4232, which consists of a NS spinning at 2.3 ms in orbit with a white dwarf companion, with an orbital period of 2 days. The mass ratio is measured to be $0.13 \pm 0,04$ [9]. According to equation (2), we find that the orbital period derivative of the system is $\dot{P}_{orb} = 2.5 \times 10^{-14} I_{45}$, where I_{45} is the moment of inertia of the NS in units of 10^{45} g cm^2. The variation in the orbital period can be derived from the measurement of the periastron time delay. The effect of an orbital period derivative on the periastron arrival

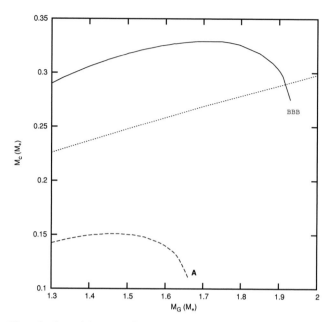

FIGURE 1. Allowed values of the mass of the primary versus the mass of the companion (both in solar masses) for the system described in the text. The two lines are for NS with EOS A (dashed line) and NS with EOS BBB (solid line). The dotted line indicates the lower limit on the companion mass obtained if the mass function is $5 \times 10^{-3} M_\odot$.

time is given by

$$\Delta T_{\text{per}} \simeq 0.5 \frac{\dot{P}}{P} \Delta T_{\text{obs}}^2. \tag{4}$$

We find that we will need $118/I_{45}^{1/2}$ years of observation to detect a delay of 1 second. However, if a pulsar with a higher spin-down energy will be discovered, this effect might become observable in a much shorter observation time.

REFERENCES

1. N. Stergioulas, *Living Reviews in Relativity* **1**, 8 (1998).
2. M. Burgay, N. D'Amico, A. Possenti, R. N. Manchester, A. G. Lyne, B. C. Joshi, M. A. McLaughlin, M. Kramer, J. M. Sarkissian, F. Camilo, V. Kalogera, C. Kim, and D. R. Lorimer, *Nature* **426**, 531 (2003).
3. J. M. Bardeen, *ApJ* **162**, 71 (1970).
4. L. W. Esposito and E. R. Harrison, *ApJ* **196**, L1 (1975).
5. D. R. Lorimer, *Living Reviews in Relativity* **4**, 5 (2001).
6. G. B. Cook, S. L. Shapiro and S. A. Teukolsky, *ApJ* **424**, 823 (1994).
7. W. D. Arnett and R. L. Bowers, *ApJS* **33**, 415 (1977).
8. M. Baldo, I. Bombaci and G. F. Burgio, *Astron. Astrophys.* **328**, 274 (1997).
9. C. G. Bassa, M. H. van Kerkwijk, and S. R. Kulkarni, *Astron. Astrophys.* **403**, 1067 (2003).

The Einstein-Vlasov system with a scalar field

Hayoung Lee

Max-Planck-Institut für Gravitationsphysik, Am Mühlenberg 1, Golm, 14476, Germany

Abstract. We study the Einstein–Vlasov system coupled to a nonlinear scalar field with a nonnegative potential in locally spatially homogeneous spacetime, as an expanding cosmological model. It is shown that solutions of this system exist globally in time. When the potential of the scalar field is of an exponential form, the cosmological model corresponds to accelerated expansion. The system with an exponential potential shows the causal geodesic completeness of the spacetime towards the future. The asymptotic behavior of solutions of this system in the future time is analysed in various aspects, which shows power-law expansion.

We consider the dynamics of expanding cosmological models, particularly accelerated expansion. A simple way to obtain accelerated expansion is to introduce a positive cosmological constant, which leads to exponential expansion. In homogeneous spacetimes, the detailed asymptotics of solutions of the Vlasov system (collisionless matter) have been analysed in [1]. Another choice for accelerated expansion is a nonlinear scalar field. In the case of an exponential potential, the models show power-law expansion.

Let G be a simply connected three-dimensional Lie group and $\{e_i\}$, a left invariant frame and $\{e^i\}$, the dual coframe. Consider the spacetime as a manifold $G \times I$, where I is an open interval and the spacetime metric has the form $ds^2 = -dt^2 + g_{ij}(t)e^i \otimes e^j$. The constraint equations are

$$R - (k_{ij}k^{ij}) + (k_{ij}g^{ij})^2 = 16\pi T_{00} + 8\pi \psi^2 + 16\pi V(\phi), \qquad \nabla^i k_{ij} = -8\pi T_{0j}. \quad (1)$$

The evolution equations are

$$\partial_t \phi = \psi, \qquad \partial_t \psi = (k_{lm}g^{lm})\psi - V'(\phi), \qquad \partial_t g_{ij} = -2k_{ij}, \quad (2)$$

$$\partial_t k_{ij} = R_{ij} + (k_{lm}g^{lm})k_{ij} - 2k_{il}k^l_j - 4\pi(2T_{ij} + T_{00}g_{ij} - (T_{lm}g^{lm})g_{ij} + 2V(\phi)g_{ij}). \quad (3)$$

These equations are written using frame components. Here k_{ij} is the second fundamental form, R is the Ricci scalar curvature and R_{ij} is the Ricci tensor of the three-dimensional metric. Moreover, ϕ is a scalar field, depending only on t, with a non-negative potential $V(\phi)$, while ψ is a function introduced by the first evolution equation above. The non-negative assumption on the potential is very natural. That is, the dominant energy condition is satisfied. The components of the energy-momentum tensor of the Vlasov matter are (here $v := (v^1, v^2, v^3)$ and $dv := dv^1\,dv^2\,dv^3$)

$$T_{00}(t) = \int f(t,v)(1 + g_{rs}v^r v^s)^{1/2}(\det g)^{1/2}\,dv,$$

$$T_{0i}(t) = \int f(t,v)v_i(\det g)^{1/2}\,dv, \qquad T_{ij}(t) = \int f(t,v)v_i v_j(1 + g_{rs}v^r v^s)^{-1/2}(\det g)^{1/2}\,dv.$$

CP751, *General Relativity and Gravitational Physics, 16th SIGRAV Conference*, edited by G. Esposito et al.
© 2005 American Institute of Physics 0-7354-0236-1/05/$22.50

The Vlasov equation reads as

$$\partial_t f + \left\{ 2k^i_j v^j - (1 + g_{rs} v^r v^s)^{-1/2} \gamma^i_{mn} v^m v^n \right\} \partial_{v^i} f = 0, \tag{4}$$

where γ^i_{mn} is the Ricci rotation coefficient which is obtained from the metric and the structure constants of the Lie algebra of G. In the discussion of expanding cosmology, the models being considered in this paper are all Bianchi types except IX.

Now we present results on this system. We refer to [2] for details and proofs.

Theorem 1 *Let $g_{ij}(t_0)$, $k_{ij}(t_0)$, $\phi(t_0)$, $\psi(t_0)$ and $f(t_0, v)$ be an initial data set for the evolution equations (2) – (3) and the Vlasov equation (4), which has Bianchi symmetry and satisfies the constraints (1). Let $f(t_0, v)$ be non-negative C^1 with compact support. Assume that the potential $V(\phi)$ is a non-negative C^2 function. Then there exists a unique C^1 solution $(g_{ij}, k_{ij}, \phi, \psi, f)$ of the Einstein–Vlasov system for all time.*

We study the asymptotics of solutions in the future time with a particular form of the potential, given by

$$V(\phi) = V_0 e^{-\lambda \kappa \phi}, \tag{5}$$

where V_0 is a positive constant, $0 < \lambda < \sqrt{2}$ and $\kappa^2 = 8\pi$. In has been shown in [3] that, in order for power-law inflation to occur, λ must be smaller than $\sqrt{2}$. Note that the case $\lambda = 0$ corresponds to the model with a positive cosmological constant which exhibits exponential expansion. This has been well understood in [1].

Theorem 2 *Suppose the hypotheses of Theorem 1 hold. Assume that the potential $V(\phi)$ has the form (5). Then the spacetime is future complete.*

We now observe asymptotic behaviors of solutions in various aspects.

Proposition 1

$$k_{ij} g^{ij} = -\frac{6}{\lambda^2} t^{-1} + \begin{cases} \mathcal{O}(t^{-2} \ln t), & \text{if } \lambda^2 \kappa \gamma / 2 = \eta \\ \mathcal{O}(t^{-(\zeta + 1)}), & \text{if } \lambda^2 \kappa \gamma / 2 > \eta \\ \mathcal{O}(t^{-2}), & \text{if } \lambda^2 \kappa \gamma / 2 < \eta \end{cases}$$

$$\phi = \frac{2}{\lambda \kappa} \ln t + C + \begin{cases} \mathcal{O}(t^{-1} \ln t), & \text{if } \lambda^2 \kappa \gamma / 2 = \eta \\ \mathcal{O}(t^{-\zeta}), & \text{if } \lambda^2 \kappa \gamma / 2 > \eta \\ \mathcal{O}(t^{-1}), & \text{if } \lambda^2 \kappa \gamma / 2 < \eta \end{cases}$$

$$g_{ij}(t) = t^{4/\lambda^2} \left(\left(\frac{\lambda^2 \kappa \gamma}{2 e^{\lambda \kappa C_1 / 2}} \right)^{4/\lambda^2} \mathcal{G}_{ij} + \begin{cases} \mathcal{O}(t^{-1} \ln t), & \text{if } \lambda^2 \kappa \gamma / 2 = \eta \\ \mathcal{O}(t^{-\zeta}), & \text{if } \lambda^2 \kappa \gamma / 2 > \eta \\ \mathcal{O}(t^{-1}), & \text{if } \lambda^2 \kappa \gamma / 2 < \eta \end{cases} \right)$$

$$g^{ij}(t) = t^{-4/\lambda^2} \left(\left(\frac{2 e^{\lambda \kappa C_1 / 2}}{\lambda^2 \kappa \gamma} \right)^{4/\lambda^2} \mathcal{G}^{ij} + \begin{cases} \mathcal{O}(t^{-1} \ln t), & \text{if } \lambda^2 \kappa \gamma / 2 = \eta \\ \mathcal{O}(t^{-\zeta}), & \text{if } \lambda^2 \kappa \gamma / 2 > \eta \\ \mathcal{O}(t^{-1}), & \text{if } \lambda^2 \kappa \gamma / 2 < \eta \end{cases} \right)$$

where C and C_1 are positive constants and $\gamma := \sqrt{2V_0}/\sqrt{6 - \lambda^2}$, $\zeta := 2\eta/\lambda^2 \kappa \gamma$ and $\eta := \min\{\frac{1}{2} \kappa \gamma (6 - \lambda^2), \varepsilon^*/2\}$. (Note that ε^* depends only on the initial condition of the spacetimes. We refer to [2] for details.) Here $\mathcal{G}_{ij}$ and $\mathcal{G}^{ij}$ are independent of t.

Let λ_i be the eigenvalues of k_{ij} with respect to g_{ij}, that is, the solutions of $\det(k^i_j - \lambda \delta^i_j) = 0$. Define *the generalized Kasner exponents* by

$$p_i := \frac{\lambda_i}{\sum_l \lambda_l} = \frac{\lambda_i}{(k_{lm}g^{lm})}.$$

Note that, in the special case of the Kasner solutions, these are the Kasner exponents. The following proposition implies that spacetime becomes isotropic at late times.

Proposition 2

$$p_i(t) = \frac{1}{3} + \mathcal{O}(t^{-\xi/2}),$$

where $\xi := 2\varepsilon^*/\lambda^2\kappa\gamma$.

Proposition 3 *Let q be the deceleration parameter.*

$$q = -1 - \frac{\lambda^2}{6} + \begin{cases} \mathcal{O}(t^{-1}\ln t), & \text{if } \lambda^2\kappa\gamma/2 = \eta \\ \mathcal{O}(t^{-\xi}), & \text{if } \lambda^2\kappa\gamma/2 > \eta \\ \mathcal{O}(t^{-1}), & \text{if } \lambda^2\kappa\gamma/2 < \eta. \end{cases}$$

Note that, in accelerated expanding models, this deceleration parameter is negative.

Proposition 4 *Let $\{\hat{e}_i\}$ be an orthonormal frame. The energy-momentum tensor is described by (here $\hat{v} := (\hat{v}_1, \hat{v}_2, \hat{v}_3)$ and $d\hat{v} = d\hat{v}_1 d\hat{v}_2 d\hat{v}_3$)*

$$\rho(t) = \int f(t,\hat{v})(1+|\hat{v}|^2)^{1/2} d\hat{v},$$

$$J_i(t) = \int f(t,\hat{v})\hat{v}_i d\hat{v}, \qquad S_{ij}(t) = \int f(t,\hat{v})\hat{v}_i\hat{v}_j(1+|\hat{v}|^2)^{-1/2} d\hat{v},$$

where $\rho := \hat{T}_{00}$ is the energy density, $J_i := \hat{T}_{0i}$ the components of the current density and $S_{ij} := \hat{T}_{ij}$ the spatial components of the energy-momentum tensor. Here the hats indicate that objects are written in the orthonormal frame.
Then $\rho(t)$, $J_i(t)$ and $S_{ij}(t)$ tend to zero as t goes to infinity. More precisely,

$$\rho(t) = \mathcal{O}(t^{-6/\lambda^2}), \qquad J_i(t) = \mathcal{O}(t^{-8/\lambda^2}), \qquad S_{ij}(t) = \mathcal{O}(t^{-10/\lambda^2}).$$

Furthermore,

$$\frac{J_i(t)}{\rho(t)} = \mathcal{O}(t^{-2/\lambda^2}), \qquad \frac{S_{ij}(t)}{\rho(t)} = \mathcal{O}(t^{-4/\lambda^2}). \tag{6}$$

By virtue of this proposition, solutions of the Einstein–Vlasov system coupled to a nonlinear scalar field with an exponential potential are approximated by vacuum Einstein solutions. At a more detailed level, the statement in (6) resembles the non-tilted dust-like solutions in which $J_i(t)$ and $S_{ij}(t)$ are identically zero.

REFERENCES

1. H. Lee, (gr-qc/0308035, AEI-2003-066), *Math. Proc. Camb. Phil. Soc.* **137**, 495–509 (2004).
2. H. Lee, (gr-qc/0404007, AEI-2004-030), *Ann. H. Poincare*, to appear.
3. J. J. Halliwell, *Phys. Lett.* **185B**, 341–344 (1987).

Gravitational wave emission from matter accretion onto a Schwarzschild black hole

Alessandro Nagar

Departament d'Astronomia i Astrofísica, Universitat de València,
Dr. Moliner 50, 46100 Burjassot (Valencia), Spain

Abstract. We discuss recent results concerning the gravitational wave (GW) emission from matter accretion onto a Schwarzschild black hole. A numerical "hybrid" procedure is adopted in which the dynamics of the accreting matter is studied by solving, in axisymmetry ($m = 0$), the non-linear general relativistic equations of fluid dynamics using a code based on high-resolution-shock-capturing (HRSC) methods; the GW signal is correspondingly obtained as a solution of the inhomogeneous odd-and even-parity equations of linearized metric perturbations of Schwarzschild spacetime (i.e., Regge–Wheeler and Zerilli–Moncrief equations). Self-gravity of the accreting matter and radiation-reaction effects are neglected. Different configurations are considered: *i)* quadrupolar dust shells; *ii)* runaway unstable thick accretion disks and *iii)* oscillations of (stable) thick disks.

INTRODUCTION

This contribution deals with ingoing work done in collaboration with O. Zanotti, J.A. Font and J.A. Pons regarding the gravitational radiation that may be generated in the last phases of gravitational collapse, once that a system formed by a black hole surrounded by accreting matter has emerged. A thorough discussion on the analytic and numerical framework as well as the main results have been published in [1]; the current results will be presented in complete form in [2]. The aim of this work is to exploit the use of hybrid methods (i.e., full non-linear hydrodynamics evolution coupled to linear perturbations of the spacetime) to study gravitational waveforms generated in accretion scenarios onto a Schwarzschild black hole. Such an approach was performed in [3] and recently revisited in [1] (see references therein for alternative approaches based on the frequency domain [4]). Our method is based on the solution of the odd- and even-parity gauge-invariant master equations for non-spherical perturbations of Schwarzschild spacetime (also known as Regge–Wheeler [5] and Zerilli–Moncrief [6, 7] equations), with general source terms to account for the dynamics of the accreting fluid (the reader is addressed to Ref. [8] for a review). It is well known that, on a Schwarzschild spacetime, the linearized Einstein equations for the three odd-parity and the seven even-parity degrees of freedom reduce to a couple of wave-like equations with potential for a couple of gauge-invariant quantities built up as suitable combinations of the metric multipoles; the odd is referred to as $\Phi^{\ell m}$ and the even as $Z^{\ell m}$. For a

CP751, *General Relativity and Gravitational Physics, 16th SIGRAV Conference*, edited by G. Esposito et al.
© 2005 American Institute of Physics 0-7354-0236-1/05/$22.50

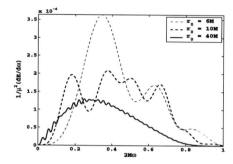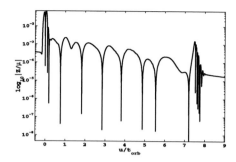

FIGURE 1. Accreting models: on the left, the energy spectra produced by quadrupolar dust shells of mass μ of one hundredth of that of the black hole, plunging into it from different distances r_0. The modulations result from interference among GW. On the right, typical signal from a runaway unstable thick accretion disk plunging into a $M = 2.5 M_\odot$ hole. The orbital period of the center (i.e. maximum density point) is $t_{orb} = 1.98$ ms. The early time waveform comes from initial data choice; then, we have fluid oscillations, followed by a small burst and the quasi-normal-mode ringing (followed by the late-time tail phase), occurring when the object accretes onto the hole.

chosen convention, the GW amplitude can be written as [8]

$$h_+ - ih_\times = \frac{1}{r}\sqrt{\frac{(\ell-2)!}{(\ell+2)!}} \sum_{\ell=2}^{\infty} \sum_{m=-\ell}^{\ell} \left\{ \frac{1}{2} Z^{\ell m} + \frac{i}{(\ell-1)(\ell+2)} \Phi^{\ell m} \right\} {}_{-2}Y^{\ell m}(\theta,\phi) , \quad (1)$$

where ${}_{-2}Y^{\ell m}(\theta,\phi)$ are the spin-weighted spherical harmonics. In our numerical code, for a given distribution of matter surrounding the hole, we update in time the hydrodynamical variables by using a state-of-the art HRSC scheme based upon approximate Riemann solvers [1, 9]; from the hydrodynamic quantities the source terms of odd-and even-parity master equations are computed, and the equations themselves are solved by using standard second-order centered methods.

RESULTS

We show first results for unstable models. Fig. 1 on the left exhibits the energy spectra due to the plunging of quadrupolar shells of dust into the hole. As the shells are accreting from a finite distance r_0, the energy spectra show typical interference modulations. The spacing between two subsequent bumps decreases as r_0 increases, to disappear for $r_0 \to \infty$ [4]. This qualitatively matches the findings of [10] in which the radial plunging of point particles was analyzed. The (normalized) emitted energies are about $2M/\mu^2 \approx 10^{-4}$ or smaller, μ and M being the mass of the shell and of the black hole, respectively. Correspondingly, Fig. 1 on the right shows a typical waveform when the accreting model is represented by a runaway unstable thick accretion disk with constant specific angular momentum (see [11] for simulations of the runaway instability of thick disks). Fluid oscillations are present before the accretion of the total bulk of matter occurs after about 7 orbital periods. As a matter of fact, in [12] it was argued that

222

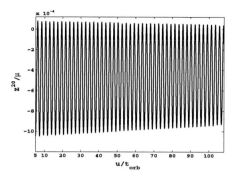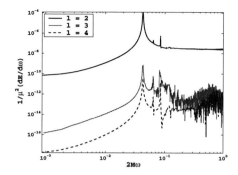

FIGURE 2. Oscillations of a stable thick disk with $t_{orb} = 1.66$ ms and $\mu = 0.02M$, where $M = 2.5M_\odot$. On the left, the $\ell = 2$ waveform; on the right we show the energy spectra for $\ell = 2$ and 4 (even multipoles) and $\ell = 3$ (odd multipole). Logarithmic scale is chosen for both axes. Just the oscillation frequencies of the disk (fundamental mode and overtones) are noticeable in the energy spectra.

oscillations of these objects are interesting sources of GW; in that case, the gravitational waveforms were computed by using a suitable expression of the quadrupole formula. We have recently revisited this scenario, but from the point of view of metric perturbations. Figure 2 on the left shows the $\ell = 2$ gravitational h_+ waveform (in geometrical units) for an oscillationg object, extracted at $r_{obs} = 400M$ and plotted versus retarded time $u = t - r^*_{obs}$. On the right, we compare the energy spectra for $\ell = 2$, 3 and $\ell = 4$ multipoles: by virtue of the very small amount of matter plunging into the hole, no evidence of spacetime quasi-normal modes excitations seems to be present, the whole GW signal being driven by fluid motions. Our results are consistent with those of [12]. A detailed comparison with the signals extracted by using the quadrupole formula will be presented elsewhere [2].

We thank L. Rezzolla for useful comments and suggestions throughout the development of this work. The stay at the University of València has been supported by an Angelo della Riccia fellowship. Numerical computations were carried out on the INFN Albert100 cluster for Numerical Relativity at the University of Parma.

REFERENCES

1. A. Nagar, G. Diaz, J.A. Pons, and J.A. Font, Phys. Rev. D **69**, 124028 (2004).
2. A. Nagar, J.A. Font, J.A. Pons, and O. Zanotti, in preparation.
3. P. Papadopoulos and J.A. Font, Phys. Rev. D **59**, 044014 (1999).
4. S.L. Shapiro and I. Wasserman, Astrophys. J., **260**, 838 (1982).
5. T. Regge and J.A. Wheeler, Phys. Rev. **108**, 1063 (1957).
6. F.J. Zerilli, Phys. Rev. D **2**, 2141 (1970).
7. V. Moncrief, Ann. Phys. (N.Y.) **88**, 323 (1974).
8. A. Nagar and L. Rezzolla, submitted to Classical and Quantum Gravity.
9. F. Banyuls, J.A. Font, J.M. Ibáñez, J.M. Martí, and J.A. Miralles, Astrophys. J. **476**, 221 (1997).
10. C.O. Lousto and R.H. Price, Phys. Rev. D **55**, 2124 (1997).
11. F. Daigne and J.A. Font, Mon. Not. R. Astron. Soc. **349**, 841 (2004).
12. O. Zanotti, L. Rezzolla, and J.A. Font, Mon. Not. R. Astron. Soc. **341**, 832 (2003).

The Cosmological Black Holes Hypothesis

Cosimo Stornaiolo

Istituto Nazionale di Fisica Nucleare, Sezione di Napoli, Complesso Universitario di Monte S. Angelo, Edificio N', via Cinthia, 45 – 80126 Napoli

Abstract. In this paper we introduce the Cosmological Black Holes hypothesis in order to explain the dark matter distribution problem and the generation of cosmological voids.

Introduction – This work was developed in order solve two important cosmological problems. The existence of large cosmological voids among the galaxy distribution and the discrepancies between the measurements of Ω_m.

The existence of voids has been evident after the discovery by Kirshner et al. [1] of a 60 Mpc large void in the Böotes constellation. Later there was a *visual* evidence of the existence of many empty regions and it was proved that voids occupy a considerable part of the volume of the universe by analyzing their *physical* extension, shape and the relative underdensity of galaxies with the code Voidfinder described in [2]. Recent analysis in the 2dF Galaxy Redshift Survey [3] estimated a total volume of about 40% and an average radius of 12.4 ± 1.9 Mpc h^{-1}, with underdensities of the internal galaxy distribution of the order of -0.95.

Recent observations indicated that in the most reliable model of universe $\Omega_{tot} = 1$ where the fraction of dark and baryonic matter is $\Omega_m \sim 1/3$ [4]. But direct measurements of Ω_m, performed with different methods, lead to a wide range of values. Bahcall and coworkers obtained $\Omega_m(M/L) = 0.16 \pm 0.05$ by extrapolating the mass-to-light ratio function of the universe [5] and recently, they improved their analysis to obtain $\Omega_m(M/L) = 0.17 \pm 0.05$ [6]. Estimates of the baryon fraction in clusters yield $\Omega_m \leq 0.3 \pm 0.05$, the evolution of cluster abundance gives $\Omega_m \simeq 0.25$ [7]. Finally Turner, comparing the measurements of the physical properties of clusters, CMB anisotropy and the power spectrum of mass inhomogeneity [8], infers that $\Omega_m = 0.33 \pm 0.035$ (see also the discussion in [9]).

In this paper we show that these two problems can be explained by identifying the voids as the "holes" of a Swiss Cheese cosmological model.

The Swiss Cheese model was introduced by Einstein and Straus [10]. It is obtained by cutting out spherical regions from a Friedmann–Lemaître–Robertson–Walker (FLRW) universe, with vanishing pressure. These regions are substituted by other ones with a spherically symmetric metric such as the Schwarzschild or the L-T solutions.

The CBH hypothesis – It is important to note that in determining the value of Ω_m with the mass-to-light ratio function, it was assumed that voids do not contribute additional dark matter [11].

But as observed by Peebles in [12], the low dispersion of the velocities of galaxies indicates, when $\Omega_m = 1$, that most of the matter must be inside the voids. Moreover, the

CP751, *General Relativity and Gravitational Physics, 16th SIGRAV Conference*, edited by G. Esposito et al.

author suggests that this is true even in the case of smaller Ω_m. This last observation motivated [13][14] us to consider an Einstein–Straus universe characterized by the following properties:

1. we shall consider, for sake of simplicity, a homogeneous and isotropic spatially flat universe dominated by matter and a cosmological constant;
2. in all voids there is a central spherical black hole with mass $M = 4\pi/3\,\Omega_{cbh}\rho_c R_v^3$, where the parameter $\Omega_{cbh} = \rho_{cbh}/\rho_c$ represents the contribution of these black holes to Ω_{tot}, and ρ_c is the critical density of the universe;
3. all voids are spherical.

If we consider voids with radius ranging from 10 Mpc to 30 Mpc, according to points 1)-3) such voids have a central black hole with masses varying from $1.15 \times 10^{15} M_\odot\, h\, \Omega_{cbh}^{-1}$ to $3.105 \times 10^{16} M_\odot\, h\, \Omega_{cbh}^{-1}$, and Schwarzschild radii ranging in between $0.1\, h\, \Omega_{cbh}^{-1}$ kpc and $2.7 \times 10\, h\, \Omega_{cbh}^{-1}$ kpc.

Therefore, the central black holes formed at densities in between $5.8 \times 10^{-14}h^{-2}\,\Omega_{cbh}^2\, g/cm^3$ and $8.0 \times 10^{-17}h^{-2}\,\Omega_{cbh}^2\, g/cm^3$. The orders of magnitude of these densities suggest that these black holes formed from a smooth collapse of very large perturbations, the voids being the result of the cosmic expansion of the rest of matter. To explain the existence of these large perturbations it is enough to suppose an inflationary epoch that lasted for a relatively long time.

By virtue of their cosmological relevance, we shall call these black holes Cosmological Black Holes (CBH).

Stability of voids – The model proposed does not need to represent the *real* voids. The matching conditions, defined by Einstein and Straus for a universe with pressure $P = 0$ can be extended to a universe with $P \neq 0$. The time evolution of the comoving radius, denoted by R, is given by the equation (see Lake [15])

$$\frac{\dot{R}_{void}^2 a^2}{1 - kR_{void}^2} = \frac{\varepsilon^2 - j}{1 - j}, \tag{1}$$

where $\varepsilon = P/\rho$. We can identify the last term with the kinetic energy of galaxies. The cosmological observations lead to the average value [16] $\langle v \rangle \sim 200 km/sec \sim 10^{-3}$ in units $c = 1$, which corresponds to $\varepsilon = 10^{-6}$. We also fix $j = 0$ as the voids are supposed to be empty along their internal border. On solving equation (1) we find that $\Delta V/V = 3\Delta R_{void}/R_{void} \leq 6 \times 10^{-4}$. Then voids are very stable structures and our evaluation of the mass of CBH is correct.

Testing the CBH hypothesis– As the voids should not have other matter than that of CBH, the best way to verify the existence of such black holes is to observe their lensing properties.

By applying the equations for a Schwarzschild lens, the Einstein angle for a void with radius R_v, expressed in radians, is $\alpha_0 = 4.727 \times 10^{-4}\Omega_{CBH}^{1/2}R_v\sqrt{R_v D_{ds}/D_d D_s}$ where D_s is the distance of the source from the observer, D_{ds} is the distance of the source from the CBH, and D_d is the distance of the CBH from the observer. All these quantities are expressed in Mpc. The corresponding characteristic length is $\xi_0 = \alpha_0 D_d$.

Conclusions – The CBH hypothesis so far proposed is able to explain some important properties of the universe.

Despite our original assumption, voids appear to be not totally empty. According to the analysis in [3] there is a an underdense distribution of galaxies. Mostly, these galaxies reside close to the edges of voids.

But, since the definition of void is model dependent, we consider here two possibilities to develop our model.

One is to perturb our model in order to explain dynamically the presence of galaxies inside the voids. We can assume, for instance, that CBH have a (small) angular momentum. In this case the Einstein–Straus model would be correct only to a first approximation. The result is that the Birkhoff theorem (see [13] and [14]) does not hold and the CBH would interact with the galaxies through gravitomagnectic effects.

The other one is instead to adapt the void analysis with the Voidfinder code to the CBH hypothesis.

Our next important task is to compare the results obtained from measurements of Ω_m with the void analysis. This should lead to a modified mass-luminosity relation to be applied to the whole universe.

The last word about the reliability of our model will follow from the cosmological observations. A deeper study of the void lensing properties applied to the galaxy background would yield us the necessary predictions for observation.

REFERENCES

1. R.P. Kirshner, A. Oemler, P.L. Schechter and S.A. Shectman, Astrophys. J. **248**, L57 (1981).
2. H. El-Ad and T. Piran, Astrophys. J. **491**, 421 (1997).
3. F. Hoyle and M. S. Vogeley, Astrophys. J. **607**, 751 (2004).
4. S. Perlmutter, M. Turner and M. White, Phys. Rev. Lett. **83**, 670 (1999).
5. N.A. Bahcall, R. Cen, R. Davé, J.P. Ostriker and Q. Yu, Astrophys. J. **541**, 1 (2000).
6. N.A. Bahcall and J.M. Comerford, Astrophys. J. **565**, L5 (2002).
7. N.A. Bahcall "Constructing the Universe with Clusters of Galaxies" IAP meeting, Paris (France) July 2000, Florence Durret & Daniel GErbal eds.
8. M. S. Turner, astro-ph/0106035.
9. S. Khalil and C. Munoz, "The enigma of the dark matter," hep-ph/0110122.
10. A. Einstein and E.G. Straus, Rev. Mod. Phys. **17**, 120 (1945) and Rev. Mod. Phys. **18**, 148 (1946).
11. N.A. Bahcall, L.M. Lubin and V. Dorman, Astrophys. J. **447**, L81 (1995).
12. P.J.E. Peebles, Astrophys. J. **557**, 495 (2001).
13. C. Stornaiolo, Gen. Rel. Grav. **34**, 2089 (2002).
14. S. Capozziello, M. Funaro and C. Stornaiolo, Astron. Astrophys. **420**, 847 (2004).
15. K. Lake, Astrophysical Journal **240**, 744 (1980).
16. C.W. Misner, K.S. Thorne, and J.A. Wheeler, 1973, Gravitation (San Francisco: W.H. Freeman).

Retro gravitational lensing for Sgr A^* with Radiastron

A.F. Zakharov* and A.A. Nucita, F. De Paolis, G. Ingrosso[†]

*Institute of Theoretical and Experimental Physics, Moscow &
Astro Space Center of Lebedev Physics Institute, Moscow, Russia
[†]Dipartimento di Fisica, Università di Lecce and INFN, Sezione di Lecce, Italy

Abstract. Recently Holz & Wheeler (2002) have considered a very attractive possibility to detect retro-MACHOs, i.e. retro-images of the Sun by a Schwarzschild black hole. We analyze the case of a Kerr black hole with an arbitrary spin for some selected positions of a distant observer with respect to the equatorial plane of a Kerr black hole. We discuss glories (mirages) formed near rapidly rotating Kerr black hole horizons and propose a procedure to measure masses and rotation parameters by analyzing these forms of mirages. In some sense, that is a manifestation of gravitational lens effect in the strong gravitational field near the black hole horizon and a generalization of the retro-gravitational lens phenomenon. We also propose to use future radio interferometer Radioastron facilities to measure shapes of mirages (glories) and to evaluate the black hole spin as a function of the position angle of a distant observer.

Recently Holz & Wheeler have suggested that a Schwarzschild black hole may form retro-images (called retro-MACHOs) if it is illuminated by the Sun [1] (astronomical applications of this idea were discussed by [2] and its generalizations for Kerr black hole are considered by [3]). We analyze a rapidly rotating Kerr black hole case for some selected positions of a distant observer with respect to the equatorial plane of the Kerr black hole. We discuss glory (mirage) formed near a rapidly rotating Kerr black hole horizon and propose a procedure to measure the mass and the rotation parameter by analyzing the mirage shapes. Since a source illuminating the black hole surroundings may be located in an arbitrary direction with respect to the observer line of sight, a generalization of the retro-gravitational lens phenomenon is needed [4]. Based on an qualitative analysis of geodesics near Kerr black holes [5, 6] and numerical analysis of photon geodesics [7, 8, 9, 10, 11, 12, 13, 14, 15, 16, 17] we considered shapes of mirages black holes and a possibility to evaluate parameters of nearby massive black holes, for example for Sgr A^* using Radioastron facilities [4] (mirage forms discuss earlier [18, 19] but a complete classification of the shapes could be given only by using a photon trajectory classification [5]). Here we summarize results of mirage shape analysis given in [4].

Let us remind parameters of the Radioastron instrument. The space radio telescope Radioastron will be launched in 2006. This project was initiated by Astro Space Center (ASC) of Lebedev Physical Institute of Russian Academy of Sciences (RAS) in collaboration with other institutions of RAS and RosAviaKosmos. This space based 10-meter radio telescope will be used for space – ground VLBI measurements. The measurements will have extraordinary high angular resolutions, i.e. about 1 – 10 microarcseconds (in

CP751, *General Relativity and Gravitational Physics, 16th SIGRAV Conference*, edited by G. Esposito et al.

TABLE 1. The fringe sizes (in micro arc seconds) for the standard and advanced apogees B_{max} ($3.5 * 10^5$ and $3.2 * 10^6$ km correspondingly).

$B_{max}(km) \backslash \lambda (cm)$	92	18	6.2	1.35
3.5×10^5	540	106	37	8
3.2×10^6	59	12	4	0.9

particular about 8 microarcseconds at the shortest wavelength 1.35 cm and a standard orbit and could be about 0.9 microarcseconds for the high orbit at the same wavelength. For observations, four wave bands will be used corresponding to $\lambda = 1.35$ cm, $\lambda = 6.2$ cm, $\lambda = 18$ cm, $\lambda = 92$ cm. An orbit for the satellite was chosen with high apogee and with period of satellite rotation around the Earth 9.5 days, which evolves as a result of weak gravitational perturbations from the Moon and the Sun. The perigee is in a band from 10 to 70 thousand kilometers, the apogee is a band from 310 to 390 thousand kilometers. The basic orbit parameters will be the following: the orbital period is p = 9.5 days, the semi-major axis is a = 189 000 km, the eccentricity is e = 0.853, the perigee is H = 29 000 km. After several years of observations, it would be possible to move the spacecraft to a much higher orbit (with apogee radius about 3.2 million km), by additional spacecraft maneuver using gravitational force of the Moon. In this case it would be necessary to use 64-70 m antennas for the spacecraft control, synchronizations and telemetry. The fringe sizes (in micro arc seconds) for the apogee of the above-mentioned orbit and for all Radioastron bands are given in Table 1. Thus, there are non-negligible chances to observe such mirages around the black hole at the Galactic Center and in nearby AGNs in the radio-band by using Radioastron facilities.

Observations of Sgr A* in radio, near-infrared and X-ray spectral bands develop very rapidly also because it harbours the closest massive black hole. The mass of this black holes is estimated to be $4 \times 10^6 \, M_\odot$. We use the length parameter $r_g = \dfrac{GM}{c^2} = 6 \times 10^{11}$ cm to calculate all values in these units as it was explained [4]. If we take into account the distance towards the Galactic Center $D_{GC} = 8$ kpc, then the length r_g corresponds to angular sizes $\sim 5\mu$as. Since the minimum arc size for the considered mirages are about $2r_g$, the standard Radioastron resolution of about 8 μas is comparable with the required precision. The resolution in the case of the higher orbit and shortest wavelength is $\sim 1\mu$as (Table 1), good enough to reconstruct the shapes. Therefore, in principle it will be possible to evaluate a and θ parameters after mirage shape reconstructions from observational data even if we will observe only the bright part of the image (the bright arc) corresponding to positive parameters α. We may summarize by saying that there are indications that the spin of the Galactic Center black hole can be very high, although this problem is not completely solved so far but there are indirect evidences for high spins of the black hole in Sgr A*. We could summarize that angular resolution of the space Radioastron interferometer will be high enough to resolve radio images around black holes, therefore by analyzing the shapes of images one might evaluate the mass and the spin a for the Kerr black hole inside the Galactic Center, and a position angle θ_0 for a distant observer. As is clear, a position angle could be determined in a simpler way for rapidly rotating black holes $a \sim 1$ (in principle, by measuring the mirage shapes we might evaluate mass, inclination angle and spin parameter provided that the distance towards the observed black hole is known [4]). Otherwise one can only evaluate the spin parameter in units of the black hole mass, since even for not very small spin $a = 0.5$ we have very weak dependence on θ_0 angle for mirage shapes and hardly ever one could determine the θ_0 angle from mirage shape analysis. Moreover, we have a chance to evaluate parameters a and θ (for rapidly rotating black holes) if we reconstruct only the bright part of mirages (bright arcs) corresponding to co-moving photons ($\alpha > 0$). However, for slowly rotating black holes $\alpha \lesssim 0.5$ it would be difficult to evaluate parameters a and θ because we have very slow dependence of mirage shapes on these parameters [4]. In spite of the difficulties of measuring the shape of images near black holes, it is such an attractive challenge to look at the "faces" of black holes because the mirages outline the "faces" and correspond to a fully general relativistic description of a region near black hole horizon without any assumption about a specific model for astrophysical processes around black holes (of course, we assume that there are sources illuminating black hole surroundings). No doubt that the rapid growth of observational facilities will give a chance to measure the mirage shapes by using not only Radioastron facilities but also other instruments and spectral bands (for example, X-ray band).

A.F. Zakharov would like to thank the Organizers of 16th SIGRAV Conference, especially prof. G. Vilasi for a kind invitation to present this contribution and the hospitality in Vietri sul Mare.

REFERENCES

1. D. Holz and J.A. Wheeler, *ApJ*, **578**, 330-334 (2002).
2. F. De Paolis, A. Geralico, G. Ingrosso, A.A. Nucita, *A&A*, **409**, 809-812 (2003).
3. F. De Paolis, A. Geralico, G. Ingrosso, A.A. Nucita, A. Qadir, *A&A*, **415**, 1-7 (2004).
4. A.F. Zakharov, A.A. Nucita, F. De Paolis, G. Ingrosso, *A&A*, (submitted).
5. A.F. Zakharov, *Sov. Phys. - Journ. Experim. and Theor. Phys.*, **64**, 1-3 (1986).
6. A.F. Zakharov, *Sov. Phys. – Journ. Experimental and Theoretical Phys.*, **68**, 217-220 (1989).
7. A.F. Zakharov, *SvA*, **35**, 30-33 (1991).
8. A.F. Zakharov, 1994, *MNRAS*, **269**, 283-287 (1994).
9. A.F. Zakharov, in *Proc. of the 17th Texas Symposium on Relativistic Astrophysics*, Ann. NY Academy of Sciences, 1995, **759**, pp. 550-553.
10. A.F. Zakharov and S.V. Repin, *Astronomy Reports*, **43**, 705-717 (1999).
11. A.F. Zakharov and S.V. Repin, *Astronomy Reports*, **46**, 360-365 (2002).
12. A.F. Zakharov and S.V. Repin, in *Proc. of the Eleventh Workshop on General Relativity and Gravitation in Japan*, eds. J. Koga, T. Nakamura, K. Maeda, K. Tomita (eds) Waseda University, Tokyo, 2002, pp. 68-72.
13. A.F. Zakharov, N.S. Kardashev, V.N. Lukash and S.V. Repin, *MNRAS*, **342**, 1325-1333 (2003).
14. A.F. Zakharov and S.V. Repin, 2004, *Advances in Space Res.* (accepted).
15. A.F. Zakharov and S.V. Repin, *A &A*, **406**, 7-13 (2003).
16. A.F. Zakharov and S.V. Repin, in *XEUS - studying the evolution of the hot universe*, eds, G. Hasinger, Th. Boller, and A.N. Parmer, MPE Report 281, 2003, pp. 339-344.
17. A.F. Zakharov and S.V. Repin, *Astron. Rep.*, **47**, 733-739 (2003).
18. P. Young, *Phys. Rev. D*, **14**, 3281-3289 (1976).
19. S. Chandrasekhar, *Mathematical Theory of Black Holes*, Clarendon Press, Oxford, 1983.

WORKSHOP—EXPERIMENTAL GRAVITY

Beyond the Standard Quantum Limit

G. Cella

Istituto Nazionale di Fisica Nucleare sez. Pisa, Largo Bruno Pontecorvo 1, Pisa

Abstract. I discuss various possibilities of using squeezing to reduce the optical noise in an interferometric detector of gravitational waves. Influences of other sources of noise on the efficiency of the proposed solutions are analyzed, trying to understand how far we are from the possibility of a concrete application of these ideas.

INTRODUCTION

In an interferometric detector of gravitational waves like Virgo or LIGO, the space-time metric perturbation is monitored by comparing the relative positions of some test masses. The basic optical scheme is a Michelson interferometer, and in the linearized regime the phase shift is proportional to a position operator $\hat{Q}$ for an effective free mass μ. The non-vanishing commutator $[\hat{Q}(t_1), \hat{Q}(t_2)] = i\hbar(t_2 - t_1)/\mu$ enforces an indetermination relation which forbids repeated measurements of $\hat{Q}$ of infinite precision. The physical interpretation of this result is very simple: a position measurement introduces an indeterminacy of the momentum, which in turn acts back on the position during the following time evolution. A rigorous and complete discussion of the accuracy achievable with continuous measurements can be found for example in [1].

In a Virgo-like interferometer the best compromise between radiation pressure and shot noise is reached at a particular frequency f_{SQL} which depends on the laser power, and the locus of all these minimum strain effective noise amplitudes is given by

$$S_h^{SQL} = \frac{2}{\omega L}\sqrt{\frac{\hbar}{m}}, \tag{1}$$

which is called *standard quantum limit*. Looking at the sensitivity curves of current interferometers we see that a reduction of thermal noise of at least two orders of magnitude is mandatory before the standard quantum limit can become an issue. In spite of that, in these years there has been a considerable activity on this subject. Many proposals for avoiding the standard quantum limit exist now, and in the following we will give a coincise overview of these. Before starting one may wonder if the standard quantum limit is a fundamental limit, as one might expect that it is strictly connected to a fundamental principle of quantum mechanics. The answer is negative, because in detecting gravitational waves we are not really trying to measure the position of test masses, but instead to get informations about a (classical) force which acts upon them. For a detailed analysis of this point see [2].

CP751, *General Relativity and Gravitational Physics, 16th SIGRAV Conference*, edited by G. Esposito et al.
© 2005 American Institute of Physics 0-7354-0236-1/05/$22.50

PROPOSALS FOR SQL EVASION

We list here some current proposals for evading SQL. As there is no place for a careful discussion we point to the literature for the details. Also, we do not pretend to achieve completeness, and the list is based mainly on personal taste and knowledge.

Modification of input and/or output

The basic starting point of this kind of approach is the observation that the interaction of the laser beam with the free masses of the mirror generates *ponderomotive squeezing*. Put in simple terms, this means that as the mirror recoils in response to a fluctuation of the laser amplitude, a fluctuation of the beam phase is generated. Phase and amplitude quantum fluctuations are no longer uncorrelated, as it occurs in a coherent state, and there is an appropriate quadrature of the relevant mode of the electromagnetic field which sees reduced quantum fluctuations. We can exploit this fact by measuring at the output the quadrature with the best SNR ratio, or injecting a squeezed state in the dark port of the interferometer (the only one which contributes to output quantum noise). These possibilities are discussed in detail in [3]: in principle it is possible to completely suppress radiation pressure noise, which means that there are no obstructions to reducing shot noise also by increasing laser power. The improvement obtainable by injecting squeezed light depends on the available degree of squeezing.

Measurement of quantum non demolition observables

The commutator at different times between the momentum operators for a free mass is zero, hence there is no indetermination limit to the precision obtainable in a repeated measurement of the mirror velocity. The velocity of a free mass is an example of a *quantum non demolition* observable: more generally, a conserved quantity is QND also (see [1] for a general discussion). There are alternative optical schemes which give, at least approximately, a measurement connected with the speed of the mirrors, like for example a Sagnac interferometer: these will not have a standard quantum noise limit [4, 5].

Modification of test mass dynamics

In a resonant cavity the cavity power is maximized, and in the linearized approximation radiation pressure does not change when the mirrors are moved. This means that information about the signal is encoded only in the phase of the output beam. Out of resonance, or in presence of a more complicated optical scheme (notably if a signal recycling mirror is present) this no longer holds: part of the signal is present also in the intensity quadrature, and a restoring force proportional to the mirrors' displacement is present ("optical spring"). The modified dynamics can be exploited to evade the standard quantum limit [6, 7]: however, not all optomechanical modes are stable, and a control strategy must be devised.

Quantum feedback

It is possible to monitor the movement of a test mass with an auxiliary cavity. On using a carefully implemented feedback strategy, which exploits the light squeezing generated by ponderomotive effects, this test mass can be controlled in principle at the quantum level, by subtracting completely radiation pressure noise [8, 9]. This subtraction is possible also for a complete cavity, using a single feedback [10].

CONCLUSION

The feasibility of an experimental study of these techniques is linked to the possibility of reducing up to a very high degree other sources of noise, technical and fundamental. An important issue is the production of states of light with high levels of squeezing. So far the maximum level of squeezing obtained is around 7 dB, and the next challenge is fixed at 10 dB. A related issue is the reduction of losses inside the apparatus, because a loss is equivalent to the introduction of a source of non squeezed quantum fluctuations.

The study of quantum noise is an active field of research. So far many proposals for avoiding the standard quantum limit have been put forward at theoretical level, and we think that there is a lot of room left for new ideas.

REFERENCES

1. V. B. Braginsky and F. Y. Khalili,"Quantum measurement", Cambridge University Press (1999).
2. C. M. Caves, K. S. Thorne, R. W. Drever, V. D. Sandberg, M. Zimmermann, "On the measurement of a weak classical force coupled to a quantum mechanical oscillator. I. Issue of principle", Rev. Mod. Phys. **52**, 341 (1980).
3. H. J. Kimble, Y. Levin, A.B. Matsko, K. S. Thorne and S. P. Vyatchanin, "Conversion of conventional gravitational-wave interferometers into QND interferometers by modifying their input and/or output optics", Phys. Rev. D **65**, 022002 (2002).
4. Y. Chen, "Sagnac Interferometer as a Speed-Meter-Type, Quantum-Nondemolition Gravitational-Wave Detector", Phys.Rev. D **67**, 122004 (2003).
5. P. Purdue and Y. Chen, "Practical speed meter designs for QND gravitational-wave interferometers", Phys. Rev. D **66**, 122004 (2002)
6. A. Buonanno and Y. Chen, "Signal recycled laser-interferometer gravitational-wave detectors as optical springs", Phys. Rev. D **65**, 042001 (2002).
7. A. Buonanno and Y. Chen, "Quantum noise in second generation, signal-recycled laser interferometric gravitational wave detectors", Phys. Rev. D **64**, 042006 (2001).
8. J.-M. Courty, A. Heidmann, M. Pinard, "Back-action cancellation in interferometers by quantum locking", Europhys. Lett. **63**, 226 (2003).
9. A. Heidmann, J.-M. Courty, M. Pinard, J. Lebars, "Beating quantum limits in interferometers with quantum locking of mirrors", Journal of Optics B **6**, S684 (2004).
10. A. Giazotto, G. Cella, "Some considerations about future interferometric GW detectors", Class. Quantum Grav. **21**, S1183 (2004).

Dynamic Matched Filter for Gravitational Wave Detection

Fausto Acernese*†, Fabrizio Barone**†, Rosario De Rosa*†, Antonio Eleuteri*†, Leopoldo Milano*† and Silvio Pardi‡†

*Dipartimento di Scienze Fisiche, Universitá di Napoli Federico II, Complesso Universitario di Monte Sant'Angelo, via Cintia, I-80126, Napoli, Italy;
†Istituto Nazionale di Fisica Nucleare, sez. Napoli, Complesso Universitario di Monte Sant'Angelo, via Cintia, I-80126, Napoli, Italy;
**Dipartimento di Scienze Farmaceutiche, Universitá di Salerno, via Ponte Don Melillo, I-84084, Salerno, Italy;
‡Dipartimento di Matematica e Applicazioni R. Caccioppoli, Universitá di Napoli Federico II, Complesso Universitario di Monte Sant'Angelo, via Cintia, I-80126, Napoli, Italy;

Abstract. The detection of gravitational waves is usually based on very complex data analysis algorithms, by virtue of the low signal-to-noise ratio. In particular the search for signals coming from coalescing binary systems can be very demanding in terms of computing power, as in the case of the well known Standard Matched Filter Technique. To overcome this problem, we have tested a Dynamic Matched Filter Technique, still based on Matched Filters, whose main advantage is the requirement of a lower computing power. In this work this technique is described, together with its possible application as a pre-data analysis algorithm.

INTRODUCTION

The procedures for detection of Gravitational Waves (GW) from coalescing binary systems usually require high performances data analysis techniques for signal extraction from the detector noise. This is due to the intrinsic weakness of gravitational waves [1] and to the low signal-to-noise ratio expected on the actual earth based interferometric antennas like VIRGO [2], LIGO [3], GEO [4], TAMA [5]. Actually, the best technique are the matched filters, that integrates the classical optimum Wiener–Komolgorov filter [6] with a static coverage of the parameter space. Unfortunately, this procedure is very computing power demanding [7], [8]. The matched filters have one weak point, that is their sensitivity to the template shape. In particular, even a slight difference between the theoretical template and the real signal largely reduces the quality of the recovered signal. An alternative solution is a hierarchical strategy, consisting in a on-line rough data analysis to select all frames which may contain a signal, followed by a refined off-line search, as described in [9]. We have decided to follow this strategy by introducing a dynamic matched filter technique, in which the static grid of templates is replaced with a dynamic one, on-line generated using a global optimization algorithm.

CP751, General Relativity and Gravitational Physics, 16th SIGRAV Conference, edited by G. Esposito et al.
© 2005 American Institute of Physics 0-7354-0236-1/05/$22.50

DYNAMIC MATCHED FILTERS TECHNIQUE

The goal of the Standard Matched Filter Technique ($SMFT$) is that of computing the SNR for all the static set of templates, characterized by the full coverage of the space of parameters using the level of confidence of the detection as requirement. In presence of a GW signal, the largest SNR corresponds to its template, in absence of a GW signal, the SNRs should lie below a threshold, evaluated from the knowledge of the detector output noise.

The Dynamic Matched Filter Technique ($DMFT$) simply changes the static template grid with a dynamic one, defined by the evolution of the global optimization algorithm used for its management. Although the $DMFT$ is still based on Matched Filters, it is clear that it cannot guarantee the level of confidence of detection as the $SMFT$. In fact, it is not possible to demonstrate that an optimization algorithm, although global, always reaches the global maximum of the objective function, while it is possible to demonstrate that it always converges toward a local maximum. Critical in the $DMFT$ is the choice of the optimization algorithm. In fact, it must be robust, constrained and global, that are very important characteristics in a difficult search like this one, especially when the number of physical and geometrical parameters is very high and the objective function is not smooth. On the other hand, the real advantage of the $DMFT$ is that it requires a number of SNR evaluations that is orders of magnitude less than the $SMFT$. This characteristic makes it possible for us to use it as an on-line pre-data analysis technique, since only small computer farms are required and the detection of candidates to be analyzed with more refined off-line techniques can be very useful.

For our tests we decided to use the Price Algorithm as global optimization algorithm [10]. This algorithm, also known as Controlled Random Search (CRS), is a very powerful extension of the simplex algorithm, that we have applied to very difficult optimization procedures in astrophysics with up to 30 degrees of freedom, developing and applying also improved versions [11].

The Price optimization technique can be easily introduced in the matched filter algorithm to search GW signal from coalescing binaries in data coming from interferometric antennas. In this case, in fact, the function to optimize is the signal-to-noise ratio between the data and a certain template. Each template can be parameterized with the star masses, their spin, and other physical parameters connected with the system properties. In this way the goal of the algorithm is to maximize the SNR with respect to these quantities.

The application of the Price optimization algorithm within the matched filters scenario is easy and effective. In fact, clear advantages of this procedure are the lower number of templates needed to build the search array and its independence of the number of parameters of a template. On the other hand, problems arise due to the lack of statistical information about the confidence level of the detection and the need of a dynamic evaluation of templates, instead of their static storage as in the $SMFT$. Finally, the choice of a suitable number of templates in the starting search array is important and critical, to avoid a poor definition of the maximum (too small search arrays) or too long computational time (too large search arrays).

We tested the Dynamic Matched Filter Technique by using a signal simulating the output of a GW interferometric detector with $SNR = 20$ [12]. The GW signal was a

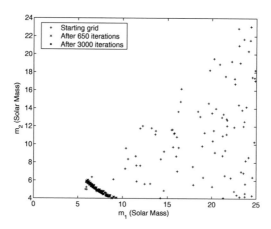

FIGURE 1. Evolution of the Price Algorithms Search Array during the Optimization Procedure.

chirp computed by using the *PN*2 approximation, simulating the coalescence of a binary system with masses equal to $5M_\odot$ and $7M_\odot$, within Gaussian White Noise. The length of the data set was 16.384 s, sampled at 4 kHz. The lower cut-off frequency was fixed at 40 Hz. The Price algorithm used an array of 100 points with a mass range from $4M_\odot$ to $25M_\odot$. Although statistical stop criteria should be used for stopping the optimization procedure [11], in the example reported here we preferred to use, for clarity, the number of iterations, fixed at 3000, as stop criterion. The evolution of the optimization procedure is shown in Figure 1. The clustering around the true solution is evident, as is the presence of a signal according to the values of the *SNR*. This clearly shows that the *DMFT* can be effectively used as an on-line pre-analysis tool and as a trigger, by checking for the presence of a GW signal and giving the possibility of reducing the amplitude of the volume of the space of parameters, making the application of the *SMFT* easier.

REFERENCES

1. C. W. Misner, K. S. Thorne, J. A. Wheeler, Gravitation (Freeman & Co., San Francisco, 1973).
2. F. Acernese et al., *Class. Quantum Grav.* **21**, S385 (2004).
3. D. Sigg, *Class. Quantum Grav.* **21**, S409 (2004).
4. B. Willke et al, *Class. Quantum Grav.* **21**, S417 (2004).
5. R. Takahashi and the TAMA Collaboration, *Class. Quantum Grav.* **21**, S403 (2004).
6. C. W. Helstrom, *Statistical Theory of Signal Detection*, 2nd ed. (Pergamon Press, London, England, 1968).
7. B. Owen, *Phys. Rev. D* **53**, 6749 (1996).
8. P. Canitrot, L. Milano, A. Viceré, VIR-NOT-PIS-1390-149 (2000).
9. K. S. Thorne, in *First International LISA symposium*, Chilton, Oxfordshire, England, 1996, edited by M.C.W. Sandford.
10. W. L. Price, *Computer J.* **20**, 367 (1976).
11. F. Barone, L. Di Fiore, L. Milano, G. Russo, *Astrophys. J* **407**, 237 (1993).
12. F. Acernese, F. Barone, R. De Rosa, A. Eleuteri, S. Pardi, G. Russo and L. Milano, *Class. Quantum Grav.* **21**, S807 (2004).

Preliminary Analysis of the July 2004 data of the Low-Frequency Facility

Angela Di Virgilio

INFN Sez. Di Pisa, Polo Fibonacci ed. C, via F. Bonarroti 2, 56100 Pisa

Abstract. The Low-Frequency Facility is an R&D experiment of the Virgo project. It measures with very high sensitivity at low frequency, around and above 10 Hz, the displacement noise of mirrors attached to a Virgo-like suspension. The preliminary analysis of measurements done in July 2004 is presented.

INTRODUCTION

The study of thermal noise, and noises in general, is a fundamental issue for the improvement of future gravitational wave antennas [1], especially at low frequency, where most of the sources of events are predicted and where noises are higher. The LFF experiment [2, 3, 4, 5, 6, 7, 8, 9, 10, 11] is an INFN Commission II experiment, supported by Firenze-Urbino, Napoli, Pisa and Roma1. The LFF is designed to study the noise of a 1 cm Fabry Perot (FP) cavity, suspended by a Virgo-like super attenuator (SA) [1], see figures 1. With the use of this suspension, the noise of a test mass attached to the SA should be dominated by the suspension thermal noise by 10 Hz, and thus this experiment is aimed at measuring off-resonance thermal noise in the 10 Hz region. In addition the LFF can measure the seismic noise transfer function of a standard SA suspension. The control philosophy used is as close as possible to that used in the Virgo SA. The LFF assembly was completed in summer 2001, with first lock of the cavity achieved in summer 2002[8]. Results taken in 2003[10, 11] showed that there were problems with non-stationary noise in the 10100 Hz region and that DAC and coil driver noise were limiting performance at 10 Hz, the achieved sensitivity was around $10^{-14}m/\sqrt{Hz}$ at 10 Hz. Various modifications have been made since then to optimise the system, including incorporation of coil drivers which can be operated in two ways, high noisehigh power and low noiselow power, moreover the control of the bandwidth has been increased up to 1 kHz, the finesse is higher than 3000 and the optical gain higher than $10^{10}V/m$.

PRELIMINARY ANALYSIS

Several runs of approximately 10 minutes long have been acquired, running in the low noise state and it has been checked that there is no seismic noise contamination above a few Hertz, and the high non-stationary noise has been removed. In fig. 2 the reconstructed displacement noise [8, 9, 10] is shown for two different runs. The low-

CP751, *General Relativity and Gravitational Physics, 16th SIGRAV Conference*, edited by G. Esposito et al.

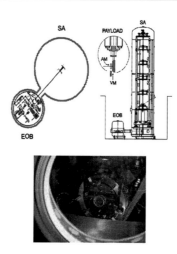

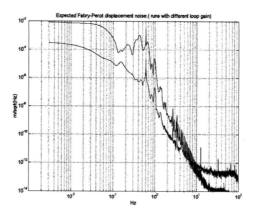

FIGURE 1. Upper: Schematic diagram of the FP cavity suspended from the superattenuator, the light comes from the optics in the EOB tank. AM is the flat auxiliary mirror, VM, Virgo-like test mass, is the curved mirror. Lower: Photo of the FP cavity in its tank.

FIGURE 2. Displacement power spectrum for two different runs with different longitudinal control loop gain; the low-frequency structure of the SA suspension is evident, but the absolute scales are inconsistent.

frequency structure clearly shows the typical peaks of the SA suspension, the 30 mHz resonance peak of the inverted pendulum is clearly visible. It is clear that there is a big mismatch between the two curves, the absolute scale looks wrong for both curves; in fact the expected relative R.M.S. displacements, evaluated by the integral of the two power spectra, are much larger than the one measured open loop. The relative comparison of signals taken in different parts of the longitudinal control loops (closed loop) shows that part of the electronics and the cavity itself exhibit transfer functions different from the one measured open loop. It will be necessary to restart the whole experiment and repeat the measurements in order to well understand the absolute calibration.

240

ACKNOWLEDGMENTS

I would like to thank G. Cella, D. Passuello, S. Solimeno and F. Vetrano.

REFERENCES

1. See papers in: Class. Quan. Grav. Vol. 21, number 5, March 2004, and the following sites: http://www.ligo.caltech.edu, http://www.virgo.infn.it, and http://tamago.mtk.nao.ac.jp
2. M.Bernardini et al., Phys. Lett. A 243 (1998) 187-194. North Holland.
3. Angela Di Virgilio et al., Optics Communication, 162 (1999) 267-279.
4. Angela Di Virgilio et al., Il Nuovo Cimento- Vol 114 B, N. 10-Ottobre 1999, pp 1197-1212.
5. G. Cella et al. Physics Letters A 266 (2000) 1-10.
6. A. Di in Gravitational Waves, edited by I. Ciufolini, V. Gorini, U. Moschella and P. Fré, Institute of Physics Publishing, Bristol and Philadelphia
7. F. Benabid et al J. Opt. B: Quantum Semiclass. Opt. 2 (2000), 1-7.
8. A. Di Virgilio et. Al, Physics Letters A 318 (2003) 199-204.
9. A. Di Virgilio et al., Physics Letters A 316 (2003) 1-9.
10. A. Di Virgilio et al, Class. Quantum Grav., 21 (2004) 1099-1106.
11. A. Di Virgilio et al., Phys. Lett. A 322 (2001) 1-9.

Adaptive Filters for Detection of Gravitational Waves from Coalescing Binaries

Fausto Acernese[*†], Fabrizio Barone[**†], Rosario De Rosa[*†], Antonio Eleuteri[*†], Lara Giordano[‡†] and Leopoldo Milano[*†]

[*]Dipartimento di Scienze Fisiche, Università di Napoli Federico II, Complesso Universitario di Monte Sant'Angelo, via Cintia, I-80126, Napoli, Italy;
[†]Istituto Nazionale di Fisica Nucleare, sez. Napoli, Complesso Universitario di Monte Sant'Angelo, via Cintia, I-80126, Napoli, Italy;
[**]Dipartimento di Scienze Farmaceutiche, Università di Salerno, via Ponte Don Melillo, I-84084, Salerno, Italy;
[‡]Dipartimento di Matematica e Applicazioni R. Caccioppoli, Università di Napoli Federico II, Complesso Universitario di Monte Sant'Angelo, via Cintia, I-80126, Napoli, Italy;

Abstract. In this work we propose a new strategy for gravitational waves detection from coalescing binaries, using Adaptive Line Enhancer filters. An analysis of the filters performance is shown in terms of the tracking capability and determination of filter parameters. Furthermore, receiver operating characteristics are derived, by extensive Monte Carlo simulation studies. Some tests have been performed on simulated Virgo noise, demonstrating that adaptive filters are promising both for the small computational power needed (with respect to the matched filter) and for the statistical robustness of the algorithms used.

INTRODUCTION

Signal detection of Gravitational Wave signals from coalescing binary systems requires high performance data analysis techniques. This is due to the low signal-to-noise ratio expected on the actual Earth-based interferometric antennas like VIRGO [3], LIGO [4], GEO [5], TAMA [6]. If the shape of the expected signal is known, the optimum Wiener–Komolgorov matched filter [7] can be used to detect the presence of a signal in the data stream.

Matched filters are sensible to the template shape. Even a slight difference between the template and the true signal reduces the quality of the recovered signal. A solution is to improve the template model, but this increases the number of parameters. The consequence is that a full on-line data analysis would require computing power not available now [8]. Therefore, there is a clear need for developing also different data analysis strategies. In this paper we propose a fast algorithm which does not use templates, as a trigger to select data frames which could contain a signal. The selected frames are then processed with matched filters for a refined analysis.

CP751, General Relativity and Gravitational Physics, 16th SIGRAV Conference, edited by G. Esposito et al.
© 2005 American Institute of Physics 0-7354-0236-1/05/$22.50

SIGNAL AND FILTER MODEL

Let the observed signal be $z_t = s_t + v_t$, where s_t is a chirp signal and v_t is the interferometer noise. For the filter operation we shall need a white noise sequence, which we can get from the observed z_t by applying a whitening operator $\mathcal{W}$, so that the input sequence to the filter can be written as

$$x_t = f_t + w_t, \quad f_t \equiv \mathcal{W} s_t, w_t \equiv \mathcal{W} v_t. \tag{1}$$

Since $f_t \equiv \mathcal{W} s_t$ is still a locally sinusoidal signal, it is convenient to use a parameterised filter with a "notch shape" in the frequency domain. Such a behavior is exhibited by Adaptive Line Enhancer (ALE) filters, that are linear predictor filters which "track" and "enhance" time-varying spectral lines.

The ALE filter model [9] is a technique designed to approximate the SNR gain obtained by the matched filter solution for a locally sinusoidal signal. The advantage of using ALE filters is that no a priori knowledge of the signal parameters is required.

Here we consider as a prototype ALE a second-order band-pass Infinite Impulse Response (IIR) filter whose (instantaneous) transfer function is [1]

$$H(z) = (1 - r^2) \left[\frac{W_t z/(r + r^2) - 1}{z^2 - W_t z + r^2} \right], \tag{2}$$

where $W_t = 2r \cos(2\pi f_t)$ is the only parameter dependent on the center-frequency f_t, while r is a fixed design parameter related to the filter bandwidth B by the relation $B = F_s(1 - r)/2$, where F_s is the sampling frequency. This filter has the advantage over FIR filters of only having one parameter, and therefore its adaptation is very fast, while at the same time providing a typically narrower filtering band. The filter tracks the signal based on the following recursion (where y_t is the filter output):

$$y_t = W_t \left(\frac{1 - r}{r} \right) x_{t-1} - (1 - r^2) x_{t-2} + W_t y_{t-1} - r^2 y_{t-2}. \tag{3}$$

Stability of the filter requires $|W_t| < 2r$.

SIGNAL DETECTION BY ALE

On using an ALE filter for detection, a delicate problem is the choice of the threshold, T, to discriminate between the alternative hypotheses of absence or presence of a signal at an assigned confidence level (we assume the Neyman–Pearson approach to detection [7]). We have used the ALE as a detector with no incoherent integration, i.e. at every sample we output a decision. The assumed output statistics is y_t^2. We have implemented an extensive Monte Carlo simulation study to get the false alarm and detection probabilities. In Fig.1 the resulting Receiver Operating Characteristic (ROC) are shown both for the IIR ALE and the matched filter. Clearly, IIR ALE performance are worse than those of the matched filter, however it should be noted that the latter was

243

operating at the best conditions of known signal shape, duration and time of arrival. It is not a realistic condition, but it allowed us to evaluate an upper bound on the IIR ALE filter performance.

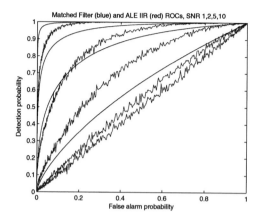

FIGURE 1. IIR ALE and matched filter ROCs evaluated at different SNRs.

An example: simulated Virgo data

To assess the performance of the filter, we have used simulated Virgo data. A 3 minute data set with software injections was whitened. The whitened data were fed to the ALE, which was set to operate at 1 false alarm/6 hours at 20kHz sampling rate. The running time of ALE on all frames is about 1 minute with a MATLAB script running on a notebook equipped with an Athlon MP 1GHz. In the first example, a Newtonian chirp with chirp mass 1.4 and starting frequency 40 Hz was injected at SNR 5. The signal is long 19s, starts at 10s and ALE first detects it at 13.5s. In the second example, a Newtonian chirp with chirp mass 7 and starting frequency 30 Hz was injected at SNR 5. The signal is long 3s, starts at 10s and ALE first detects it at 10.4s.

REFERENCES

1. F. Acernese, F. Barone, R. De Rosa, A. Eleuteri, L. Milano, *Class. Quantum Grav.* **21**, S781 (2004).
2. C. W. Misner, K. S. Thorne, J. A. Wheeler, Gravitation (Freeman & Co., San Francisco, 1973).
3. F. Acernese at al, *Class. Quantum Grav.* **21**, S385 (2004).
4. D. Sigg, *Class. Quantum Grav.* **21**, S409 (2004).
5. B. Willke et al, *Class. Quantum Grav.* **21**, S417 (2004).
6. R. Takahashi and the TAMA Collaboration, *Class. Quantum Grav.* **21**, S403 (2004).
7. C. W. Helstrom, *Statistical Theory of Signal Detection*, 2nd ed. (Pergamon Press, London, England, 1968).
8. B. S. Sathyaprakash, *"Problems of Searching for Spinning Black Hole Binaries"* in Proceedings of the *XXXVIII Rencontres de Moriond*, 24-29 March 2003 (in press).
9. B. Widrow, and S. Stearns, *Adaptive Signal Processing*, Prentice Hall, 1985.

The BepiColombo Radio Science Experiment and the Non–Gravitational Perturbations to the Mercury Planetary Orbiter orbit: key role of the Italian Spring Accelerometer

Valerio Iafolla*, David M. Lucchesi* and Sergio Nozzoli*

*Istituto di Fisica dello Spazio Interplanetario (IFSI/CNR), Via Fosso del Cavaliere, 100, 00133 Roma, Italy

Abstract. The advantages of an on-board accelerometer are outlined in the case of the Bepi-Colombo mission to Mercury with respect to the modeling of the non-gravitational perturbations at work in the strong radiation environment of Mercury. The readings from the Italian Spring Accelerometer guarantees a very significant reduction of the non-gravitational accelerations impact on the space mission accuracy, especially of the strong direct solar radiation pressure.

INTRODUCTION

The space mission BepiColombo (BC) aims at performing a detailed study of the planet Mercury and its environment [1], and to test Einstein's General Relativity (GR) to an unprecedented accuracy. The Radio Science Experiment (RSE) is one of the main experiments considered for BC. For such measurements it is necessary to allocate on-board the Mercury Planetary Orbiter (MPO) a set of dedicated instrumentation. These instruments allow the determination of the MPO orbit around the planet and the accurate determination of Mercury's center-of-mass around the Sun, by which it is possible to estimate several parameters related to the planet structure and verify the theory of GR [2, 3, 4]. In particular, the scientific goals to achieve by very precise measurements are: i) the rotation state of Mercury; ii) the global gravity field of Mercury and its temporal variations resulting from solar tides; iii) the local gravity anomalies; iv) the orbit of Mercury's center-of-mass around the Sun and the propagation of electromagnetic waves between the Earth and the spacecraft. One of the instruments involved in the RSE is a very high sensitivity accelerometer, able to measure the inertial accelerations acting on the MPO, i.e., able to remove the perturbing non-gravitational accelerations from the list of unknowns. The Italian Spring Accelerometer (ISA), a three-axis instrument with a sensitivity of $10^{-9} g_{\oplus}/\sqrt{Hz}$ ($g_{\oplus} \simeq 9.8$ m/s^2) in the frequency band of 10^{-4}—10^{-1} Hz, has been considered to fly on–board the MPO [5]. In the following we prove the superiority of the readings from an on-board accelerometer with respect to the modeling of the non-gravitational perturbations (NGP).

CP751, *General Relativity and Gravitational Physics, 16th SIGRAV Conference*, edited by G. Esposito et al.
© 2005 American Institute of Physics 0-7354-0236-1/05/$22.50

DIRECT SOLAR RADIATION EFFECTS

We analyze the perturbing effects of direct solar radiation pressure (SRP), the strongest NGP, on the MPO satellite and we compare such effects with the readings from the ISA accelerometer. We assumed, for the sake of simplicity, the MPO spherical in shape and with the same area–to–mass ratio of the planned spacecraft (about $1.9 \cdot 10^{-2}$ m²/kg, corresponding to a mass of approximately 357 kg). The resulting acceleration produced by the Sun radiation pressure is

$$\vec{a}_\odot = -\frac{A\Phi_\odot}{mc} \left(\frac{\bar{R}_\odot}{R_\odot}\right)^2 \hat{S}, \tag{1}$$

where A/m represents the area-to-mass ratio of the satellite, c the speed of light, $\Phi_\odot$ the solar irradiance at the average distance $\bar{R}_\odot$ of Mercury from the Sun, and $\hat{S}$ represents the unit vector towards the Sun. We neglected the diffusive term of the reflected radiation, while the penumbra effects have been considered in our numerical simulations because of the vicinity of the Sun. The radiation environment around Mercury is quite strong. The solar irradiance varies between 14,448 W/m² at Mercury's perihelion, to 6272 W/m² at Mercury's aphelion (i.e., with a variation of about 80% at half of the sidereal period of Mercury around the Sun, to be compared with the corresponding 0.37% variation on Earth). Therefore the perturbing acceleration due to direct SRP vary in the range between $4 \cdot 10^{-7}$ m/s² and $1 \cdot 10^{-6}$ m/s². In the following we give the results of the numerical analysis performed over a characteristic arc length of about 7 hours (used for the recovery of Mercury's gravity field and for the GR experiment), corresponding to three orbital periods of the MPO (about 2.3 h) around Mercury. In Fig. 1 the results are shown of the numerical simulation in the case of transversal component of the direct SRP acceleration for an orbital configuration of the MPO characterized by the presence of eclipses from Mercury. The larger peaks are about 10^{-6} m/s². That is, the perturbing effect at the MPO orbital period is two orders-of-magnitude larger than the accelerometer accuracy ($10^{-9} g_\oplus \simeq 10^{-8}$ m/s²), e.g., the orbit propagation error is just controlled by the accelerometer measurement error.

CONCLUSIONS

With the use of an on–board accelerometer the MPO spacecraft can be considered as an a-posteriori drag-free satellite, without the necessity of modeling the NGP. Indeed, the SRP is very difficult to model because of the complex shape of the MPO (not spherical in shape, plus solar panels and the High Gain Antenna) and also because the spacecraft reflects (in the visible) and reradiates (in the infrared) in a very complex way. In the present study we have assumed a spherical in shape spacecraft just to look at the main features of the visible radiation effects on the MPO. Indeed, with a refined model of the NGP on the "true" MPO we cannot reach a modeling better than 10%, in accuracy, of their subtle and complex effects. However, in order to reach such 10% modeling, we need a not so complex in shape spacecraft (because of the shadowing effects), with homogeneous surfaces of known optical properties and, very important, how these

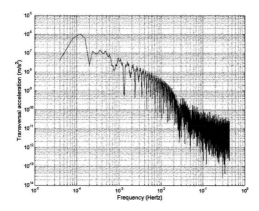

FIGURE 1. Fourier analysis of the Transversal acceleration in the accelerometer bandwidth (10^{-4} – 10^{-1} Hz). The 10^{-6} m/s^2 peak at the MPO orbital period (2.3 h) is clearly visible in the spectrum.

optical properties fade away in the strong radiation environment of Mercury during the MPO mission. All this represents a quite difficult task, and the use of the accelerometer is enough to justify the complexity of the problem of the NGP modeling. Anyway, even in the case of a 10% modeling of the direct SRP, with the use of the ISA accelerometer we gain about one order-of-magnitude with respect to this best modeling. In reality, if we also consider the variations of the SRP (as during the MPO eclipses, because of the penumbra effects) we gain even more than a factor 10 with the use of the accelerometer which is able to detect such variations.

REFERENCES

1. Grard, R., Balogh, A., Returns to Mercury: science and mission objectives, Plan. Space Science, 49, 1395–1407 (2001).
2. Milani, A., Rossi, A., Vokrouhlický, D. Villani, D., Bonanno, C., Gravity field and rotation state of Mercury from the BepiColombo Radio Science Experiments, Plan. Space Science, 49, 1579–1596 (2001).
3. Milani, A, Vokrouhlicky, D., Villani, D., Bonanno, C., Rossi, A., Testing general relativity with the Bepicolombo radio science experiment, Phs. Rev. D 66, 082001 (2002).
4. Iess, L., Boscagli, G., Advanced radio science instrumentation for the mission BepiColombo to Mercury, Plan. Space Science, 49, 1597–1608 (2001).
5. Iafolla, V., Nozzoli, S., Italian spring accelerometer (ISA): a high sensitive accelerometer for "Bepi-Colombo" ESA CORNERSTONE, Plan. Space Science, 49, 1609–1617 (2001).

Orbital analysis of LAGEOS and LAGEOS II laser ranged satellites: relativistic effects and geophysical issues

Roberto Peron

Dipartimento di Fisica, Università di Lecce
Via Arnesano, 73100 Lecce, Italy
E-mail: peron@ifsi.rm.cnr.it

Abstract. We present here the results of a recent analysis of LAGEOS and LAGEOS II laser range data. The higher accuracy in determining the orbits of these satellites makes it possible to see very tiny relativistic effects like frame-dragging and a wide variety of other phenomena at work. In particular, it is apparent the need of better understanding some effects of non-gravitational origin. The importance of these orbital fits as a geophysical probe is also stressed with a particular example. The analysis is carried out with GEODYN II software, whose broad structure and use is described.

INTRODUCTION

Among the various techniques to track satellites, *Satellite Laser Ranging* (SLR) is the most precise. Basically, a laser pulse is shot from an Earth station towards the satellite, and reflected back by *Corner Cube Retroreflectors* (CCR) mounted on it. The measurement of the pulse travel time gives the station-satellite range. The precision that can be obtained with this technique is below 1 centimeter in single-shot ranges, and reaches the sub-millimeter level for normal points. This has to be compared with the order of magnitude of station-satellite distance, ~ 10000 Km.

SLR ranges can be used to reconstruct the orbit of the satellite, by means of a fitting procedure. In order to achieve a good estimate of the satellite orbit, it is of fundamental importance to model as accurately as possible the dynamics of the satellite. This means having at our disposal models for all forces acting on the satellite, as well as any other information useful to correctly process the data (like a model for atmospheric delay of laser pulse). As we will see below, the analysis of LAGEOS and LAGEOS II data yields an accuracy $\sim$ cm.

LAGEOS satellites are in the class of "geodetic satellites". They are very close to the ideal concept of "test mass", and are suited to test a wide class of phenomena which involve very small accelerations [5, 1]. In particular, they can be used to study geophysical and geodynamical effects, as well as the relativistic ones. We focused our analysis on the study of an important prediction of general relativity, the so-called **Lense–Thirring effect** (also known as **Frame Dragging**) [2], which is caused by a rotating body (as the Earth) curving spacetime in a peculiar way so as to perturb the orbit of every object around it.

CP751, *General Relativity and Gravitational Physics, 16th SIGRAV Conference*, edited by G. Esposito et al.
© 2005 American Institute of Physics 0-7354-0236-1/05/$22.50

In 1918 J. Lense and H. Thirring [4] first obtained the formula for the secular precession of the nodal longitude of a satellite orbiting around a rotating body

$$\dot{\Omega}^{L-T} = \frac{2GJ}{c^2 a^3 (1-e^2)^{3/2}}, \tag{1}$$

where J is the angular momentum of the central body, a the semimajor axis of the satellite orbit and e its eccentricity, respectively. An analogous formula holds for the argument of perigee ω. This effect is very small — ~ 40 milliseconds of arc (mas) for the LAGEOS satellites — nevertheless it is fully in the range of SLR precision. The real problem is to pick it out among all other effects that perturb the satellite orbit; i.e., to precisely model its dynamics.

MODELING AND ERROR SOURCES

Fitting range data with an accuracy sufficient to sense frame-dragging requires a detailed knowledge of all processes that affect the motion of the satellite ("perturbations"). These can be roughly divided into two main classes:

• Gravitational perturbations
• Non-gravitational perturbations

Among the gravitational ones, the only critical point for our analysis is related to the Earth quadrupole (C_{20} term in the expansion of gravitational geopotential in spherical harmonics). This problem is solved by a suitable combination of orbital elements of the two satellites.

The rôle of non-gravitational perturbations is not so critical for the node as it is for the perigee. There are nevertheless some effects related to the thermal behavior of satellites and to their attitude, which produce accelerations of order of 10^{-12} m/s^2, and whose modeling is not easy. Work is in progress to better understand these forces and to assess their importance in lowering the fit accuracy.

DATA ANALYSIS

We analyzed about ten years of range data of LAGEOS and LAGEOS II satellites with **Geodyn II** software (NASA GSFC) [6], by using time chunks (*arcs*) of fifteen days for the data. Geodyn did a least-squares fit of data using a detailed set of models to describe the satellite dynamics and the observation conditions. For each arc the program adjusted the initial conditions ($\vec{x}$, $\dot{\vec{x}}$) and a number of other parameters, such as the reflection coefficient C_R. The post-fit RMS is of the order of few centimeters in range, throughout the entire period of analysis.

In Fig. 1 we can see a plot of the combined nodal residuals of LAGEOS and LAGEOS II, for the period of analysis. It is apparent the secular trend resulting from the relativistic frame-dragging effect, apart from observational errors and errors in the modeling. This trend is in agreement with the prediction of general relativity.

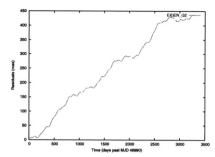

FIGURE 1. Combined integrated node residuals.

As a useful by-product of our analysis, we observed, starting from 1999, a change in the trend for the separate integrated residuals of both satellites. This is an independent confirmation of the observed abrupt change of the quadrupole moment of Earth in recent times [3], and a nice example of how varied are the scientific products of such an analysis.

ACKNOWLEDGMENTS

We are pleased to thank Erricos C. Pavlis (JCET-UMBC, NASA-GSFC) and David M. Lucchesi (IFSI-INAF), with whom this work was done, and Gaetano Chionchio (IFSI-INAF) for his invaluable computer support.

REFERENCES

1. B. Bertotti, P. Farinella P. and D. Vokrouhlick, *Physics of the Solar System - Dynamics and Evolution, Space Physics, and Spacetime Structure*, Kluwer Academic Publishers, Dordrecht (2003).
2. I. Ciufolini and J. A. Wheeler, *Gravitation and Inertia*, Princeton University Press, Princeton, NJ (1995).
3. C. M. Cox and B. F. Chao, Science **297**, 831 (2002).
4. B. Mashhoon, F. W. Hehl and D. S. Theiss, Gen. Rel. Grav. **16**, 711 (1984).
5. A. Milani, A. M. Nobili and P. Farinella, *Non-Gravitational Perturbations and Satellite Geodesy*, Adam Hilger, Bristol (1987).
6. D. E. Pavlis, *GEODYN II Operations Manual*, NASA GSFC technical report (1998).

Rotation Effects and The Gravito-Magnetic Approach

Matteo Luca Ruggiero

Dipartimento di Fisica, Politecnico di Torino and INFN, Sezione di Torino
matteo.ruggiero@polito.it

Abstract. Gravito-electromagnetism is somewhat ubiquitous in relativity. In fact, there are many situations where the effects of gravitation can be described by formally introducing "gravito-electric" and "gravito-magnetic" fields, starting from the corresponding potentials, in analogy with the electromagnetic theory [1],[2] (see also A. Tartaglia's contribution to these proceedings). The "many faces of gravito-electromagnetism" [3] are related to rotation effects in both approximate and full theory approaches. Here we show that, by using a 1+3 splitting, relativistic dynamics can be described in terms of gravito-electromagnetic (GEM) fields in full theory. On the basis of this formalism, we introduce a "gravito-magnetic Aharonov-Bohm effect", which makes it possible to interpret some rotation effects as gravito-magnetic effects. Finally, we suggest a way for measuring the angular momentum of celestial bodies by studying the gravito-magnetic effects on the propagation of electromagnetic signals.

GRAVITO-MAGNETIC AHARONOV-BOHM EFFECT

Let a physical reference frame (PRF) be defined by a time-like congruence Γ of world-lines of particles constituting the 3-dimensional reference frame.[4] The motion of free particles relative to Γ can be described by projecting the equation of motion onto the 3-space of the PRF by means of the "natural splitting".[1] The projected equation can be written in the form [4]

$$\frac{\hat{D}\tilde{p}_i}{dT} = m\tilde{E}_i^G + m\gamma_0 \left(\frac{\tilde{v}}{c} \times \tilde{B}_G\right)_i , \tag{1}$$

in terms of the GEM fields $\tilde{E}_G, \tilde{B}_G$. In other words, the dynamics of particles, relative to the reference frame Γ, is described in terms of motion under the action of a GEM Lorentz force, which is obtained without any approximations on fields and velocities. On the basis of the analogy between the "classical" Lorentz force and the GEM Lorentz force (1), we may formally introduce the gravito-magnetic Aharonov-Bohm phase shift and the corresponding time delay [4]

$$\Delta T = \frac{2\gamma_0}{c^3} \oint_C \tilde{A}_G \cdot d\tilde{x} = \frac{2\gamma_0}{c^3} \int_S \tilde{B}_G \cdot d\tilde{S}. \tag{2}$$

---

[1] We refer to the projection technique developed by Cattaneo: see [4] and references therein.

CP751, *General Relativity and Gravitational Physics, 16th SIGRAV Conference*, edited by G. Esposito et al.
© 2005 American Institute of Physics 0-7354-0236-1/05/$22.50

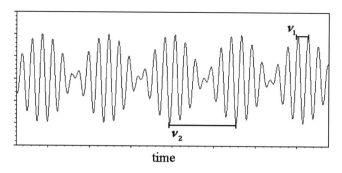

FIGURE 1. The basic signal frequency v_1 depends on the mass of the rotating body only, while the amplitude modulation frequency v_2 depends on its angular momentum only.

Equation (2) describes the time delay for two massive or massless beams, which propagate in opposite directions along a closed path, with the same velocity in absolute value. Because of the gravito-magnetic field, the two beams take different times (as measured by a clock at rest in Γ) for a round trip, and (2) expresses this time delay. In particular, some rotation effects can be thought of as gravito-magnetic Aharonov-Bohm effects: this is the case, for instance, of the Sagnac effect, both in flat and curved-space time [5], [6].

MEASURING THE ANGULAR MOMENTUM OF CELESTIAL BODIES

Gravito-magnetic effects originating from masses rotation can be exploited for measuring the angular momentum of celestial bodies. Shapiro *et al.* [7],[8],[9] verified the influence of the solar mass on signals propagation, however, in weak-field approximation, the time of flight does depend also on the angular momentum, and its contribution has the form [10]

$$t_J = \pm \frac{2GJ}{c^4 b} f, \tag{3}$$

where J is the magnitude of the angular momentum, whose direction is supposed to be perpendicular to the plane of propagation of the signals; b is the impact parameter and f is a geometrical factor. The double sign means that the propagation time is shortened on one side of the rotating mass and lengthened on the other side. The interplay of motions of the source of electromagnetic signals, the rotating mass and the observer results in a change of the impact parameter in (3): this induces a frequency shift [10]. To fix the ideas, let us consider electromagnetic signals coming from a distant source (such as a star, or a pulsar); these signals, after propagating in the field of the Sun, are received by an Earth-based observer. For the sake of simplicity, let us suppose that the source, the Sun and the observer are in the same plane. Since the source is much more distant from the Sun than the receiver, to first approximation the time variation of the impact parameter is due to the motion of the receiver only. Around the occultation of the source by the Sun, the impact parameter depends linearly on time and its time variation is approximately given by $|\dot{b}| = v_0$, where v_0 is the apparent velocity of the source in the sky. If the signal is emitted with a frequency ν, it is received with a (slowly varying) frequency $\nu + \delta\nu$, such that [10]

$$\frac{\delta\nu}{\nu} = 4\frac{GM}{c^2 b}\frac{v_0}{c} - \frac{GMb}{c^2 R^2}\frac{v_0}{c} \mp 4\frac{GJ}{c^3 b^2}\frac{v_0}{c}, \tag{4}$$

where R is the distance of the observer from the rotating body whose mass is M. The order of magnitude of this frequency shift, for a Solar System experiment, is [10]

$$\frac{GM}{c^2 b}\frac{v_0}{c} \sim 10^{-10}, \quad \frac{GMb}{c^2 R^2}\frac{v_0}{c} \sim 10^{-14}, \quad \frac{GJ}{c^3 b^2}\frac{v_0}{c} \sim 10^{-16}. \tag{5}$$

The frequency shift due to the mass has been recently measured by Bertotti *et al.* [11], using radio pulses from Cassini spacecraft. The angular momentum contribution could be measured by producing interference between the received signals. In other words, if we record the signal before and after the occultation, and then superimpose them, we get a beat function (see Figure 1). The beat has a basic frequency [10]

$$\nu_1 = \nu\left(4\frac{GM}{c^2 b}\frac{v_0}{c} - \frac{GMb}{c^2 R^2}\frac{v_0}{c}\right), \tag{6}$$

which depends on the mass only, and a frequency of amplitude modulation

$$\nu_2 = \nu\left(4\frac{GJ}{c^3 b^2}\frac{v_0}{c}\right), \tag{7}$$

which depends on the angular momentum only: the two effects are then decoupled. Though small, this rotation effect is, in principle, detectable, and it would provide a way for measuring the angular momentum of celestial bodies. More favourable conditions for observing this and other gravito-magnetic effects could be attained by studying the recently observed double pulsar system J0737-3039 [12].

REFERENCES

1. Mashhoon B., (2003), gr-qc/0311030.
2. Ruggiero M.L., Tartaglia A., *Nuovo Cimento B* **117**, 743 (2002), gr-qc/0207065.
3. Jantzen R.T., Carini P., Bini D., *Annals of Physics* **215**, 1 (1992), gr-qc/0106043.
4. Rizzi G., Ruggiero M.L., in *Relativity in Rotating Frames*, eds. Rizzi G. and Ruggiero M.L., in the series "Fundamental Theories of Physics", ed. A. Van der Merwe, Kluwer Academic Publishers, Dordrecht (2004), gr-qc/0305084.
5. Rizzi G., Ruggiero M.L., *Gen. Rel. Grav.* **35**, 1743 (2003), gr-qc/0305046.
6. Ruggiero M.L., to appear in Proceedings of *Analysis, Manifolds and Geometric Structures in Physics, International Conference in Honor of Y. Choquet-Bruhat, Isola d'Elba, June 2004*.
7. Shapiro I.I. *et al.*, *Phys.Rev. Lett.* **26**, 1132 (1971).
8. Shapiro I. I., Reasenberg R. D. *et al.*, *J. Geophys. Res.* **82**, 4329 (1977).

9. Reasenberg R. D., Shapiro I.I. *et al.*, *Astrophys. J.* **234**, L219 (1979).
10. Tartaglia A., Ruggiero M.L., *Gen. Rel. Grav.* **36**, 293 (2004), gr-qc/0305093.
11. Bertotti B., Iess L., Tortora P., *Nature* **425**, 374 (2003).
12. Lyne A., *Science* **303**, 1153 (2004), astro-ph/0401086.

The Weak Equivalence Principle (WEP) and the General Relativity Accuracy Test (GReAT) with an Einstein Elevator

V. Iafolla*, D.M. Lucchesi*, S. Nozzoli*, M. Ravenna*, F. Santoli*, I.I. Shapiro[†], E.C. Lorenzini[†], M.L. Cosmo[†], J. Ashenberg[†], P.N. Cheimets[†] and S. Glashow[**]

*Istituto di Fisica dello Spazio Interplanetario (IFSI/CNR; Roma, Italy)
[†]Harvard–Smithsonian Center for Astrophysics (CfA; Cambridge, MA)
[**]Boston University (Boston, MA)

Abstract. The experiment GReAT (General Relativity Accuracy Test) aims at testing the Weak Equivalence Principle with a precision of several parts in 10^{15}. The test will be performed by using a cryogenic differential accelerometer consisting of two test masses of different materials, released from a stratospheric altitude to free fall inside an evacuated capsule. The detector is spun about its horizontal axis to modulate the possible violation signal. As a result, the signal will appear at the spin frequency while the effects due to the diagonal components of the gravity gradient tensor in case of imperfect coincidence of the two proof masses centers of mass will appear at twice the rotation frequency.

INTRODUCTION

The Weak Equivalence Principle (WEP) is at the basis of the Einstein Equivalence Principle (EEP) and states that the ratio of the inertial mass to the (passive) gravitational mass is the same for all bodies. The WEP can be stated also in terms of Universality of Free Fall (UFF) of a test body: the motion of any freely falling test body is independent of its internal structure and composition. Therefore, a possible violation of the UFF can be evidenced through a differential acceleration measurement. A differential accelerometer is then an ideal probe to test for a possible violation of the WEP. This detector must have a very high sensitivity in measuring the possible differential acceleration on a pair of proof masses of different materials. At the same time, the detector must be capable of rejecting common–mode external accelerations and gravity gradients. In the present paper we report the progress in the development of a differential accelerometer to be used in the GReAT experiment [1, 2, 3], that aims at reducing the uncertainty of the WEP to a level of several parts in 10^{15}. The detector will be released to free fall inside an evacuated capsule which has been previously dropped from a stratospheric balloon. In order to reach the accuracy goal of the experiment, the accelerometer must attain a sensitivity close to $10^{-14} g_{\oplus}/\sqrt{Hz}$ in a 25 s integration time (with $g_{\oplus}$ the Earth's gravity acceleration). The free–fall duration is determined by the time that the detector takes to span the co–falling capsule. The detector will be slowly rotated about a horizontal axis to modulate a possible signal of WEP violation in order to distinguish it from background noise.

CP751, *General Relativity and Gravitational Physics, 16th SIGRAV Conference*, edited by G. Esposito et al.

We first describe briefly the overall experiment. Then, we present a new configuration of the differential accelerometer which is less sensitive to higher–order mass moments generated by nearby masses. This configuration of the differential accelerometer is an evolution of a high–sensitivity accelerometer developed in the IFSI/CNR laboratory over many years.

DESCRIPTION OF THE OVERAL EXPERIMENT AND THE DIFFERENTIAL ACCELEROMETER

An instrument package containing the differential accelerometer detector, free falls inside a shielding capsule, with an internal length of 2 m and a diameter of 1 m, released from a stratospheric balloon at an altitude of about 40 km. The capsule is slightly decelerated by the rarefied atmosphere so that, in 25 s the co–falling experiment reaches the bottom of the capsule while the non–propelled capsule falls by 3 km and attains a maximum Mach number of 0.8. At the end of the fall, the capsule is decelerated by a parachute for retrieval and re–flight. The cryogenic refrigeration of the capsule provides low thermal noise, high thermal stability, low thermal gradients, and a high Q–factor of the acceleration detector. These are necessary conditions to achieve the desired measurement accuracy. The overall mass of the capsule inclusive of the gondola is less than 1500 kg, which is an average load for high–altitude balloons. The elements of the experimental apparatus are as follows: helium balloon; the gondola attached to the balloon with the mechanism to release the capsule and other house–keeping equipment; shielding capsule with large–vacuum chamber/cryostat; free–falling instrument package which houses the differential acceleration detector, batteries and electronics; spin/release mechanism for spinning the instrument package and releasing it into the capsule at the start of the fall; two video cameras at the top of the capsule for monitoring the motion of the instrument package; telemetry system for the down link to the ground; supersonic parachute for decelerating the capsule at the end of the fall. The experimental package moves, at low speed, inside the capsule where the residual gas produces a small force on the free–falling package that can be expressed in terms of acceleration noise. Moreover, the high vacuum ($< 10^{-6}$ mbar) inside the capsule attenuates strongly the propagation of disturbances from the wall to the free–floating detector. Consequently, the free–falling capsule reduces the acceleration noise to values unmatched by any other Earth–based drop facility and comparable to values achieved on board drag–free satellites. The acceleration noises produced by the residual gas are common–mode–type (i.e., they affect equally both test masses) accelerations which are further reduced by the common–mode rejection factor (CMRF) of the differential accelerometer. For a conservative value of 10^{-4} for the CMRF, the influence of these two accelerations on the differential measurement is made negligible. The detector is a differential accelerometer with a noise spectral density smaller than $10^{-14} g_{\oplus}/\sqrt{Hz}$, able to measure the differential accelerations acting on two sensing masses of different materials by measuring the variation of their position, with two capacitive transducers. The detector is based on the experimental heritage acquired by the IFSI laboratory in developing highly sensitive accelerometers. Its further development requires technology within the current state–

of–the–art. Two cylindrical masses are attached to a frame by means of two pairs of flexural suspension that restraint the two proof masses to rotate about the same pivot axis. The accelerations acting along the sensing axis are detected by means of two capacitive bridges formed by two double–faced condensers, and two external capacitors. The double–faced capacitors are formed by a sensing mass and two fixed plates. Two additional capacitors are used for feed–back and system control. The bridges are pumped by a quartz oscillator at a stable frequency of about 20 kHz in order to reduce the relevant noise temperature of the preamplifier. After amplification the signal is sent to a lock–in amplifier for phase–detection and signal acquisition. The two sensing masses are right solid hollow cylinders which, in the flight model, will have spherical ellipsoids of inertia in order to cancel the 2nd–order gravity–gradient torques (quadrupole moments). The design of the capacitive detector proposed can accommodate a variety of sensing masses with different dimensions. Moreover, the centers–of–mass of the sensing masses are as close as technically possible to one another, to minimize the effect of gravity–gradient forces, rotational motion, and linear accelerations upon the differential output signal. The two sensing masses are made of different materials: a high–density material for mass 1 and a low–density material for mass 2. We also plan to build a (zero–signal) differential detector with sensing masses of the same material to be flown in a test balloon flight to bound the noise environment during free fall.

CONCLUSIONS

The GReAT experiment, which is sponsored by ASI (Agenzia Spaziale Italiana) and NASA, aims to test the WEP with an accuracy about two orders–of–magnitude better than the present state of art. A prototype of the differential accelerometer has been built also based on the knowledge acquired through analyses and simulation of the flight experiment conducted at SAO (Smithsonian Astrophysical Observatory), which is the US partner (sponsored by NASA) on this project. The laboratory experiments carried out on the differential accelerometer prototype have demonstrated its performance. Specifically, we have studied the damping of transient effects after instrument release, the rejection of the common–mode effects due to perturbation external to the differential accelerometer, and the suitable shape of the sensing masses for reducing the perturbing effects of higher–order mass moments (this work has been supported by ASI contract I/R/098/02 and NASA grants NAG8–1780 and NAG3–2881).

REFERENCES

1. Iafolla, V., Nozzoli, S., Lorenzini, E.C., and Milyukov, V., Methodology and instrumentation for testing the weak equivalence principle in stratospheric free fall, Review of Scientific Instruments, Vol. 69, No. 12, 4146–4151 (1998).
2. Lorenzini E.C., Shapiro I.I., Fuligni F., Iafolla V., Cosmo M.L., Grossi M.D., Cheimets P.N. and Zielinski J.B., Test of the Weak Equivalence Principle in an Einstein Elevator, Il Nuovo Cimento, Vol. 109B, No. 12, 1195–1209 (1994).
3. Shapiro, I.I., et al., Test of the Equivalence Principle in an Einstein Elevator, Final Report, NASA Grant NAG8–1780, April (2004).

Conference Program

17:00 - 19:30   *registration*

9:30   **welcome address**

10:00 – 10:45   Carlo Rovelli: *Gravità quantistica: sviluppi recenti*

10:50 – 11:20   *coffee break*

11:20 – 12:05   Andrea Possenti: *The double pulsar binary J0737-3039: a two-clocks relativistic system*

12:10 – 12:55   Giovanni Losurdo: *Virgo and the worldwide search for gravitational waves*

13:00   *lunch*

15:00 – 16:30   *SIGRAV Prizes*

Carlo Angelantonj: *Open strings and supersymmetry breaking*

Giovanni Miniutti: *Flux and energy modulation of iron emission due to relativistic effects in NGC 3516G*

16:30   *Transfer to Salerno - Palazzo della Provincia*

17:00   *"Amaldi Medal" Ceremony*

Professor Roger Penrose: *Galilei-Einstein equivalence principle and quantum state reduction*

20:00   *dinner*

9:00 – 9:45   Laurent Friedel: *An overview of group fiel theory in quantum gravity*

9:55 – 10:40   Vincent Pasquier: *The Chaplyng gas: a model for dark energy in cosmology"*

10:45 – 11:15   *coffee break*

11:15 – 12.00   Guglielmo Tino: *Atomic clocks and atom interferometers for gravitational physics experiments*

12:10 – 12:55   Giulio Magli: *Analytical modelling of gravitational collapse*

13:00   *lunch*

Workshop A: CLASSICAL AND QUANTUM GRAVITY

15:00 – 15:20   Fabrizio Canfora: *Singular PP waves, junctions conditions and BPS states*

15:25 – 15:45   Sergio Zerbini: *Asimptotics of quasinormal modes for Schwarzschild-de Sitter black holes*

15:50 – 16:10   Marcella Palese: *Bergman-Bianchi Identities in Field Theories*

16:15 – 16:35   *coffee break*

16:40 – 17:00   Antonio Feoli: *Some properties of the de Broglie gravitational waves*

17:05 – 17:30   Andrzej Borowiec: *Three-dimentional Chern-Simons and BF theories*

17:35 – 18:00	Alfio Bonanno: *Proper-Time regulators and RG flow in QEG*
18:05 – 18:30	Daniele Malafarina: *Static axially symmetric sources of the gravitational field*

Workshop B: ASTROPHYSICS AND COSMOLOGY

15:00 - 15:20	Marco Bruni: *Newman-Penrose quantities as valuable tools in astrophysical relativity*
15:20 - 15:40	Antonio Capolupo: *Neutrino mixing and cosmological constant*
15:40 - 16:00	Viviana Acquaviva: *Weak lensing and cosmological investigation*
16:00 - 16:20	Alexander Zakharov: *Retro gravitational lensing for Sgr A* with Radiastron*
16:20 - 16:50	*coffe break*
16:50 - 17:10	Alfio Bonanno: *Pulsar magnetism and dynamo actions*
17:10 - 17:30	Leonardo Gualtieri: *Gravitational waves from neutron stars described by modern EOSO*
17:30 - 17:50	Giuseppe Lavagetto: *Constraining the equation of state of neutron stars with general relativity*
17:50 - 18:10	Alessandro Nagar: *Gravitational waves emission from matter accretion onto a Schwarzschild black hole*
18:10 - 18:30	Cosimo Stornaiolo: *The cosmological black hole hypothesis*
18:30 - 18:50	Hayoung Lee: *The Einstein-Vlasov system with a scalar field*
18:50 - 19:00	*discussion of the poster session*
18:00 – 19:00	*SIGRAV Members Meeting*
20:00	*dinner*

WEDNESDAY 15 SEPTEMBER 2004

9:00 – 9:45	Alessandro Bottino: *Particle dark matter*
9:55 – 10:40	Salvatore Capozziello: *Higher order curvature theories matched with observations: a bridge between dark energy and dark matter problems*
10:40 – 11:00	*coffee break*
11:00 – 11:45	Massimo Visco: *The explorer and nautilus gravitational wave detectors and beyond*
11:50 – 12:35	Massimo Bianchi: *Higher spins and stringy ads5xs5*
12:40 – 13:00	Gaetano Lambiase: *Propagation of neutrinos and photons in gravitational field*
13:00	*lunch*
15:00 – 15:45	Donato Bini: *Inertial effects of an accelerating black hole*
15:55 – 16:40	Roberto Cianci: *Gravity and Yang-Mills fields: geometrical approaches*
16:45 – 17:15	*coffee break*
17:15 – 18:00	Livia Conti: *Interferometric readout for acoustic gravitational wave detectors*
18:10 – 18:55	Gianluca Allemandi: *Gravitational energies and generalized geometric entropy*
20:00	*Social dinner*

21:30	**Concert**

THURSDAY 16 SEPTEMBER 2004

9:00 – 9:45	Luciano Iess: *A new test of general relativity with the Cassini spacecraft*
9:55 – 10:40	Gianluca Israel: *Taking the pulse of the shortest orbital period double degenerate binary rxj0806.3+1527*
10:45 – 11:15	*coffee break*
11:15 – 12:00	Angelo Tartaglia: *Rotation effects and the gravito-magnetic approach*
12:10 – 12:55	Raffaella Schneider: *L'alba cosmica: dalle prime stelle all'universo osservato*
13:00	*lunch*

Workshop A: CLASSICAL AND QUANTUM GRAVITY

15:15 – 15:40	Alessandro Tanzini: *Penrose limit and duality between string and gauge theories*
15:45 – 16:10	Valeria Gili: *Simplicial aspects of string dualities*
16:15 – 16:45	*coffee break*
16:45 – 17:10	Laurent Freidel: *Coupling spinning particles to three dimensional quantum gravity*
17:15 – 17:40	Stefano Ansoldi: *Minisuperspace, WKB, quantum states of general relativistic extended objects*
17:45 – 18:10	Claudio Dappiaggi: *Holography and BMS Field Theory*
18:15 – 18:40	Luisa Doplicher: *D-particles and the localization limit in quantum gravity*

Workshop C: EXPERIMENTAL GRAVITY

15:00 - 15:20	David Lucchesi: *The BepiColombo radio science experiment and the non-gravitational perturbations to the Mercury planetary orbiter orbit: key role of the Italian spring accelerometer*
15:20 - 15:40	Roberto Peron: *Orbital analysis of LAGEOS and LAGEOS II laser ranged satellites: relativistic effects and geophysical issues*
15:40 - 16:00	Valerio Iafolla: *The weak equivalence principle (WEP) and the general relativity accuracy test (GReAT) with an Einstein elevator*
16:00 - 16-20	Giancarlo Cella: *Beyond the standard quantum limit*
16:20 - 17:00	*coffee break*
17:00 - 17:20	Angela Di Virgilio: *Preliminary analysis of the July 2004 data of the low frequency facility*
17:20 - 17:40	Matteo Luca Ruggiero: *Rotation effects and gravito-magnetic approach*
17:40 - 18:00	Rosario De Rosa: *Dynamic matched filter for gravitational wave detection*
18:00 - 18:20	Antonio Eleuteri: *Adaptive filters for detection of gravitational waves from coalescing binaries*
20:00	*dinner*

PARTICIPANTS

Acquaviva Viviana, SISSA/ISAS, Via Beirut 4, 34014 Trieste, Italy; Ph: 040 3787475; Fax: 040 3787528; e-mail: acqua@sissa.it.

Allemandi Gianluca, Dipartimento di Matematica, Università di Torino, Via Carlo Alberto 10, 10123 Torino; Ph: 349-2694243; Fax: 0116702878; e-mail: allemandi@dm.unito.it.

Angelantoni Carlo, INAF/Osservatorio Astronomico di Roma.

Ansoldi Stefano, Dipartimento di Matematica e Informatica Universita` di Udine & INFN Sezione di Trieste; Ph: 347 8077047; Fax: 0432 558499; e-mail: ansoldi@trieste.infn.it.

Baiotti Luca, SISSA, Via Beirut 4, 34014 Trieste, Italy; Ph: 0403787525; Fax: 0403787528; e-mail: baiotti@sissa.it.

Bassan Massimo, Dipartimento di Fisica, Università Tor Vergata and INFN, Sezione Roma2; Ph: 06 7259 456; Fax: 06 202 3507; e-mail: bassan@roma2.infn.it.

Bianchi Massimo, Dipartimento di Fisica, Università "La Sapienza", Roma.

Bibbona Enrico, Università di Torino; Ph: 011-6702933; Fax: 011-6702878; e-mail: bibbona@dm.unito.it.

Bini Donato, Istituto per le Applicazioni del Calcolo, "M. Picone", CNR, via del Policlinico, 137, 00161 ROMA; Ph: + 39 06 88470255; Fax: +39 06 4404306.

Bonanno Alfio, INAF - Osservatorio Astrofisico di Catania; Ph: +39 0957332319; e-mail: abo@ct.astro.it.

Borowiec Andrzej, Institute of Theoretical Physics, Universit of Wroclaw, Pl. Maxa Borna 9, 50-204, WROCLAW, POLAND; Ph: +48-71-3759406; Fax: +48-71-3214454; e-mail: borow@ift.uni.wroc.pl.

Bottino Alessandro, Dipartimento Fisica Teorica, Universita` di Torino and INFN; Ph: 011 670 7202; e-mail: bottino@to.infn.it.

Bruni Marco, Institute of Cosmology and Gravitation, Mercantile House, Portsmouth University, PO1 2EG, Portsmouth, United Kingdom; Ph: +44 (23) 9284 3136; Fax: +44 (23) 9284 5626; e-mail: marco.bruni@port.ac.uk.

Canfora Fabrizio, Dipartimento di Fisica "E. R. Caianiello", Università degli Studi di Salerno , Via S. Allende, I-84081 Baronissi (SA) Italy; Ph: 089965228; Fax: 089965275; e-mail: canfora@sa.infn.it.

Capolupo Antonio, Dipartimento di Fisica "E. R. Caianiello", Università degli Studi di Salerno , Via S. Allende, I-84081 Baronissi (SA) Italy; Ph: +39 089 965360; Fax: +39 089 965275; e-mail: capolupo@sa.infn.it.

Capone Monica, Dipartimento di Fisica "E. R. Caianiello", Università degli Studi di Salerno , Via S. Allende, I-84081 Baronissi (SA) Italy; Ph: 3290728528; Fax: 0817775963; e-mail: agravaine@inwind.it.

Capozziello Salvatore, Dipartimento di Fisica "E. R. Caianiello", Università degli Studi di Salerno , Via S. Allende, I-84081 Baronissi (SA) Italy; Ph: 089/965238; Fax: 089/965275; e-mail: capozziello@sa.infn.it.

Cardone Vincenzo Fabrizio, Dipartimento di Fisica "E. R. Caianiello", Università degli Studi di Salerno , Via S. Allende, I-84081 Baronissi (SA) Italy; Ph: 089965256; Fax: 089965275; e-mail: winny@na.infn.it.

Carfora Mauro, Dipartimento di Fisica Nucleare e Teorica dell'Universita` di Pavia, Via Bassi 6, 27100 Pavia; Ph: 0382 507 443; e-mail: mauro.carfora@pv.infn.it.

Cavaliere Alfonso, Dipartimento di Fisica, Università di Tor Vergata, Roma; Ph: 06 7259 4432; Fax: 06 202 3507; e-mail: alfonso.cavaliere@roma2.infn.it.

Cella Giancarlo, Universita di Pisa and INFN sez. Pisa; Ph: 0583724515; Fax: 0583710542; e-mail: giancarlo.cella@pi.infn.it.

Cerdonio Massimo, INFN Section and Physics Department, Università di Padova, Italia.

Cianci Roberto, DIPEM, Università di Genova; Ph: 0103536005; Fax: 0103536003; e-mail: cianci@dipem.unige.it.

Coccia Eugenio, Dipartimento di Fisica, Università degli Studi di Roma "Tor Vergata", Via della Ricerca Scientifica, 1 - 00133 Roma; Ph: +39 06 72594588; Fax: +39 06 2023507; e-mail: eugenio.coccia@roma2.infn.it.

Colpi Monica, Dipartimento di Fisica "G. Occhialini", Università di Milano Bicocca, Piazza della Scienza 3, I-20126 Milan, Italy

Conti Livia, Univ & INFN of Padova; Ph: 049 8068826; Fax: 049 8068824; e-mail: conti@lnl.infn.it.

Dappiaggi Claudio, Dipartimento di Fisica Nucleare e Teorica - Università di Pavia; Ph: 0382507447; Fax: nessuno; e-mail: claudio.dappiaggi@pv.infn.it.

De Giuli Enrico, Università degli Studi di Trento, Dipartimento di Fisica; Ph: 0461 881575; Fax: 0461 882014; e-mail: degiuli@science.unitn.it.

De Pietri Roberto, Parma University; Ph: 0521 906068; Fax: 0521 905223; e-mail: depietri@fis.unipr.it.

De Rosa Rosario, Università di Napoli "Federico II"; Ph: +39 081 676456; Fax: +39 081 676346; e-mail: rosario.derosa@na.infn.it.

Di Marino Vincenzo, Dipartimento di Fisica "E. R. Caianiello", Università degli Studi di Salerno , Via S. Allende, I-84081 Baronissi (SA) Italy; Ph: 089965316; Fax: 089965275; e-mail: dimarino@sa.infn.it.

Di Virgilio Angela, INFN - Sezione di Pisa, Edificio C - Polo Fibonacci - Via F. Buonarroti, 2 PISA; Ph: 39 050 2214343; Fax: +39 - 050 2214 317; e-mail: ANGELA.DIVIRGILIO@pi.infn.it.

Doplicher Luisa, Università di Roma "La Sapienza", Piazzale Aldo Moro, 2 - 00185 Roma, Italy; Ph: +390649914350; e-mail: doplicher@roma1.infn.it.

Eleuteri Antonio, Dipartimento di Scienze Fisiche, Università degli Studi di Napoli "Federico II" and INFN Sez. Napoli; Ph: 081676882; Fax: 081676346; e-mail: eleuteri@na.infn.it.

Esposito Giampiero, INFN - Sezione di Napoli; Ph: 081-676470; Fax: 081-676346; e-mail: giampiero.esposito@na.infn.it.

Fatibene Lorenzo, Dipartimento di Matematica, University of Torino; Ph: 011 670 2933; Fax: 011 670 2878; e-mail: fatibene@dm.unito.it.

Feoli Antonio, Dipartimento di Ingegneria dell'Università degli Studi del Sannio - Benevento; Ph: 0825/37490; Fax: 089/965275; e-mail: feoli@unisannio.it.

Francaviglia Mauro, Dipartimento di Matematica, University of Torino; Ph: 011 670 2932; Fax: 011 670 2878; e-mail: francaviglia@dm.unito.it.

Freidel Laurent, Laboratoire de Physique, ENS-Lyon, France and Perimeter institute, Waterloo, Ontario, Canada.; Ph: 4 7272 8578; e-mail: lfreidel@perimeterinstitute.ca.

Giazotto Adalberto, INFN - Sezione di Pisa, Edificio C - Polo Fibonacci - Via F. Buonarroti, 2 PISA; e-mail: adalberto.giazotto@pi.infn.it.

Gili Valeria, Dipartimento di Fisica Nucleare e Teorica, Università degli Studi di Pavia; Ph: +390382507447; e-mail: valeria.gili@pv.infn.it.

Gorini Vittorio, Dipartimento di Fisica e Matematica - Facoltà di Scienze MM.FF.NN., Sede di Como; Fax: 031 238 6119; e-mail: vittorio.gorini@uninsubria.it.

Gualtieri Leonardo, Dipartimento di Fisica "G. Marconi", Università degli Studi di Roma "La Sapienza", P.le A. Moro 2, 00185 Roma, Italy.

Iafolla Lorenzo, Student of Tor Vergata Univ. (Rome); Ph: 3495095361; e-mail: iafol@yahoo.com.

Iess Luciano, Dipartimento di Ingegneria Aerospaziale ed Astronautica Università La Sapienza; Ph: 06-44585336; Fax: 06-44585952; e-mail: iess@hermes.diaa.uniroma1.it.

Israel Gianluca, INAF - OA Roma; Ph: 0694286437; Fax: 069447243; e-mail: gianluca@mporzio.astro.it.

Lambiase Gaetano, Dipartimento di Fisica "E. R. Caianiello", Università degli Studi di Salerno , Via S. Allende, I-84081 Baronissi (SA) Italy; Ph: 089965422; Fax: 089965275; e-mail: lambiase@sa.infn.it.

Lavagetto Giuseppe, DSFA - Università di Palermo; Ph: +390916234289; Fax: +390916234281; e-mail: lavaget@fisica.unipa.it.

Lee Hayoung, Max Planck Institute for gravitational physics, Golm bei Potsdam, Germany; Ph: +49 (0)331 567 7127; Fax: +49 (0)331 567 7298; e-mail: hayoung@aei.mpg.de.

Losurdo Giovanni, INFN Firenze; Ph: 050 752317; Fax: 050 752550; e-mail: losurdo@fi.infn.it.

Lucchesi David M., Istituto di Fisica dello Spazio Interplanetario (IFSI/CNR); Ph: 0649934367; Fax: 0649934383; e-mail: lucchesi@ifsi.rm.cnr.it.

Lusanna Luca, INFN - Sezione di Firenze; Ph: 0554572334; Fax: 0554572121; e-mail: lusanna@fi.infn.it.

Magli Giulio, Politecnico di Milano, Dipartimento di Matematica, Via Bonardi 9, 20133, MILANO; Ph: 02-23994597; Fax: 02-23994568; e-mail: magli@mate.polimi.it.

Malafarina Daniele, Dipartimento di Matematica, Politecnico di Milano; Ph: 3493720434; Fax: 0223994513; e-mail: malafarina@mate.polimi.it.

Marmo Giuseppe, Dipartimento di Scienze Fisiche, Università Federico II di Napoli; Ph: 081 676492; e-mail: marmo@na.infn.it.

Marzuoli Annalisa, Dipartimento di Fisica Nucleare e Teorica, Università di Pavia via Bassi 6, 27100 PAVIA; Ph: 0382 507442; Fax: 0382 526938; e-mail: annalisa.marzuoli@pv.infn.it.

Menotti Pietro, Dipartimento di Fisica, Università di Pisa; Ph: 050 2214 885; Fax: 050 2214 887; e-mail: menotti@df.unipi.it.

Miniutti Giovanni, Institute of Astronomy, University of Cambridge; Ph: 0044 1223 339281; Fax: 0044 1223 337523; e-mail: miniutti@ast.cam.ac.uk.

Moschella Ugo, Universita dell'Insubria; Ph: 0312386229; Fax: 0312386119; e-mail: ugo.moschella@uninsubria.it.

Nagar Alessandro, Department of Astronomy and Astrophysics University of Valencia; Ph: 00390115097213; e-mail: alex@nagarsoft.com.

Nobili Anna, Dipartimento di Matematica, Via F. Buonarroti, 2 - 56127 Pisa; e-mail: nobili@dm.unipi.it.

Palese Marcella, Department of Mathematics, University of Torino, via C. Alberto 10, I-10123 Torino, Italy; Ph: +39 011 6702889; Fax: +39 011 6702878; e-mail: marcella.palese@unito.it.

Parisi Luca, Dipartimento di Fisica "E. R. Caianiello", Università degli Studi di Salerno, Via S. Allende, I-84081 Baronissi (SA) Italy; Ph: 089383646; Fax: +39 089 965275; e-mail: luca-parisi@libero.it.

Pasquier Vincent, spht cea Saclay; Ph: 0033169088125; Fax: 0033169088120.

Penrose Roger, Mathematical Institute, University of Oxford, 24-29 St Giles', Oxford, OX1 3LB England; Ph: +44 (0)1865 273550; Fax: +44 (0)1865 273583; e-mail: rouse@maths.ox.ac.uk.

Peron Roberto, Università di Lecce, Dipartimento di Fisica; Ph: 06-49934367; Fax: 06-49934383; e-mail: peron@ifsi.rm.cnr.it.

Possenti Andrea, INAF-Osservatorio Astronomico di Cagliari; Ph: +39-070-71180249; Fax: +39-070-71180244; e-mail: possenti@ca.astro.it.

Rosquist Kjell, Dept of Physics, Stockholm University; Ph: +46707557649; e-mail: kr@physto.se.

Rovelli Carlo, Università del Mediterraneo, Marsiglia; Ph: +33614593885; e-mail: rovelli@cpt.univ-mrs.fr.

Rubano Claudio, Dept. of Physical Sciences
Univ. "Federico II" Naples; Ph: +39 081 676497; e-mail: rubano@na.infn.it.

Ruggiero Matteo Luca, Dipartimento di Fisica, Politecnico di Torino, 10129 Torino - Italy; Ph: 00390115647380; Fax: 00390115647399; e-mail: matteo.ruggiero@polito.it.

Scarpetta Gaetano, Dipartimento di Fisica "E. R. Caianiello", Università degli Studi di Salerno , Via S. Allende, I-84081 Baronissi (SA) Italy; Ph: 089 965210; e-mail: scarpetta@sa.infn.it.

Schneider Raffaella, Centro Enrico Fermi, INAF/Osservatorio Astrofisico Arcetri; Ph: 055 2752250; Fax: 055 220039; e-mail: raffa@arcetri.astro.it.

Scudellaro Paolo, Dipartimento di Scienze Fisiche, Università Federico II di Napoli; Ph: 081 676498; Fax: 081-676346; e-mail: scud@na.infn.it.

Stornaiolo Cosimo, INFN sezione di Napoli; Ph: 081676471; e-mail: cosmo@na.infn.it.

Tanzini Alessandro, SISSA, Via Beirut 4, 34014 Trieste, Italy; Ph: +33-1-44277397; Fax: +33-1-44277088; e-mail: tanzini@lpthe.jussieu.fr.

Tartaglia Angelo, Dipartimento di Fisica, Politecnico di Torino; Ph: 011-5647328; Fax: 011-5647399; e-mail: angelo.tartaglia@polito.it.

Tino Guglielmo M., Dipartimento di Fisica and Laboratorio LENS,INFN, Università degli Studi di Firenze, Polo Scientifico, I-50019 Sesto Fiorentino (Firenze), Italy; Ph: 055 4572034; Fax: 055 4572121; e-mail: guglielmo.tino@fi.infn.it.

Troisi Antonio, Dipartimento di Fisica "E. R. Caianiello", Università degli Studi di Salerno , Via S. Allende, I-84081 Baronissi (SA) Italy; Ph: +39 089 965266; Fax: +39 089 965275; e-mail: antro@sa.infn.it.

Tzenov Stephan, Dipartimento di Fisica "E.R. Caianiello", Università degli Studi di Salerno, Via S. Allende, I 84081 Baronissi (SA), Italy; Ph: +39 089 965242; Fax: +39 089 965275; e-mail: tzenov@sa.infn.it.

Vignolo Stefano, Dipartimento di Ingegneria della Produzione e Modelli Matematici dell'Università di Genova; Ph: 0103536043; Fax: 0103536003; e-mail: vignolo@dipem.unige.it.

Vilasi Gaetano, Dipartimento di Fisica "E. R. Caianiello", Università degli Studi di Salerno , Via S. Allende, I-84081 Baronissi (SA) Italy; Ph: 0039-089965317; Fax: 0039-089-965275; e-mail: vilasi@sa.infn.it.

Visco Massimo, CNR IFSI Roma; Ph: 0649934366; Fax: 0649934383; e-mail: massimo.visco@ifsi.rm.cnr.it.

Vitagliano Luca, Dipartimento di Matematica e Informatica, Università degli Studi di Salerno; Ph: 089 963329; e-mail: vitagliano@icra.it.

Zakharov Alexander, State Scientific Centre -- Institute of Theoretical and Experimental Physics, Moscow, Russia; Ph: 007-095-1507540; Fax: 007-095-8839601; e-mail: zakharov@itep.ru.

Zerbini Sergio, Department of Physics, Trento University; Ph: 0461-881527; Fax: 0461-881696; e-mail: zerbini@science.unitn.it.

AUTHOR INDEX

A

Acernese, F., 92, 236, 242
Acquaviva, V., 199
Allemandi, G., 25
Amico, P., 92
Angelantonj, C., 3
Ansoldi, S., 159
Aoudia, S., 92
Arcioni, G., 176
Arnaud, N., 92
Ashenberg, J., 255
Avino, S., 92

B

Babusci, D., 92
Ballardin, G., 92
Barillé, R., 92
Barone, F., 92, 236, 242
Barsotti, L., 92
Barsuglia, M., 92
Beauville, F., 92
Belvedere, G., 202
Benhar, O., 211
Bini, D., 37
Bizouard, M. A., 92
Blasone, M., 208
Boccara, C., 92
Bonanno, A., 162, 202
Bondu, F., 92
Borowiec, A., 165
Bosi, L., 92
Bottino, A., 46
Braccini, S., 92
Bradaschia, C., 92
Brillet, A., 92
Brisson, V., 92
Brocco, L., 92
Bruni, M., 205
Burderi, L., 214
Burgay, M., 126
Buskulic, D., 92

C

Calloni, E., 92
Camilo, F., 126
Campagna, E., 92
Canfora, F., 168
Capolupo, A., 208
Capozziello, S., 54, 208
Cardone, V. F., 54
Carfora, M., 182
Carloni, S., 54, 208
Cavalier, F., 92
Cavalieri, R., 92
Cella, G., 92, 233
Cerdonio, M., 75
Chassande-Mottin, E., 92
Cheimets, P. N., 255
Cherubini, C., 37
Cianci, R., 64
Cleva, F., 92
Conti, L., 75
Cosmo, M. L., 255
Coulon, J.-P., 92
Cuoco, E., 92

D

D'Amico, N., 126
D'Antona, F., 214
Dappiaggi, C., 176, 182
Dattilo, V., 92
Davier, M., 92
De Paolis, F., 227
De Rosa, M., 75
De Rosa, R., 92, 236, 242
Di Fiore, L., 92
Di Salvo, T., 214
Di Virgilio, A., 92, 239
Dujardin, B., 92

E

Eleuteri, A., 92, 236, 242
Enard, D., 92

269

W

White, F., 205
Winterroth, E., A., 188

Y

Yvert, M., 92

Z

Zakharov, A. F., 227
Zerbini, S., 194
Zhang, Z., 92

767